THE INTERNET SCIENCE, RESEARCH, AND TECHNOLOGY

YELLOW PAGES

Rick Stout
and
Morgan Davis

Osborne McGraw-Hill

Berkeley   New York   St. Louis   San Francisco   Auckland   Bogotá   Hamburg
London   Madrid   Mexico City   Milan   Montreal   New Delhi   Panama City   Paris
São Paulo   Singapore   Sydney   Tokyo   Toronto

THE INTERNET SCIENCE, RESEARCH, AND TECHNOLOGY YELLOW PAGES

**OSBORNE MCGRAW-HILL
2600 TENTH STREET
BERKELEY, CALIFORNIA 94710
U.S.A.**

For information on translations or book distributors outside the U.S.A., or to arrange bulk purchase discounts for sales promotions, premiums, or fundraisers, please contact Osborne McGraw-Hill at the above address.

The Internet Science, Research, and Technology Yellow Pages

234567890 SEM 99876

ISBN 0-07-882187-8

Acquisitions Editor
  Scott Rogers

Project Editor
  Emily Rader

Copy Editor
  Linda Medoff

Proofreaders
  Pat Mannion
  Linda Medoff

Computer Designer
  Peter F. Hancik
  Richard Whitaker

Illustrators
  Leslee Bassin
  Rosemarie Ellis

Series Design
  Peter F. Hancik

Quality Control Specialist
  Joe Scuderi

To my loving wife, Dawn, for all her patience and understanding.

—Rick

To *my* loving wife, Dawn, for even more patience and understanding.

—Morgan

ABOUT THE AUTHORS...

Rick Stout is a CPA and the author of a number of computer books, including
The Peter Norton Introduction to Computers (a textbook from Glencoe/Macmillan
McGraw-Hill). Stout is also the coauthor of Osborne's best-selling *The Internet Complete
Reference,* First Edition, *The Internet Yellow Pages,* First Edition, *The Internet Yellow Pages,*
Second Edition, and most recently, the author of the acclaimed *The World Wide Web
Complete Reference.*

Morgan Davis is a writer, speaker, and director of one of the nation's largest Internet
providers, and is among the foremost authorities on the Internet today. With his unique
ability to explain the Internet's most complex issues to anyone, he is often invited to radio
and TV shows, conferences, and user groups.

Table of Contents

Introduction

There are two kinds of people who use the Internet: those who have cruised the Net before and those who are just pulling away from the curb and are heading for the onramp.

If you're new to the Internet, thank you for purchasing this book. Now place it carefully on your bookshelf and get the original *Internet Yellow Pages* (or the *Internet Golden Directory,* for those of you in the U.K.). Spend some time at *biancaTroll's Smut Shack*, abuse yourself on the *Profanity and Insult Server*, and shop for a hot date at the *Virtual MeetMarket*. It's important to purge yourself of the urges that attract you to the *Out-of-Body Experiences* site, and cause you to check on the fluid levels and temperatures of coffee pots around the Net. You must get it out of your system. Go on. The rest of us will wait here for you.

Now, after you've had your fun experiencing the silly and seamy side of the Net from a tourist's perspective, you'll soon realize that there's a lot more to it than summaries of every episode of *The Simpsons* along with biographies of the characters and broadcast schedules.

Remember: you, your employer, or your school are paying good money to access this stuff. Now it's time to roll up your sleeves and do something useful with all this technology and information, because the Internet is filled with amazing material that can help you in your job, school, and hobbies. That's where this wonderful book comes in. That's why you bought a copy even if you're new to the Net—because you're just pretty darn smart.

Inside you'll find valuable links and descriptions to some of the best science, research, and technology resources available on the Internet. However, unlike a dull textbook, they're presented in a way that's entertaining and downright fun. (Face it, we must satisfy that silly side in all of us or be tempted to run for political office). And yet, the entries in this book are only the tip of this iceberg. On the web pages listed here, you'll find links to hundreds of thousands of other web pages and Internet resources.

Will this book change your life? Probably not. But whether you're an amateur photographer, nuclear physicist, educator, student, history buff, or armchair scientist, this book will open your eyes to the rich world of scientific-, research-, and technology-related resources that await your discovery.

HOW TO USE THIS BOOK

We learned a lot from this book's predecessors. And answering six to ten email messages every day is a great way to learn of the problems people experience when they begin to use a book like this.

One of the areas we've improved upon in this book is the consistency of the addresses. When Rick began work on the original book over two years ago, the World Wide Web was still in its infancy and hadn't yet really taken off. As a result, there were few web pages in the book compared with the many more FTP sites, Gophers, and sites to finger for information of all sorts. Now, of course, this has changed dramatically.

Today, the World Wide Web is very much the graphical user interface (or GUI) for the entire Internet. From the comfort of your Web browser, you can access the best of everything the Internet has to offer. In fact, the few types of resources that you can't access directly from your Web browser, you can access with a couple of other types of programs—an email program and a telnet client.

There are, by far, more addresses for web pages in this book than any other type of Internet resource. In fact, about 88 percent of the resources are web pages. The remaining resources are (in descending order of prevalence):

Usenet newsgroups

Mailing lists

Anonymous FTP sites or files

Gophers

Telnet resources

Email addresses

Of course, with your reasonably up-to-date Web browser, you can view web pages, Usenet newsgroups, FTP sites, and Gophers. All of the addresses in the book, with the exception of email addresses and mailing lists, are in the standard Universal Resource Locator (or URL) format. We did this to provide an additional level of consistency throughout. What this means to you is that you can type into your Web browser the text of the address exactly as you see it in the entry. If there's a file that you want on an FTP server, you don't need to know how to use an FTP program or anything else about it. For example, if an entry in the book shows the address for a file like this:

URL:
   ftp://ds.internic.net/rfc/rfc1320.txt

all you have to do is type the text **ftp://ds.internic.net/rfc/rfc1320.txt** into your Web browser's *Location* or *Open Location* edit box. Your browser will fetch the file and display it in its window.

In some cases, we list FTP sites and reference a directory instead of an actual file. You don't have to do anything different for these. For example, you might see an entry like this:

URL:
   ftp://cert.org/pub/papers/

When you type this URL into your browser's Location box, you'll see a list of the files and subdirectories within that directory. Here's how it will look:

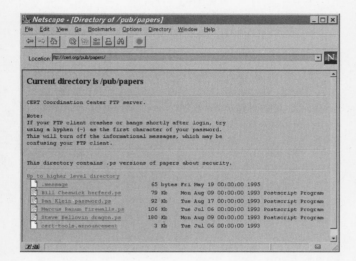

We include URLs for directories, since everything in a particular directory might be of interest. For example, if the FTP site houses a collection of picture files, you might want to see or look at more than just one of them. By giving you the URL for the directory, you can see each file and subdirectory in that directory and choose to look at any you like.

Similarly, if a resource you're interested in happens to be an item on a Gopher server, you can just type the URL into your Web browser. For example, if the address listed is:

URL:
   gopher://cell-relay.indiana.edu/11/docs/rfc

simply type that URL into your browser's Location box, and here's what you'll see:

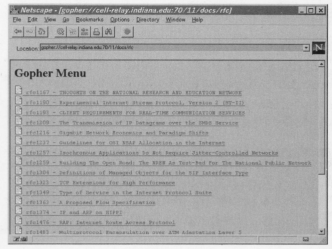

The types of resources you can't immediately access directly through your Web browser are telnet resources and email—which includes both a regular email address to mail a message to someone and mailing lists. However, with a little ingenuity, you can even use these resources seamlessly.

The developers of some of the newest Web browsers are including full-featured email functionality in their products. Most notably—Netscape. Some Netscape users swear by the power and ease of use of Netscape's built-in email system. But on the other hand, it's as easy to get both commercial and shareware email packages as it is to get hold of a Web browser.

Telnet clients are also easy to find. Just check the resources in the "Software" section. And once you get one, simply integrate it with your Web browser by going into its Preferences dialog and specifying the path and filename of your telnet client.

CONTACTING US

Early in the process of compiling this book, Morgan (being the programming wizard that he is) wrote a fantastic program that takes every address in our database and automatically checks each one for validity. Among the many tests this program performs, it checks to see if a trailing slash is really necessary at the end of a URL. It also determines the most efficient way to represent an address so that you don't have to type **index.html** or enter a port number if you don't have to. This tool has proven to be invaluable in helping us produce a book that is both accurate and as up to date as possible.

As we go to press, we can confidently assure you that this is the most current and up-to-date directory of Internet resources available. However, you may, from time to time, find an address that has changed or gone bad altogether. If you do, we'd appreciate hearing from you about it. We've come up with a couple of ways for you to let us know if you find a bad address or let us know of an upcoming change.

The best way to get the information to us is through our web page. The address is **http://www.iypsrt.com**. We can automatically classify your suggestions and corrections and act on them much more efficiently. Also, we may be able to give you some up-to-date information through our web pages that we can't email to every person who sends us email.

If you would like to submit a new science-, research-, or technology-oriented URL for possible inclusion in the next edition, please do so on our web site. Just fill out the online form and we'll check it out.

Another way to reach us is by sending email to **iyp@iypsrt.com**. We'll both get a copy of your mail, and one of us will respond to your message personally. We do get a lot of email, so please be patient while awaiting a reply.

Even if you don't have corrections to existing entries or suggestions for new ones, but would like to send us a note anyway, we'd love to hear from you.

Acknowledgments

Writing a book like this is a lot of fun. It's also a tremendous amount of work. And we couldn't possibly have completed this project without the help of a great many talented people.

First, we would like to thank our primary researcher on the project—Jim Hall. Jim worked tirelessly researching subjects of interest, collecting entries, and he even did some of the ads in the book. Our hats are off to Jim as we thank him for his time and endless effort.

Ronda Stout did a great job working with the copy editor and proofreader to make sure that every i was dotted and every t crossed, and that every one of these corrections got reliably back into our master database. Ronda also did some research, and a few ads as well. Thanks very much, Ronda.

The rest of our outstanding research team is a diverse collection of multi-talented people with expertise and interests in a wide range of topics. The unity of their diverse interests and voices have contributed significantly to make this book interesting and fun to read. They are:

Dawn Davis	San Diego, California
Bill Hubbard	Athens, Georgia
Linda Koons	San Diego, California
Charley Marshall	Yale University
Frances Marshall	Northwestern University
Mike Peirce	Trinity College, Dublin, Ireland
Ben Plouganou	El Paso, Texas
Robert I. Tilling	Palo Alto, California

Scott Rogers has done it again. Even in his new position at Osborne McGraw-Hill as Executive Editor, Scott continues to come up with the best ideas and books in the entire computer book industry. We just hope that the corporate folks at McGraw-Hill appreciate Scott as much as we do.

As good as he is, working with Scott would be impossible without the calming influence of his assistant, Daniela Dell'Orco. The soft-spoken Daniela can take the craziest of situations and make it all make sense. We don't even mind our Monday morning beatings for not getting more work done over the weekend, when Daniela is administering. Talking with Daniela on any day is a sheer delight.

Emily Rader, Project Editor Extraordinaire, faced down the daunting task of pulling this book together. This involved coordinating between us, the copy editor, the proofreader, the artists, and the production department. You might not realize it if you haven't thought about it, but producing a book like this is a long and complex process. Not only did Emily pull it all together, but she did it with style and unrivaled professionalism. Thanks, Emily.

At one time or another, everyone at Osborne had a hand in making this book the success that it is. Here are just a few of them:

Linda Medoff, Freelance Copy Editor
Pat Mannion, Freelance Proofreader
Deborah Wilson, Manufacturing and
    Production
Leslee Bassin and Rosemarie Ellis, Art
Peter Hancik and Richard Whitaker,
    Production
Cindy Brown, Managing Editor
Jodi Forrest, Sales Administrative Manager
Polly Fusco, Special Sales
Anne Ellingsen, Public Relations

We thank each of them and want them all to know how much we appreciate their hard work and professionalism.

Finally, we couldn't have produced this book without superior connections to the Internet. A special thanks goes to CTS Network Services, the most awesome Internet service provider this side of the prime meridian. The hardworking folks at CTS made sure the Net would be there while we spent tremendous amounts of time researching more pages than we could possibly include in the book.

—Rick Stout

—Morgan Davis

AEROSPACE AND SPACE TECHNOLOGY

Apollo 11

"That's one small step for man, one giant leap for mankind." Experience the drama and awe of the first manned lunar mission with images, movies, recollections of the astronauts, key White House documents, and a retrospective analysis.

URL:
http://www.gsfc.nasa.gov/hqpao/apollo_11.html

Area 51/Groom Lake

This is a page only aliens could have created. For a combination of Air Force technology, secrecy, and whimsy, beam your browser over to Area 51. This gigantic page is devoted to a U.S. military installation in the Nevada desert that officially does not exist. For accounts of top-secret aircraft, UFO sightings and contact, and general conspiracy theories, Groom Lake has it all.

URL:
http://www.cris.com/~Psyspy/area51/

Conspiracy
Your government is secretly housing, studying, and flying extraterrestrial aircraft, er, spaceships. It's true! Groom Lake is what they call the secret base where they keep these crafts and other top-secret projects. Area 51 is where they keep the most secret of them—the alien vehicles. Read up on what your government is doing behind your back on the Area 51 / Groom Lake page.

Canadian Space Guide

Canucks in space! Blastoff with the Canadian Space Guide for news items on space organizations, businesses, conferences, events, job index, space education resources, and briefs of projects covering aspects of the Canadian space program. Out of this world, eh?

URL:
http://www.conveyor.com/space.html

Computers in Outer Space

This is really a features and specs page for the products of an aerospace company, but it's fascinating. Read about some of the standard characteristics computers must have to travel in space. These computers employ fully redundant designs, have double-sided component cards, are radiation-proof, and have many more features that you may one day find on your desktop.

URL:
http://www.isso.org/Industry/Silver/spaceFlight.html

Johnson Space Center General Aviation Web

There's lots to keep enthusiasts high on aviation at this great resource: piloting tips and tricks; newsgroups and FAQs; information on the Civil Air Patrol, aircraft simulators, FAA, and NASA; and weather links. Keep your computer flying, too, with software for planning flights and private pilot practice. There's even a section on model airplanes.

URL:
http://aviation.jsc.nasa.gov

Langley Research Center

If computers could fly, they'd have a hangar at the Langley Research Center. Dedicated to high-performance aerospace technology, the center's page is filled with studies and projects that assist researchers in developing space vehicles and missions.

URL:
http://www.larc.nasa.gov/larc.cgi

NASA Newsroom Information System

In the wake of the Challenger disaster, Hubble trouble, delayed shuttle launches, and ever-decreasing federal funding, NASA needs all the good press it can get. So it has launched its own online newsroom for press releases and announcements on current and future mission events, research, activities, status reports, press kits, fact sheets, official statements, and required reports submitted to Congress and the president. Search a database by keyword and subject term.

URL:
http://www.gsfc.nasa.gov/hqpao/newsroom.html

A B C D E F G H I J K L M N O P Q R S T U V W X Y Z

NASA's Dynamics and Control Branch

The Wright brothers had no idea it would come to this. Check out this page from NASA's Dynamics and Control Branch, which concerns itself with high-speed aircraft control systems, mathematical modeling, and other research in high-performance aircraft technology.

URL:
   http://agcbwww.larc.nasa.gov/DCBVision.html

Numerical Aerodynamic Simulation

Why build complex models and huge wind tunnels when you can simulate them in the computer and save millions of dollars, and possibly the lives of countless test pilots? The NAS program's research in computational fluid dynamics and related aerospace disciplines provide a great data tool for NASA's Office of Aeronautics.

URL:
   http://www.nas.nasa.gov/NAS/GenInfo/

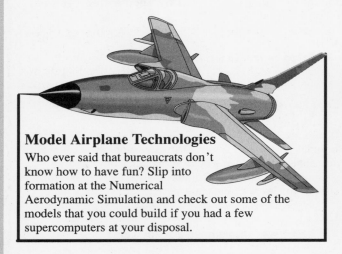

Model Airplane Technologies
Who ever said that bureaucrats don't know how to have fun? Slip into formation at the Numerical Aerodynamic Simulation and check out some of the models that you could build if you had a few supercomputers at your disposal.

Planetary Image Finders

Looking for just the right image of Mars? . . . or the perfect close-up of Phobos, Deimos, or Epimetheus? Look no further than NASA's Planetary Image Finders. This site features images from telescopes, satellites, the space shuttle, and the Viking orbiters. The linked National Space Science Data Center photo gallery also offers numerous images of planets, moons, solar systems, asteroids, and comets.

URL:
   http://ic-www.arc.nasa.gov/fia/projects/
   bayes-group/Atlas/

Planetary Rings Node

A cooperative project between NASA's Ames Research Center and Stanford's Center for Radar Astronomy, the Planetary Rings Node archives and distributes scientific data sets relating to planetary ring systems. Featured planets include Jupiter, Saturn, Uranus, and Neptune. Don't miss the stunning images from Voyager or the animated sequences.

URL:
   http://ringside.arc.nasa.gov

Russian Space Research Institute

IKI is the leading organization of the Russian Academy of Sciences in investigating outer space, the solar system, the planets, and other objects of the universe. Access their High Energy Astrophysics Department, the Space Monitoring Information Support Laboratory, archives of solar system data, solar-terrestrial data, plus information on current and future space missions. They also offer links to "our partners" home pages, both of which are NASA!

URL:
   http://www.iki.rssi.ru

AGRICULTURE

Agricultural Engineering at UIUC

Who said only supercomputers and Web browsers come out of the University of Illinois at Urbana-Champaign? The UIUC trains aggies, too! This page describes the agricultural-engineering programs offered here, including course catalogs, curriculums, schedules, faculty, and information about the campus and facilities.

URL:
   http://www.age.uiuc.edu

Agricultural Engineering—Aggie Style

If it's agricultural technology you're looking for, why go anywhere but to one of the preeminent agricultural schools in the world. Texas A&M University offers a wealth of information on its current research projects. Among these projects are volumes of information on topics such as food engineering, bioacoustics, agricultural and environmental safety, and using neural networking technology to solve food and agriculture problems.

URL:
   http://ageninfo.tamu.edu

Agriculture Online

A hayseed's haven in electronic format, *Agriculture Online* is a Web magazine with some of the best agricultural information sources in the world. Among dozens of resources, you'll find news on farming, weather, market research, and more.

URL:
    http://www.agriculture.com

AgWorld Reports

Pages of reports and statistics on world agriculture await you on the AgWorld Gopher server from the University of Michigan.

URL:
    gopher://una.hh.lib.umich.edu/11/ebb/agworld

Alternative Farming

If you think alternative farming is today's hip version of the 4-H club—where midwestern kids wear baggy overalls with the crotches at the ankles, listen to grunge bands like *The Dead Beets*, and attend raves out in the middle of cornfields at night—you're way off. Please visit the Alternative Farming Systems Information Center before you make a fool of yourself in front of a young person.

Alternative Farming Systems Information Center

One of ten information centers at the National Agricultural Library of the USDA, the Alternative Farming Systems Information Center includes links to sustainable agriculture-related sites and documents on the Net.

URL:
    http://www.inform.umd.edu:8080/EdRes/
    Topic/AgrEnv/AltFarm

The Net is humanity's greatest achievement.

American Society of Agricultural Engineers

The ASAE is a professional and technical organization of members worldwide who are interested in engineering knowledge and technology as it relates to food and agricultural industries. This page is provided by the University of Georgia's student branch of the ASAE. It offers information about the projects and engineering programs at the University of Georgia.

URL:
    http://ice.bae.uga.edu/asae/asae.html

Biological and Agricultural Engineering at NCSU

The Department of Biological and Agricultural Engineering at North Carolina State University prides itself in providing education, research, and services to the community. Take a tour of the department and see what NCSU has to offer you.

URL:
    http://www.bae.ncsu.edu/bae/

Biological and Agricultural Engineering at UGA

There's a lot of agricultural activity in the southern portion of the U.S., so it's no surprise that the faculty and students at the University of Georgia have much to share.

URL:
    http://ice.bae.uga.edu

Biotechnology Information Centers

This page offers links to many resources involving agricultural biotechnology. Categories include Gophers, publications, newsletters, patents, documents, and software.

URL:
    http://www.inform.umd.edu:8080/EdRes/Topic/
    AgrEnv/Biotech

Breeds of Livestock

Stop horsing around and check out Oklahoma State University's extensive reference work containing pages of information on various breeds of livestock, complete with photos and descriptions. The Animal Science department addresses the business of the production of beef cattle, dairy cattle, horses, poultry, sheep, and swine.

URL:
    http://www.okstate.edu/~animsci/breeds

A B C D E F G H I J K L M N O P Q R S T U V W X Y Z

Emus and Ratites

Is it an eclectic opportunist fad, or a viable farming alternative? Don't bury your head in the sand—you make the call. The rapidly growing industry of ratite farmers is attempting to get the USDA to place their seal of inspection on the meat of the ostrich, emu, and rhea, birds that have sold over the Internet as pets for more than $10,000 a beak. However, for meat production, the economics may favor cattle production. Who knows? Maybe tomorrow you'll pull into the drive-through and order an Emu McMuffin and a side-order of giblets.

URL:

http://www.axs.net/~ektor/ratite.html

Global Agricultural Biotechnology Association

GABA is a nonprofit organization dedicated to promoting agricultural biotechnology around the world through international information exchange.

URL:

http://www.lights.com/gaba/

How to Prune Trees

Do you have a sick tree or perhaps one that doesn't bear much fruit? Perhaps you should visit the USDA Forest Services' How to Prune Trees pages. You will learn everything you need to know about when, why, and how to prune your trees.

URL:

http://willow.ncfes.umn.edu/HT_prune/
prun002.htm

Integrated Pest Management

If your garden is bugged by pests and weeds, turn to the Nuclear Physics section in this book for solutions. Just kidding. Better for both you and the environment is the Center for Integrated Pest Management. Whether you grow wheat for government subsidies or just a few tomato plants on your window sill, check out this page to keep on growin'.

URL:

http://ipmwww.ncsu.edu/cipm/
Virtual_Center.html

Livestock Virtual Library

Where's the beefalo? It's right here on this frequently updated library on everything you ever wanted to know about horses, swine, and other livestock. It also provides a list of academic resources, departments, centers, fairs, and livestock exhibits.

URL:

http://www.okstate.edu/~animsci/library

Yeeehaaa! Here's more skinny on farm critters then you kin shake a stick at. Mosey on down to the Livestock Virtual Library and lasso yerself some real data from one of them newfangled computer databases.

Master Gardener's Notebook

Thinking about starting a garden or improving the one that you have? Browse through the Master Gardener's Notebook offered by the University of Florida. This large pamphlet touches on many of the important subjects in home gardening—including landscaping, fertilizers, fruit trees, vegetables, and even houseplants and flowers.

URL:

http://hammock.ifas.ufl.edu/txt/fairs/mg/
19977.html

National Pork Producers Council

As one of the largest livestock commodity organizations with over 85,000 producer members, the NPPC is in a unique position to offer the most in-depth information resource on pigs and pork in the world. The NPPC's web site offers insights into the pork industry, information on how the pork industry has changed throughout its history, and a hog trivia page with tidbits such as the origins of terms like "barbecue," "pork barrel politics," and "going whole hog."

URL:

http://www.nppc.org/porkindustry.html

PIGVISION

This page offers information about various research programs on agriculture and animal husbandry related to pigs. PIGVISION is an artistic as well as scientific collaboration that promotes education about pigs. Included on this page are literary resources; documentation on porcine research projects; images of artwork; and links to other art, science, and agriculture sites.

URL:

http://toolshed.artschool.utas.edu.au/
PigVision/pigvision.html

Purdue University Agronomy Department

Get the dirt on PU's Agronomy page, with updated lists of literary resources and activities in soil, crop, and environmental sciences. This page also offers departmental information on courses and research, a student and staff directory, and campus maps.

URL:
http://info.aes.purdue.edu/agronomy/agny.html

Terminal Market Prices

The rapidly changing price of fruits and vegetables keeps you up at night. So the Terminal Market Prices page was created by the University of Florida to help keep bean counters current with the price of beans. Now you can sleep soundly.

URL:
http://gnv.ifas.ufl.edu/~MARKETING/
tmpcmenu.html

U.S. Department of Agriculture

When the evening chores are done, and the kids are in the sack, take a load off and read about how this gigantic department has helped you today. Proudly listed first among the jobs they do is "Help American farmers and ranchers earn a good living." Read about the projects, missions, and agencies of the USDA. If you like, send a note to the secretary himself. (His email address is **agsec@usda.gov**.)

URLs:
http://web.fie.com/web/fed/agr/
http://www.usda.gov

USDA Research Database

The USDA has accumulated thousands of project summaries relating to human nutrition, nutrition research projects, progress reports, and lists of recent publications coming out of research. This is a searchable database. In fact, that's the only way to use it.

URL:
http://medoc.gdb.org/best/stc/usda-best.html

Anyone can browse the Web.

Weeds of the World

Researchers at Oxford University are kicking around the idea of publishing on the Web the volumes of information contained in the book *A Geographical Atlas to World Weeds*. The resulting database will be searchable by country, plant, genus, or family. They've already done extensive work on this project. If you think such a resource would be valuable, check out these pages and join the ongoing discussions of this project.

URL:
http://ifs.plants.ox.ac.uk/wwd/wwd.htm

Weeds
A weed is any plant you don't want growing in a particular place. But what are weeds to you may be dinner to your neighbor! So before you mow that lawn or fire up the ol' Deere, check the World Weeds Database at the Weeds of the World project.

ANTHROPOLOGY

Aboriginal Studies Electronic Data Archive

ASEDA provides a service to researchers in the field of Aboriginal and Torres Strait Islander Studies. By accessing information in electronic form, researchers easily locate references not available by keyword searching of traditional catalogs. And that's what the Net is all about.

URL:
http://coombs.anu.edu.au/SpecialProj/ASEDA/
ASEDA.html

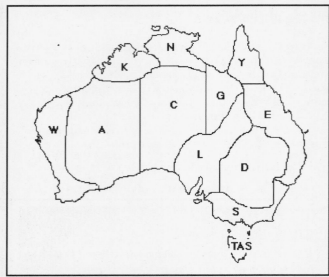

A clickable map helps you navigate regions of Australia in the Aboriginal Studies Electronic Data Archive.

Anthro-l

You may not receive mail from your relatives, but you can talk about your ancestors in Anthro-l, the mailing list for all things anthropological.

Mailing List:
Address: **listserv@ubvm.cc.buffalo.edu**
Body of Message: **subscribe anthro-l** *<your name>*

Anthropoetics

This electronic journal of Generative Anthropology contains essays and projects of scientists around the world. Find out why "the first word of human language was an aborted gesture of appropriation." This must be true. Some guy did that to me on the freeway on the way to work today.

URL:
http://www.humnet.ucla.edu/humnet/anthropoetics/home.html

Anthropology Information and Software

Dig into this FTP site for FAQs, articles, resources, software, and a brief essay on culture.

URL:
ftp://ftp.neosoft.com/pub/users/claird/sci.anthropology

Aquatic Ape Theory

Did humans descend from apes? Many people believe so. Of course, many people also have relatives they can point to as living proof. This page poses the question and theory, "Did humans evolve from aquatic apes?" The aquatic ape theory explains that many of our features can *only* be explained in the light of an aquatic stage of evolution. The author debates contrasting theories, such as the Creation, Savannah, and Mosaic theories. (Unfortunately, there is no mention of the Netscape theory.)

URL:
http://www.brad.ac.uk/~dmorgan/aat.html

Arctic Circle

The Arctic Circle home page presents a virtual classroom with photo exhibitions and audio and video recordings. Discussions on the history, culture, natural resources, social equity, and environmental justice are the central themes throughout.

URL:
http://www.lib.uconn.edu/ArcticCircle/

Center for Anthropology and Journalism

A resource center for anthropologists, students, journalists, and science writers, this page includes a worldwide listing of anthropologists, related materials, and a pointer to other online anthro resources.

URL:
http://pegasus.acs.ttu.edu/~wurlr/

Center for Social Anthropology and Computing

Discover multidisciplinary ventures in anthropology encompassing research in visual ethnography, marine biology, geography, history, and psychology.

URL:
http://lucy.ukc.ac.uk

Dead Man's Party

If it's fossils you're interested in, these groups are for you: lively discussions by living people about dead people, plants, and animals, and how they got that way. *'Dem bones, 'dem bones, if 'dey could only talk.*

URLs:
news:sci.anthropology
news:sci.anthropology.paleo

Fourth World Documentation Project

The Fourth World is defined as "Nations forcefully incorporated into states that maintain a distinct political culture but are internationally unrecognized." The mission of the Documentation Project and the Center for World Indigenous Studies is to gather information relating to the anthropological, social, political, strategic, economic, and human rights issues faced by Fourth World nations and make this information available to tribal governments, researchers, and organizations. These documents are collected and processed into electronic text for distribution on the Internet.

URL:
http://www.halcyon.com/FWDP/fwdp.html

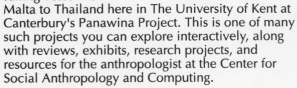

Need a Vacation?

How about a virtual journey through the circuits of the Center for Social Anthropology and Computing? Begin your 6,000-mile odyssey by sailing vessel from Malta to Thailand here in The University of Kent at Canterbury's Panawina Project. This is one of many such projects you can explore interactively, along with reviews, exhibits, research projects, and resources for the anthropologist at the Center for Social Anthropology and Computing.

Total recall is guaranteed.

Future Artifact

Using the Web as an interactive tool for an anthropology of the future, one consisting of multiple-media forms, is an intriguing concept not lost on this page. This vast collection of links to institutions, specialized fields, jobs, news, events, and more, is a great place to get started.

URL:
http://www.usc.edu/dept/v-lib/
anthropology.html

Indigenous Peoples of Mexico

Your gateway to Central America, this popular page contains a history of the people of Mexico and their poetry, songs, dances, native languages, and stories. A section on Chiapas and the current war and information on the indigenous nations of Mexico accompany this page.

URL:
http://kuhttp.cc.ukans.edu/~marc/geography/
latinam/mexico/mex_main.html

Images of artifacts are among the many images available at the University of Kansas Indigenous Peoples of Mexico page.

Native American information

Enter **soc.culture.native** in Usenet and discuss issues relating to native populations throughout the world. Other topics include education, sovereignty, religion, philosophy, human rights bulletins, and population reports on native indigenous peoples.

URL:
ftp://ftp.cit.cornell.edu/pub/special/
NativeProfs/usenet

A B C D E F G H I J K L M N O P Q R S T U V W X Y Z

NativeWeb

A home for all things Native American and beyond, including resources on the Net for native regional studies, culture, language, literature, journals, bibliographies, and organizations.

URL:

http://ukanaix.cc.ukans.edu/~marc/
native_main.html

Paleo Talk

Add your voice to the Paleo Talk page, a Web-based bulletin board system for the unmoderated discussion of Paleo Information Systems.

URL:

http://sunrae.uel.ac.uk/palaeo/talk/talk2.html

Social and Cultural Anthropology at Oxford

What anthropologist at Oxford isn't social? This organization of U.K. anthropological theses, abstracts, multimedia, and the Oxford Union list of periodicals is available to social anthropologists everywhere.

URL:

http://www.rsl.ox.ac.uk/isca/

Theoretical Anthropology

This is a newsletter devoted to anthropological theory and epistemology. It contains articles, discussions, announcements, symposia, workshops, round tables, reports, reviews, book announcements, and even gossip.

URL:

http://www.univie.ac.at/voelkerkunde/
theoretical-anthropology/

UCSB Student Projects

Check out the digs at UCSB! This handsome page introduces you to the University of California at Santa Barbara's programs, showcasing recent projects such as an interactive map of the coastline of Europe as it may have appeared 10,000 years ago, and a three-dimensional reconstruction of a Great Kiva, an architectural feature of many prehistoric Anasazi communities in the southwestern United States.

URL:

http://www.sscf.ucsb.edu/anth/

Yale Peabody Collection

The Yale Peabody Museum maintains this anthropology archive filled with photographs, field notes, correspondence, and manuscripts. Caribbean, Mesoamerican, and Intermediate Area archaeological collections are available online. You'll also find ethnological holdings from Southeast Asia, Oceania, and significant archaeological collections from the Old World.

URL:

gopher://george.peabody.yale.edu:70/11/main/
Anthropology

ARCHAEOLOGY

ABZU

ABZU is an experimental guide for the study and public presentation of the ancient Near East. Turn to ABZU for online indexes of library catalogs, journals, museum collections, projects, institutional affiliations, publishers, book dealers, and pointers to archaeological sites.

URL:

http://www.oi.uchicago.edu/OI/DEPT/RA/
ABZU/ABZU.HTML

Excerpt from the Net...

(from the Commission on Theoretical Anthropology)
In the long run of the history of anthropology, the science split into various subdisciplines, as for instance social anthropology, medical anthropology, visual anthropology, anthropology of religion, material culture studies, aesthetic anthropology, etc.

Furthermore, our science is going through different theoretical and methodological developments, questions, and focal points, aspects and orientations which are also related to the specific geographical regions where it is practiced.

The Anasazi

Who were the Anasazi, and why did they vanish? Many theories abound for their disappearance: overpopulation, starvation, internal warfare, even the plague. Explore the ruins and romance of their long-ago civilization in Arizona at this attractive page, featuring Wupatki, the Grand Canyon, and Montezuma's Castle.

URL:

http://www.primenet.com/~smudger/

Ancient City of Athens

Located at Indiana University, the Ancient City of Athens is a photographic archive of the archaeological and architectural remains of ancient Athens, Greece. It is intended primarily as a resource for students of classical languages, civilization, art, archaeology, and history who wish to take a virtual tour of excavated regions and monuments. But they'll let you in, too.

URL:

http://www.indiana.edu/~kglowack/Athens/
Athens.html

Ancient Metallurgy

If you just can't get enough early heavy-metal, move your axe over to The Crift Farm Project, a pre-sixteenth century tin smelting site. This University of Bradford page also includes links to The Ancient Metallurgy Research Group, and The Newstead Research Project, a study of the Roman fort of Trimontium near Newstead in the Borders region of southern Scotland.

URL:

http://www.brad.ac.uk/acad/archsci/
homepage.html

Ancient Monuments Laboratory

The Ancient Monuments Laboratory (AML) offers reports of investigations in archaeological science and conservation. Items of interest range from short notes to substantial papers on excavations, along with articles on conservation and geophysics.

URL:

http://robin.eng-h.gov.uk

Archaeological Dialogues

This semiannual language journal is published by researchers with innovative approaches to debating traditional issues in archaeology. Short papers, provocative statements, inquiring interviews, and critical reviews of recently published studies are included in each issue.

URL:

http://archweb.LeidenUniv.nl/ad/home_ad.html

Archaeological Fieldwork Server

If you're sifting for archaeological fieldwork, excavate this page to uncover opportunities available around the world. Browse postings submitted by those who have projects and positions for volunteers and paid workers in field schools and contract jobs.

URL:

http://durendal.cit.cornell.edu/TestPit.html

Excerpt from the Net...

(from the Archaeological Fieldwork Server)

Field School in the Bahamas

The LSU archaeology field school will be taking place in the Bahamas. The class will be excavating at two British Loyalists cotton plantations (1783-1834) on Crooked Island. Both plantations contain standing ruins, including planter houses, military barracks, slave cabins, manager's houses, and other outer buildings.

Students will earn six hours of credit for ANTH 2016 Field Methods in archaeology. This course is intended to introduce the student to basic excavation techniques. ANTH 4021 Advanced Field Methods in Archaeology is also available for undergraduates with prior archaeological field experience and graduate students.

A B C D E F G H I J K L M N O P Q R S T U V W X Y Z

Archaeology in Scotland

Weel, ye couldna found a better page of links to Scottish archaeology than the Royal Commission for Ancient and Historic Monuments. Put down the whiskey and shortbread, laddie, and check out the Glasgow University Archaeology Research Division, Edinburgh University's archaeological program, and the Council for Scottish Archaeology.

URL:
> http://www.gla.ac.uk/Acad/Archaeology/
> scotland/scotland.html

ArchNet

Dig here to reach the World Wide Web's Virtual Library for Archaeology. With a handsome visual interface, this page categorizes archaeological resources available on the Internet by geographic region and subject.

URL:
> http://spirit.lib.uconn.edu/ArchNet/ArchNet.html

Barbarians Dissertation

Marvelous Celtic objects adorn the home page of Constanze Witt's dissertation proposal in the McIntire Department of Art at the University of Virginia. The proposal examines the material evidence for the great stylistic change that took place in fifth-century Celtic art, the highly complex and distinctive cultures that produced it, and their relationships with the Mediterranean. It cites recent alternative anthropological reconstructions of the history of the period, and reevaluates historical interpretations of artistic, cultural, and stylistic change in Celtic Europe.

URL:
> http://faraday.clas.virginia.edu/~umw8f/
> Cze/HomePage.html

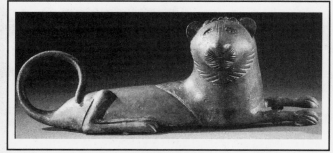

A Celtic lion from Constanze Witt's Barbarians dissertation at the University of Virginia

British Archaeology

Margaret Thatcher isn't the only relic in the U.K. Discover all things British and proper in the way of fieldwork opportunities, archaeological conferences, events, exhibits, courses, grants, and awards. This page also includes a notice board (what we Yanks call a *bulletin board*) and listings of new books on archaeology. (Just kidding, Maggie, really. John Major told us to write that. Honest.)

URLs:
> gopher://britac3.britac.ac.uk:70/11/cba

> http://britac3.britac.ac.uk/cba/

Careers in Archaeology

On your way to a career as an archaeologist? Stop here for answers to frequently asked questions, available jobs, educational requirements, college or university recommendations, general introductory books on archaeology, how to volunteer for a dig, and how to get more information on archaeology.

URL:
> http://www.museum.state.il.us/0/ismdepts/
> anthro/dlcfaq.html

Çatalhöyük

It's hard to recall what you did last week, let alone what humanity was doing 9,000 years ago. Çatalhöyük, the first urban center in the world, is perhaps the most important archaeological site in Turkey. Its wall paintings and sculptures open a direct window into ancient life. Enter the dig house and begin your own excavation now.

URL:
> http://club.eng.cam.ac.uk/~vsb1001/catal.html

Centre National de la Prehistoire

The National Center for Prehistory in France is the clearinghouse for all things Paleolithic in the southern Dordogne region of France, including the famous Lascaux Cave, and many other wonders.

Note: You need to be able to read French or enjoy the warmth and comfort of someone who can translate for you.

URL:
> http://dufy.aquarel.fr:8001/html/cnp.html

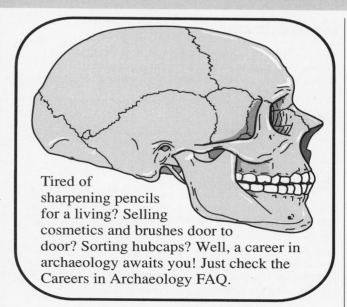

Tired of sharpening pencils for a living? Selling cosmetics and brushes door to door? Sorting hubcaps? Well, a career in archaeology awaits you! Just check the Careers in Archaeology FAQ.

Classics and Mediterranean Archaeology

A great place to continue your archaeological expedition on the Internet, with hundreds of links to projects, journals, exhibits, documents, and images that pertain to classical and Mediterranean archaeology.

URL:
    http://rome.classics.lsa.umich.edu/welcome.html

Computing in Archaeology

Mobilize human and technical resources for archaeological reasons, and you've defined the goal of the International Association of Computing in Archaeology. Featured applications include Geographic Information Systems (GIS) and remote sensing, multimedia, image processing, computer graphics, virtual reality, and networking in archaeology.

URL:
    http://www.cineca.it/visclab/aiace/aiace.htm

Dead Sea Scrolls

This online exhibit from the Library of Congress includes a selection from the ancient Dead Sea Scrolls, which have been the subject of intense public interest. The presentation describes the historical context of the scrolls and the Qumran community from whence they may have originated; it also relates the story of their discovery 2,000 years later.

URL:
    http://sunsite.unc.edu/expo/
    deadsea.scrolls.exhibit/intro.html

GIS and Remote Sensing

When digging in the dirt isn't panning out, archaeologists take to the skies. The integration of advanced airborne and satellite remote sensing data within the context of Geographic Information Systems (GIS) has significant applications for regional archaeological research. For nearly two decades, an American team has been conducting GIS research in the Arroux River Valley region of Burgundy, France. Find the results of their study on the Web.

URL:
    http://deathstar.rutgers.edu/projects/france/
    france.html

Great Zimbabwe

Archaeologists and historians believe that from the thirteenth to fifteenth centuries, Great Zimbabwe was the capital for a large area in southern Africa. A wealth of information is available on this important site, plus several photographs of the unique architecture of this culture.

URL:
    http://wn.apc.org/mediatech/VRZ10011.HTM

Dead Sea Scrolls

Since their discovery nearly half a century ago, the scrolls and the identity of the nearby settlement have been the object of great scholarly and public interest, as well as heated debate and controversy. To paraphrase Mark Twain, 'Everybody talks about the Dead Sea Scrolls, but no one ever does anything about them.' Ever since their discovery in 1947, they have been squirreled away to be studied by a select few who have not always been forthcoming with timely information. That has changed in the last several years, thanks in part to the Internet and the Huntington Museum. Visit this online exhibit to gain an overview of the scrolls, their historical significance, and a taste of the controversy surrounding them.

A B C D E F G H I J K L M N O P Q R S T U V W X Y Z

Indian Archaeology Repository

Southern Utah University curates Anasazi and Fremont Indian artifacts from its Archaeology Field School, the BLM, the Forest Service, and other groups from the Southwestern Utah region. The server has information and images regarding its Field School and explorations of a Pueblo II site on the Utah-Arizona border. There are links to the Utah State Archaeological Society; various scholarly publications; a bibliography; and, of course, an online gift shop.

URL:
    http://www.suu.edu/WebPages/MuseumGaller/
    archaeol.html

Lahav Research Project

Encounter hundreds of photos of figurine objects collected from the dig site in Lahav, Israel, in the Digmaster Figurine Database, part of the Lahav Research Project sponsored by the Cobb Institute of Archaeology.

URL:
    http://www.cobb.msstate.edu

Leptiminus Archaeological Project

Leptiminus was an important port town in Tunisia that flourished under Roman rule. The earliest remains date from the Punic period; however, it also functioned as an Arab, Byzantine, and Vandal settlement. As a result of Vandal activities, a decline in property values eventually lead the town into ruins. See for yourself.

URL:
    http://rome.classics.lsa.umich.edu/
    projects/lepti/lepti.html

Little Salt Spring

An exceptionally well-documented web page for underwater archaeology, the Little Salt Spring in southwestern Florida is a unique type of spring-fed sinkhole that may have served as an oasis in the peninsula during early prehistoric time.

URL:
    http://www.rsmas.miami.edu/groups/lss.html

If you're not on the Web, you don't exist.

Mayan Astronomy

The Maya were fascinating people, and this page is devoted not just to their astronomy, but also to their mathematics, calendar system, and writing. Also contains Maya-related web links.

URL:
    http://www.astro.uva.nl/michielb/maya/
    astro.html

Megaliths of Morbihan

Sure, it may sound like a new level in Doom, but do you know a dolmen from a menhir? You will after exploring the Megaliths of Morbihan. Located in the Brittany region near Carnac, France, these 6,500-year-old amazing rock structures are believed to be of a religious or ceremonial nature. Excellent black-and-white photographs of the stones accompany the exhibit.

URL:
    http://www.om.com.au/mkzdk/carnac/
    guiden.html

Megiddo Excavations

An ancient city in Israel of massive fortifications, Megiddo's impressive architecture, engineering, lavish palaces, and important temples are captured here in words and images.

URL:
    http://cac.psu.edu/~rlg7/hist/proj/megiddo.html

Quick! Before it's too late and the world erupts in a ball of roiling fire, check out Megiddo, the site of the New Testament's Armageddon. Megiddo is perhaps the richest archaeological site in Israel, and one of the most important in the entire Near East. Megiddo is on a bottleneck of the land route from Egypt to points north, including Phoenicia, Anatolia, Syria, and Mesopotamia. From the Neolithic (5000 B.C.) to the Persian era (500 B.C.), it governed international trade. So when the locusts invade your home during the last days, at least you'll know where it all started.

Minnesota Archaeology Week

Since we're all dying to know what's happening in the fine state of Minnesota, this compendium of archaeological happenings includes a newsletter, graphics, and a list of dig-related sources. If you dig deep enough, perhaps you'll reach some interesting petroglyphs.

URL:
> http://www.umn.edu/nlhome/g075/mnshpo/
> arcweek/index1.html

National Archaeological Database

Watch where you step. The NADB inventories over 120,000 archeological investigations, guidance on compliance with the Native American Grave Protection and Repatriation Act, and maps displaying archaeological and environmental data at the state and county level.

URLs:
> telnet:cast.uark.edu
>
> http://cast.uark.edu/d.cast/nadb.html

Nubian Wallpaintings

The Handlist of Nubian Wallpaintings is a documentation system that serves up details on the wallpaintings from Christian Nubia. It includes the paintings saved during the UNESCO campaign for the salvage of Nubian monuments, those still *in situ*, and lost paintings for which any documentation is available.

URL:
> http://wwwlet.LeidenUniv.nl/www.let.data/
> Arthis/Nubian/intro.htm

Old World Archaeology Newsletter

At night in your tent, after a long hot dig, curl up with your laptop and this URL to read the *Old World Archaeology Newsletter*. You'll enjoy its contents, including information on research reports, grants, student and summer projects, fieldwork opportunities, conferences, lectures, overseas antiquities, administration news, new publications, and articles.

URL:
> gopher://www.wesleyan.edu/11/Classics

Online Archaeology

Is our understanding of Neolithic architecture in mainland Britain the result of convenient speculation? Does the scant evidence of permanent dwellings indicate that Neolithic age people were nomadic, or that they simply had no desire to build stone flats? This and other speculative theories of archaeology are challenged in the University of Southampton's Online Archaeology page.

URL:
> http://avebury.arch.soton.ac.uk/Journal/
> journal.html

Oriental Institute at the University of Chicago

The Oriental Institute is a museum and research organization devoted to the study of the ancient Near East. Founded in 1919 by James Henry Breasted, the Institute is part of the University of Chicago. It is an internationally recognized pioneer in the archaeology, philology, and history of early Near Eastern civilizations. Here you'll find excellent descriptions and photographs of Near Eastern treasures that the University and Institute have made available online.

URLs:
> ftp://oi.uchicago.edu/pub
>
> http://www-oi.uchicago.edu/OI/

Mailing List:
> Address: **majordomo@oi.uchicago.edu**
> Body of Message: **subscribe ane-digest**

King "Tut" Tutankhamen rearly lost his head when he found out how easy it is to explore the Ancient Near East on the World Wide Web.

Palace of Ashurnasirpal II

Sit back with an urn of popcorn and enjoy QuickTime and MPEG movie clips of a 3-D animated fly-through of the Palace of Ashurnasirpal. The video was filmed by Information Systems and Computing at the University of Pennsylvania.

URL:

http://ccat.sas.upenn.edu/arth/asrnsrpl.html

Palaeoanthropology

This sophisticated, eye-popping page originates down under in New England. Australia, that is. Job offerings, brochures, course listings, available fieldwork opportunities, Museum of Antiquities from around the world, departmental publications, seminars, newsletters, software, plus links to other related servers make this a great expedition for anyone interested in Australian archaeology.

URL:

http://www.une.edu.au/~Arch/ArchHome.html

Palaeolithic Cave Paintings at Vallon-Pont-d'Arc

Today we peck at keyboards and create HTML. But 20,000 years ago, the World Wide Web had humble beginnings on cave walls, such as this site recently discovered in France. Along with the Lascaux Cave in the Dordogne region, Vallon contains several hundred meters of galleries depicting a particularly large and unusual variety of animals, symbols, panels filled with dots, and stenciled hands of the humans who created these images. An inspirational page of how individuals documented and celebrated life thousands of years before the Internet renewed the enthusiasm.

Note: English and French versions available.

URLs:

http://www.culture.fr/culture/gvpda-d.htm
http://www.culture.fr/culture/gvpda-en.htm

Papyrology

The word "paper" comes from the Greek *papuros* and the Latin plant classification *Cyperus papyrus*. All that from a tall, aquatic sedge whose pith and stems were crushed, dried, and spread out to form one of the most enduring writing materials in history. The University of Michigan has an impressive online collection here. Read all about it!

URL:

http://www.umich.edu/~jmucci/papyrology/
home.html

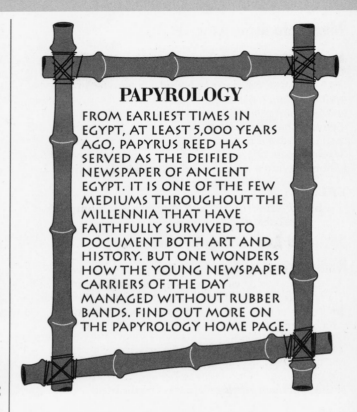

PAPYROLOGY

FROM EARLIEST TIMES IN EGYPT, AT LEAST 5,000 YEARS AGO, PAPYRUS REED HAS SERVED AS THE DEIFIED NEWSPAPER OF ANCIENT EGYPT. IT IS ONE OF THE FEW MEDIUMS THROUGHOUT THE MILLENNIA THAT HAVE FAITHFULLY SURVIVED TO DOCUMENT BOTH ART AND HISTORY. BUT ONE WONDERS HOW THE YOUNG NEWSPAPER CARRIERS OF THE DAY MANAGED WITHOUT RUBBER BANDS. FIND OUT MORE ON THE PAPYROLOGY HOME PAGE.

Petra, the Drama of History

One of the most fascinating archaeological destinations in the world is the ancient site of Petra in Jordan. Dubbed "the rose-red city, half as old as time," Petra's beautiful buildings were carved out of solid rock 2,000 years ago. Save an airline ticket and a camel ride by sojourning to this richly illustrated and well-written page.

URL:

http://www.mit.edu:8001/activities/jordanians/
jordan/petra.html

Polish Farming

Research in Oslonki, Poland, focuses on the study of the earliest farmers of the North European Plain. Excavations by a team of Polish and American archaeologists reveal a large Neolithic village about 4000 B.C., predating the well-publicized body of a prehistoric man found in the Italo-Austrian Alps in 1991 by nearly 1,000 years. This page features excellent documentation on archaeological methodology, subsistence strategy, and landscape theory.

URL:

http://www.princeton.edu/~bogucki/
oslonki.html

Pompeii Forum Project

A funny thing happened on the way to the forum . . . the Pompeii Forum. The urban center of any Roman town, the forum was the place to gather for religious, civic, and commercial institutions. This collaborative venture focuses on the urban center of Pompeii. Online discussions include documentation of standing remains, archaeological analysis, and interpretation of the developments at Pompeii in the broader context of urban history.

URL:
> http://jefferson.village.virginia.edu/pompeii/
> page-1.html

Radiocarbon Dating

Archaeologists will date any old thing, since their lives are often in ruins. For all aficionados of Carbon 14, *Radiocarbon* is the primary international journal for research articles, date lists, radioisotopes, and more. You'll also find information on radiocarbon-related resources, labs, World Wide Web servers, software, databases, and helpful dating techniques from skilled—though perhaps lonely—archaeologists.

URL:
> http://packrat.aml.arizona.edu

Roman Archaeology

A great starting point for classical archaeological interactive studies, the ROMARCH page presents resources on the art and archaeology of Italy and the Roman provinces, from circa 1000 B.C. to A.D. 600. ROMARCH is also an Internet mailing list sponsored by the Interdepartmental Program in Classical Art and Archaeology (IPCAA) at the University of Michigan.

Mailing List:
> Address: **majordomo@rome.classics.lsa.umich.edu**
> Body of Message: **subscribe romarch**

URL:
> http://www.umich.edu/~pfoss/ROMARCH.html

Want to see some cool resources? Check out Internet Resources.

Ruins at Rione Terra in Pozzuoli

When in Pozzuoli, do as the Romans do—build extensive ruins. (Well, they were perfectly fine buildings to start with, of course, until time came along and knocked them down.) Nevertheless, the extensive ruins at this town near Naples are based on Roman civic layout, which sealed their fate as an inevitable pile of marble pillars, crumbling coliseums, and broken statues. Excellent historical documentation of temples and buildings, along with many photographs of recovered artifacts, provide keys to understanding everyday life 2,000 years ago in this Italian city.

URL:
> http://www.mimesys.iunet.it/RioneTerra/
> rioneterraing.html

Society for American Archaeology Bulletin

The *SAA Bulletin* is for professional archaeologists only. Okay, so they'll let anyone browse their pages. With five issues yearly, the *SAA Bulletin* publishes contributions from field archaeologists from around the world. Even a budding Indiana Jones can uncover articles on Hopi oral history, preservation techniques, debates on training directions for future archaeologists, and more.

URL:
> http://www.sscf.ucsb.edu/SAABulletin/

Southwestern Archaeology

¡Que bueno! This Arizona State University clearinghouse for Southwest archaeological activities includes a bulletin board on topics related to the Southwest and Mexico, plus information about the Pecos Conference.

URL:
> http://seamonkey.ed.asu.edu/swa/

World Archaeological Congress

Do we really learn from history? The World Archaeological Congress (WAC) hopes so. WAC is an international forum for those concerned with recognizing the historical, social, and political context of archaeology. WAC strives to make archaeological studies relevant to the wider community.

URL:
> http://avebury.arch.soton.ac.uk/wac/

A B C D E F G H I J K L M N O P Q R S T U V W X Y Z

Southwestern Archaeology

Archaeology is one of those things you just have to love to do. How else could anyone put up with all the dirt, sweat, heat, and weather? Not to mention the snakes, tarantulas, and marauding Nazis out to get your artifacts. Nevertheless, these fearless adventurers continue to compile information and evidence on ancient civilizations, and they've put together a web site where you can read about their adventures in the southwest of the U.S. Come check out the digs at the Southwestern Archaeology Open Forum.

World of the Vikings

The World of the Vikings project, an archive of thousands of pictures and sounds, is a unique multimedia resource. The clearinghouse for everything *Viking* on the Net, the descriptive passages of text and interactive CD-ROM online tell the rest of the story.

URL:
http://www.demon.co.uk/history/vikings/vikhome.html

ARCHITECTURE

AEC ACCESS

The Architectural, Engineering, and Construction communities server contains a wealth of information of interest to architectural firms and private individuals. Find current data on public forums, news, publications, discussion areas, and an image archive. You also have access to databases on urban planning, building, and interior and landscape architecture.

URL:
http://www.interlog.com/~bhewlitt/

Alvar Aalto Museum

Finland is an amazing country full of sunlight, snow, saunas, and serendipity. Among its population are artists, composers, designers, and architects; the foremost of the latter is Alvar Aalto—known for the museums he designed. Find a wealth of information on this brilliant artist here online.

URL:
http://jkl21.jkl.fi/aalto/

American Institute of Architects

The AIA maintains an appropriately designed web page offering selections on careers in architecture, the art and science of architecture, K–12 classroom resources, a link to the American Architectural Foundation, and public interest features such as how to select an architect.

URL:
http://www.aia.org

American Institute of Architecture Students

The AIAS is a national organization for architectural students. This, the University of Washington chapter, offers UW architecture students representation at the University and the national organization. Read up on what the current crop of architectural students are up to. They solicit input from anyone.

URL:
http://www.caup.washington.edu/html/arch/people/orgns/AIAS-UW/

Architect and Interior Designer Index

Search for architects and interior designers using this WorldNet online database.

URL:
http://www.goworldnet.com/intdib.htm

Architect's Catalog

Just how much weight should that load-bearing beam take? This server's architectural catalog contains downloadable files for construction product manufacturers' specifications regarding building materials and features.

URL:
http://ideanet1.ideanet.com/~arcat/

Architectural Archives

The architectural archives of the University of Pennsylvania preserve the work of more than 250 designers from the eighteenth century to the present.

URL:

http://dolphin.upenn.edu/~gsfa/archives/archives.html

Architectural Visualization

There are a wide variety of documented in-house design, construction, and visualization projects at this site at the University of Aukland, New Zealand. Included are virtual studios, student projects, Eco House, a model of Ti House, the Bayswater Marina Studio, and a still and flyby rendering of the Aukland campus.

URL:

http://archpropplan.auckland.ac.nz/Archivis/archivis.html

Architecture and Urbanism

This page contains an abstract on the changing roles and challenges facing architects as we approach the new millennium. The page also offers the complete text of a lecture entitled *Responding to Change—Architecture and Urbanism*. The author solicits comments and suggestions.

URL:

http://www.austria.eu.net/gv95/HuwThomasText.html

Architecture Virtual Library

Build it right the first time by checking out this page offered by the University of Toronto. Search this compendium of online architectural information for articles, jobs, newsgroups, construction projects, and models—plus upcoming conference, events, and collaborative projects.

URL:

http://www.clr.toronto.edu:1080/VIRTUALLIB/arch.html

Having trouble sleeping? See Business Payment Systems.

Arcosanti

Begun in 1970, Arcosanti is an ongoing architectural experiment 70 miles north of Phoenix, Arizona, that combines architecture and ecology as an integral process to produce new urban habitats. The web site features areas regarding the philosophy of the project, updates on its progress, information on workshops and seminars, and photographs of Arcosanti.

URL:

http://www.getnet.com/~nkoren/arcosanti/

Arcosanti
Not to be outdone by the Biosphere projects, Paolo Soleri created the Arcosanti Project. Like the Biosphere projects, Arcosanti is in Arizona. (What is it about Arizona that makes people want to design ecological habitats there? Wouldn't Minnesota be better?) But the Arcosanti complex serves as a research and study center for the social, economic, and ecological implications of its architectural framework. (Palms up, now. "Ommmmmmm.")

Art and Architecture Archive

If you think art and architecture go together, join others here who agree with you. This page contains links to many texts and graphics that explore the relationship between art and architecture.

URL:

http://english-www.hss.cmu.edu/Art.html

Art Nouveau Artists

Photos and excellent text accompany this page devoted to the designers of the Paris Metro, La Samaritaine, the Amsterdam stock exchange, and the Glasgow School of Art. European Art Nouveau architects are profiled, with encouragement for other contributors to add their favorite nouveau architects. C'est cool.

URL:

http://www-stud.enst.fr/~derville/AN/authors.html

A B C D E F G H I J K L M N O P Q R S T U V W X Y Z

ArtServe

This massive Australian National University server offers access to over 16,000 images. The many gigabytes of data chronicle the history of art and architecture from the Mediterranean Basin.

URL:

http://rubens.anu.edu.au

Build a Solar House

Here is extensive information on the building of a solar home in Maine. Everything you would want to know about building a solar home is included—from optimal land surveying and equipment to efficient architecture.

URL:

http://solstice.crest.org/renewables/wlord/

Building Products Library

The AEC InfoCenter offers this online catalog of building products for architects, engineers, or anyone interested. The catalog is organized by category—for example, Appliances, Concrete, Equipment, Electrical, Metals, Plastics, Wood, and so on. These category links lead to lists of links to companies on the Web that produce these products. For example, under "Tile, Granite and Marble," links to three companies are listed.

URL:

http://www.aec-info.com/~aec/bpl.htm

Christo

Buildings are just buildings until Christo decides to wrap them, and then they become *events*. Take a long look at what he's accomplished with that creepy old Berlin State Office structure, the Reichstag, in a series of stills and MPEG movies. Wrap, wrap, wrap—they call him the wrapper. Little known rumor: Christo was fired from his day job at Macy's in the gift wrap section for wrapping a swing shift manager, four customers, and the entire sporting goods department.

URL:

http://www.cs.tu-berlin.de/~phade/christo/
christo.html

Classical Architecture Database

This is a worldwide architectural database searchable by country, date, feature, style, and title of work. Try searching all countries to find those with ancient aqueducts.

URL:

http://rubens.anu.edu.au/
architecture_form.html

Cyclorama Building

The Cyclorama Building, located on Franklin Street in Buffalo's Theater District, dates back to 1888 and, contrary to popular belief, was not used for bicycle races. It's difficult to imagine the excitement that the word *panorama* engendered among the public a century ago, but in the pre-radio, TV, movie, and Net days, any usually made-up word ending in "rama" was enough to charge-up crowds from miles around. And the panorama in the Buffalo, N.Y. Cyclorama was the cat's meow. Over a thousand Buffalonians a day packed the round building to view its 360-degree murals—first, Jerusalem on the day of Christ's crucifixion, and later, the battle of Gettysburg painted in the round. Like most fads, however, the wonderama of panoramas faded away, and the city was left with a big, round building. Scheduled twice for the wrecking ball, the saga of the Cyclorama has a happy ending. See and learn more about its architecture, history, and resurrection from some of its new, exuberant tenants at this excellent commercial web page.

URL:

http://www.grasmick.com/ourhome.htm

d-c quarterly

The d-c quarterly, an Austrian publication, features material from key Austrian design architecture and real-estate projects. Here, you'll find models, sketches, information about CAD and rendering systems, videos, plans, photos, and even some interactive walk-throughs.

URL:

http://www.ping.at/users/dc.co.at/

De Architectura

De Architectura is dedicated to collecting and dispersing European architectural research and information.

Note: Much of this information is in Italian.

URL:

http://www.dea.polimi.it/dea/dea.htm

ECO Design and Landscaping

If you thought that grass huts and thatched roofs were passé, you're out of touch! Visit the sustainable architecture (that's right—sustainable architecture) page from Hawaii. This is an archive for information about ecological planning; design; and integrating architecture and landscaping for tropical, subtropical, or temperate climates.

URL:

http://www.aloha.net/~laumana/

Frank Lloyd Wright

The work of Frank Lloyd Wright, perhaps the greatest architect of the twentieth century, represents some of the world's most important architectural treasures. Divided into two sections, this page first includes photographs, maps, and tour descriptions of his Wisconsin buildings, and the second section is a multimedia gallery with movies and stills.

URL:

http://flw.badgernet.com:2080

ARCHITECTURE 2000
EXPERTS AGREE THAT THE NEXT TREND IN AFFORDABLE AND ECO-FRIENDLY HOUSING WILL BE GRASS AND MUD HUTS WITH THATCHED ROOFS. THAT'S RIGHT! THESE ENVIRONMENTALLY SOUND HABITATS WILL COME COMPLETE WITH OUTDOOR PLUMBING, CENTRAL BONFIRE HEATING, AND BLOW-THROUGH AIR CONDITIONING. YOU'LL FINALLY BE ABLE TO SLEEP AT NIGHT KNOWING THAT YOU AREN'T RESPONSIBLE FOR THE UNTIMELY DEATH OF EVEN ONE DOUGLAS FIR. ACT NOW BY POINTING YOUR BROWSER AT **THE ECO DESIGN AND LANDSCAPING PAGE**.

Images and Sculpture

Turkey has some of the finest and most pristine ancient Roman monuments in the world, and this page features a plethora of photos dedicated to that architecture. Subject groups include Roman funerary antiquities, second- and third-century sarcophagi, sculptured theater *frons scenae* at the Roman theater at Perge, and an overview of the legendary architecture from the city of Pergamum.

URL:

http://www.ncsa.uiuc.edu/SDG/Experimental/
anu-art-history/architecture.html

Introduction to AutoCAD

What is AutoCAD? Link here and see for yourself. This hypertext description of the popular architectural design software outlines the general characteristics of AutoCAD and provides introductions to AutoCAD commands, drawing and editing controls, manipulation and grouping entities, and an index to AutoCAD commands.

URL:

http://www.arch.unsw.edu.au/helpdesk/software/
autocad/cadnotes/intro.htm

Iron One

Take a short tour of a steel house constructed in Vancouver, Canada. Read quotes from famous architects on relevant themes regarding the home's design concepts.

URL:

http://www.wimsey.com/~tom0/works/io/i1.html

Lighthouses Over the World

Oh, buoy! This paragon of illumination guides you to the best lighthouse architecture information on the Web. Sections include lighthouses near coasts and lakes, technical details regarding lighting techniques, assorted lightships, books, and tall tales about lighthouses. There are links to other lighthouse pages as well.

URL:

http://www.noord.bart.nl/~derks/

Los Angeles: Revisiting the Four Ecologies

Using the structure of Reyner Banham's landmark 1971 book *Los Angeles: The Architecture of Four Ecologies*, Matt Jones documents and illustrates each separate section on the wonders of Surfurbia, Downtown, Foothills, and Autopia.

URL:

http://www.cf.ac.uk/uwcc/archi/jonesmd/la/

Metropolis Magazine

Fritz Lang's *Metropolis* was a landmark movie, and *Metropolis* the magazine is a landmark in its own right. Published in print and online ten times a year, it offers points of view on modern architecture. Each issue contains brief synopses of articles; several full-length stories, such as I. M. Pei's design for the Rock 'n' Roll Hall of Fame; and updates on world architecture.

URL:

http://virtumall.com/newsstand/metropolis/
main.html

Los Angeles

It might be a godforsaken overpopulated wilderness of cops and cars, topless bars, movie stars, and O.J. Simpson lawyers, but L.A. does have its merits as evidenced in *Los Angeles: Revisiting the Four Ecologies*, a heavily illustrated study of the contemporary Los Angeleno architecture. The best thing about these pages: you can visit without being smogged-out on the Santa Ana freeway at rush hour, dude.

MIT School of Architecture and Planning

This overview of MIT's curriculum, facilities, faculty, programs, studies, and workshops includes the design studio of the future, an operating room of the future, women in architecture, and a virtual walk-through video of recent school construction and additions.

URL:

http://alberti.mit.edu/ap/

Modern Japanese-style Architecture

This exhibit traces the Nagoya's heritage in text and photos of 1868–1941 European-themed architecture with Japanese design accents.

Note: Versions available in Japanese and English.

URL:

http://www.tcp-ip.or.jp/~csakao/index-eg.html

New Hampshire Covered Bridges

Move over Madison County. Among America's most endearing architectural treasures are its covered bridges. New Hampshire has more of them than any other New England state, and they are proudly shown and described on this page. Included are synopses of the many bridges' different architectural styles and directions on how to get to each one the next time you're in the Granite State.

URL:

http://vintagedb.com/guides/covered.htm

Old Covered Bridges

This guide to covered bridges includes FAQs and catalogs organized by county, structural type, driving tour, and season. Images, photos, and portfolios accompany this comprehensive collection.

URL:

http://william-king.www.drexel.edu/top/bridge/CB1.html

PLAN NET Professional Online Service

Read the Daily Plan Net for the latest in the world of architecture and design. This online journal contains information from news agencies and Internet sites. Enter The Studio for open discussions on topics such as architectural styles, city planning, and building types.

URL:

http://www.plannet.com

Southeastern Architectural Archive

Can you imagine the charm of New Orleans 150 years ago? In this archive, over 200 architects from 1830 to the present are represented by a large collection of their drawings and project records. Representations range from small-scale residential projects to major public buildings, and they detail the historic importance now lost.

URL:

http://www.tulane.edu/~lmiller/SEAAHome.html

U.C. Berkeley Architecture Library

This is a compendium of architectural material that includes the Americans with Disabilities Act, Book Review Sources in Architecture and Landscape Architecture, CD-ROM Titles in the Environmental Design Library, Recent Acquisitions in Architecture, Reference Sources on Architecture (Guide), and Finding a Job in Architecture.

URL:

gopher://infolib.lib.berkeley.edu/11/resdbs/arch

Planning an outing? Check the weather in Meteorology.

Excerpt from the Net...

(from the Southeastern Architectural Archive)

The focus of the collection is the architectural and urban history of New Orleans and the Gulf South, from the 1830s through the 1980s, with significant holdings for other regions of the country. Over 200 architects are represented, a large number by drawings and project records comprising the architect's complete built work. Works represented range from small-scale residential projects to major public buildings, and include many buildings of historic importance now lost. The depth and range of the collection documents the diversity of architectural practice in New Orleans over a period of 150 years.

Unauthorised Glasgow 1999

Glasgow was awarded the title of United Kingdom City of Architecture and Design 1999 by the Arts Council of Great Britain. To celebrate, they installed this web page to keep the world informed of all the planned festivities. These include the Glasgow 1999 Information Centre, a complete listing of architectural projects proposed in Glasgow's bid, and the Designer's Habitats online catalog of the Collins Gallery.

URL:

http://www.colloquium.co.uk/www/eponym/glasgow_1999/intro.html

Unit 19

This permanent online exhibit at the Bartlett School of Architecture at University College, London proposes that, in the future, architecture will be fluid, event based, user molded, and "unsightly." Text and photos describe such off-beat issues as machine self-replication and intelligence, shamanism, celestial architectures, the Renaissance Magi and cyberspace, emergence and complex systems, virtual space and sexuality, and alchemic tech gnosis. Ommmm. Ommmmm.

URL:

http://doric.bart.ucl.ac.uk/web/unit19/unit19.html

University of Miami School of Architecture

Learn about the University of Miami School of Architecture—its faculty, facilities, current and online courses, current events, lecture series, student web pages, faculty work, a slide library, research, publications, maps, drawings, examples of local architecture, and news and projects from the current digital design studio.

URL:

http://rossi.arc.miami.edu/home.htm

Virtual City

This online lecture series focuses on future and cyber-architecture presented by the Rice University Design Alliance. Featured speakers and essayists include architects, futurists, cyberheads, and science fiction authors. A movie synopsis of accompanying *cinemarchitecture* films is included in the lecture series. Speakers include acclaimed cyberpunk author Bruce Sterling, and Deyan Sudjic, author and editor of *Blueprint*.

URL:

http://riceinfo.rice.edu/projects/RDA/VirtualCity/

Virtual Study Tour

Here's a collection of computer-modeled reconstructions of various historic and fanciful architectural works in New Zealand.

URL:

http://archpropplan.auckland.ac.nz/misc/virtual_tour.html

ARTIFICIAL INTELLIGENCE

Alan Turing

An online biography of Alan Turing by J. A. N. Lee provides a look at one of the earliest contributors to the field of artificial intelligence. Turing is most notable for the test of intelligence that we now know as the "Turing Test."

URL:

http://ei.cs.vt.edu/~history/Turing.html

A B C D E F G H I J K L M N O P Q R S T U V W X Y Z

CMU Artificial Intelligence Repository

CMU Artificial Intelligence Repository contains files, programs, and publications of interest to AI researchers, educators, and students.

URLs:

ftp://ftp.cs.cmu.edu/user/ai

http://www.cs.cmu.edu/afs/cs.cmu.edu/project/ ai-repository/ai/html/air.html

Expert Systems FAQs

Learn all about Expert Systems, software procedures, and algorithms that guide users intelligently through problem-solving sequences. This collection includes most common questions asked about Expert Systems, plus those all-important answers.

URL:

http://www.cs.cmu.edu/Web/Groups/AI/html/ faqs/ai/expert/part1/faq.html

Fuzzy Shower

Explore fuzzy logic while mastering the dangerous everyday task of using the shower. With the Fuzzy Shower, you control the volume of water flowing from both hot and cold spigots to maintain a constant temperature and flow rate. It's scaldingly clean fun!

URL:

http://ai.iit.nrc.ca/fuzzy/shower/title.html

Journal of Artificial Intelligence Research

Artificial intelligence is better than none at all. That's why JAIR covers all areas of AI, including automated reasoning, cognitive modeling, knowledge representation, learning, natural language, perception, and robotics. Papers from notable institutions such as AT&T Bell Labs, MIT, and Stanford are submitted and published here, allowing readers to offer critical feedback.

URLs:

news:comp.ai.jair.announce
news:comp.ai.jair.papers

http://www.cs.washington.edu/research/jair/ home.html

Excerpt from the Net...

(from the Alan Turing page)

"The whole thinking process is still rather mysterious to us, but I believe that a thinking machine will help us greatly in finding out how we think ourselves." --Alan Turing, May 1951

ASTRONOMY

Astro News

You'll need to read at the speed of light to follow the numerous astronomy-related newsgroups on Usenet.

URLs:

news:alt.sci.astro.aips
news:alt.sci.astro.figaro
news:sci.astro
news:sci.astro.amateur
news:sci.astro.fits
news:sci.astro.hubble
news:sci.astro.planetarium
news:sci.astro.research

Astronomical Museum in Bologna

Photos and descriptions of this eighteenth century building in Bologna, Italy, give us a greater appreciation for early celestial observers, their instruments, and the classic architecture of the time. The museum is an historic restoration by the Department of Astronomy of the University of Bologna.

URL:

http://boas3.bo.astro.it/dip/Museum/ MuseumHome.html

Astronomy and Astrophysics at NSSDC

Never mind that they have a CD collection larger than yours. Much of the National Space Science Data Center's (NSSDC) most popular NASA astronomy and astrophysics data is accessible on the Net. In addition to actual space flight mission data, NSSDC archives hundreds of astronomy and astrophysics catalogs. NSSDC also archives and distributes over 500 CD-ROMs, of which one-third hold astronomy and astrophysics data. This is also a great place to find other astronomy and astrophysics resources.

URL:

http://nssdc.gsfc.nasa.gov/astro/astro_home.html

AstroWeb

Directories of astronomical resources abound on the Internet. What makes this directory unique, according to its creators, is that each link is tested three times a day to "verify aliveness." After all, there's nothing worse than stale links. Stale links stink.

URL:

http://fits.cv.nrao.edu/www/yp_astronomy.html

Shhhhh! Don't tell anyone you read this here (some senators might be listening). But there is a place on the Internet where you can download gobs of hot pictures of some of the most exotic, heavenly bodies you'd ever want to see! There are even photos of famous stars—in color, untouched, and unclothed! Is your pulse quickening? Your heart racing? Well, comb your hair and straighten that tie! Get ready for the night of your life in **alt.binaries.pictures.astro**!

Black Hole Physics

Become absorbed in lots of calculations, formulas, and physics of the incredible phenomenon of a black hole. Size and mass are calculated in terms of a googolplex. If you go gaga for big numbers, hang on tight while getting sucked into this page.

URL:

http://www.uni-frankfurt.de/~fp/Tools/
GetAGoogol.html

CCD Images of Galaxies

Conceived in 1970 at Bell Labs, charge-coupled devices (CCDs) have been moving closer to becoming ideal detectors for many types of astronomical phenomena. View the results of the University of Oregon's experiments with CCDs in graphics and MPEG movies of comet crashes, galactic nebulae, Messier objects, interacting galaxies, supernovae, and perhaps even UFOs.

URL:

http://zebu.uoregon.edu/galaxy.html

Earth Viewer

Galileo and Copernicus would have flipped over this great interactive program. Observe the Earth from the Sun, moon, satellites, or from any altitude at any location on the planet! Choose between daylight or night views, add clouds for authenticity, and you have a first-rate astronomical and geophysical tool. There are also a number of public-domain Windows and X programs for downloading that were written by the page's author himself. These include the above-described program, an Excel-based catalog of the Palomar Observatory Sky Survey, and several outer space screen savers.

URL:

http://www.fourmilab.ch/earthview/
vplanet.html

Exploration In Education

Exploration In Education (ExInEd) is a NASA-supported program of the Special Studies Office at the Space Telescope Science Institute. Using technology to overcome the time lag and information barriers that keep fresh research results from reaching wide audiences, ExInEd explores with the public new ways to derive social benefit from space research. Take a look at their latest electronic offerings of Macintosh, Windows, and DOS software available for downloading.

URL:

http://marvel.stsci.edu/exined-html/
exined-home.html

Extragalactic Database

A veritable hitchhiker's guide to the galaxy, the NASA/IPAC Extragalactic Database (NED) contains positions, basic data, and over half a million names for extragalactic objects, as well as bibliographic references to published papers, notes from catalogs, and other publications. But don't panic! NED includes a powerful search facility to help locate tons of useful information for professional astronomers and star gazers alike.

Note: You can reach NED from its web address, but you'll ultimately interface with it via a telnet connection. Fortunately, it has extensive online help along the way so you won't be lost in space.

URL:

http://www.ipac.caltech.edu/ned/ned.html

Like lizards? Check out Herpetology.

A B C D E F G H I J K L M N O P Q R S T U V W X Y Z

Extreme Ultraviolet Explorer

Before linking here, don those blue-blocker sunglasses your brother-in-law gave you for your birthday last year! Launched in June 1992, the Extreme Ultraviolet Explorer (EUVE) is finding stars hidden to us on Earth because of our atmosphere. The extreme ultraviolet radiation emitted by white-hot dwarf stars can only be detected by the EUVE from space. The EUVE Guest Observer Center provides information, software, movies, and data to visitors.

URL:
http://www.cea.berkeley.edu

Grand Challenge Cosmology Consortium

GC3's site contains movies of gaseous structure formations, hierarchical clustering in a scale-free universe, and star formation during galaxy mergers. There are online publications, reports, and a data and software archive.

URL:
http://zeus.ncsa.uiuc.edu:8080/
GC3_Home_Page.html

Haystack Observatory at MIT

Finding a star in the farthest reaches of space is like looking for a needle in a . . . you knew that was coming. With the four separate facilities that make up MIT's Haystack Observatory, you're likely to find that needle using radio telescopes, high-power radar antennas, and Very-Long-Baseline Interferometry corellators.

URL:
http://hyperion.haystack.edu/homepage.html

Hubble Space Telescope Astrometry Science Team

See how astrometry, one of the oldest branches of astronomy, uses the Hubble Space Telescope (HST), one of the newest tools of astronomy. These pages are well written, easy to navigate and understand, and provide a clear understanding of what the HST is used for, including current projects such as the search for planets near Proxima Centauri. You'll find Pickles here, free for the downloading.

URL:
http://dorrit.as.utexas.edu

Isaac Newton Group of Telescopes

Most people look *through* telescopes. Here's your chance to look *into* three that make up the Isaac Newton Group of Telescopes on the island of La Palma in the Canary Islands. Study photos, beautiful cutaway views, schematic drawings, and *user documentation* for the William Herschel, Isaac Newton, and Jacobus Kapteyn telescopes.

URL:
http://www.ast.cam.ac.uk/~lpinfo/

Jet Propulsion Laboratory

Did you know that NASA's JPL spacecraft have visited every known planet except Pluto? (A Pluto mission is currently under study for the late 1990s.) JPL's home page at CalTech is just what you'd expect it to be—understated and professional, but with a wealth of information behind every click. Check out many fascinating NASA facts—from current NASA mission news updates and schedules, to a guide on the basics of space flight and how spacecraft missions operate—JPL's got it all.

URL:
http://www.jpl.nasa.gov

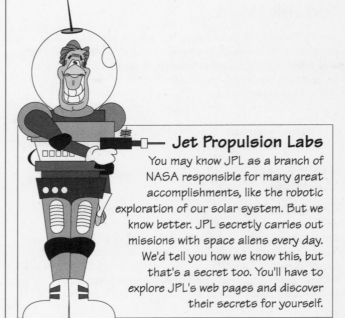

Jet Propulsion Labs

You may know JPL as a branch of NASA responsible for many great accomplishments, like the robotic exploration of our solar system. But we know better. JPL secretly carries out missions with space aliens every day. We'd tell you how we know this, but that's a secret too. You'll have to explore JPL's web pages and discover their secrets for yourself.

La Plata Observatory

One of the premier observatories in the world, La Plata in Argentina was established in 1883. This page, written in both English and Spanish, outlines the observatory's history and its current research projects in astronomy and geophysics, and lists upcoming scientific meetings and conferences.

URL:

http://www.fcaglp.unlp.edu.ar

Messier Database

Enjoy these marvelous celestial objects Monsieur Messier discovered in the eighteenth century. A complete image and information guide helps in navigating each of the 110 star clusters, nebulae, and galaxies. Data is enhanced with beautiful photographs of each object, including celestial position, visual magnitude, and approximate distance in light-years.

URL:

http://seds.lpl.arizona.edu/messier/Messier.html

NASA

What picture does NASA have on their home page? The shuttle? Astronauts on the moon? A gaseous nebula? Would you believe something more down to earth, like a map of the U. S.? Ah, but this is no ordinary map. Click a location and blast off to your favorite NASA research center! No solid rocket boosters or external fuel tank required!

URL:

http://www.gsfc.nasa.gov/NASA_homepage.html

Saturn's Ring System

"Still as, while Saturn whirls, his steadfast shade Sleeps on his luminous ring." —Alfred Lord Tennyson. And luminous Saturn is illuminated further by a wealth of images and QuickTime animations. Included on this page are Voyager 2 flybys, Hubble Space Telescope views of a storm on Saturn, and those radiant rings. References and links to relevant web pages round out this site.

URL:

http://ringside.arc.nasa.gov/www_root/000000/saturn/saturn.html

Sky & Telescope Magazine

Sky & Telescope Online is a good source of astronomical news and information, star atlases, slide sets, globes, commercial products and services, weekly news bulletins, and notices on upcoming celestial sights and how to see them. You'll enjoy tips on backyard astronomy for beginners, plus reviews from experts on telescopes and accessories, and software for your computer. The online magazine also includes a calendar of star parties, astronomical gatherings, and links to other celestial information on the Net.

URL:

http://www.skypub.com

SkyView

SkyView is the Web's interactive virtual observatory. SkyView generates images of any part of the sky at any wavelength in all spectra from radio to gamma-ray.

URL:

http://skview.gsfc.nasa.gov/skyview.html

Solar Issues

"The sun . . . In dim eclipse disastrous twilight sheds on half the nations, and with fear of change Perplexes monarchs." —John Milton. Even if you're not a monarch, you, too, can be perplexed by this MPEG of a solar eclipse captured from Kirkwood Observatory in Indiana. This site also has other eclipse images and weather satellite images of the moon's shadow on Earth.

URL:

http://www.astro.indiana.edu/solar/

Space Environment Laboratory

Heading to the beach? Although the scientists at the Space Environment Laboratory (SEL) may not be able to suggest the appropriate SPF lotion for your skin, they do provide real-time monitoring and forecasting of solar and geophysical events. Concerned about the environment between the Sun and Earth, they also conduct research in solar-terrestrial physics; develop techniques for forecasting solar and geophysical disturbances; and post current images of the Sun, solar flare-ups, and charts of X-ray flux. Are you sure you really want to go to the beach today? Better check here first.

URL:

http://www.sel.bldrdoc.gov

A
B
C
D
E
F
G
H
I
J
K
L
M
N
O
P
Q
R
S
T
U
V
W
X
Y
Z

Space News

Flash! Aliens Invade Grover's Mill! Well, perhaps it's not quite as sensational as that, but you'll still find interesting news and announcements about space programs, science, and related technologies in the **sci.space** newsgroups. They're out of this world!

URLs:
news:sci.space.news
news:sci.space.policy
news:sci.space.science
news:sci.space.tech

Space Science Education

A few orbits around this page is a science teacher's dream. Complete with NASA facts, activities, lessons in astrophysics, planetary science, space physics, and related curricula, students will plead for extended missions before having to return to Earth.

URL:
http://www.gsfc.nasa.gov/education/education_home.html

Space Telescope Electronic Information Service

The Hubble Space Telescope was really in trouble until NASA performed some spectacular repairs on it. Now it's working just fine, and the page contains pictures in several formats, as well as movies and animation of the wonders of the universe. You can also access information on the Hubble's scientific instruments, its weekly observation schedule, and an archive of past observations. Hobble to the Hubble on the double.

URL:
http://marvel.stsci.edu/top.html

Space Weather

Visit this Rice University page for stunning photography, graphics, and educational videos of auroral displays and other space weather. Don't forget your space umbrella.

URL:
http://rigel.rice.edu/~dmb/spwea.html

The Space Weather Channel

Contrary to popular assumption, space is not empty. It is filled with low-energy charged particles, photons, electric and magnetic fields, dust, and cosmic rays. (You'll also find tiles from the space shuttle, golf balls, half-eaten space food sticks, and empty jars of Tang, but that's another story.) The densities of these things are relatively low, though high enough to affect spacecraft, humans in space, and even human activities on Earth. Sometimes there are violent changes in the space environment. These are seen in dramatic auroral displays.

We call this *space weather*.

SpaceViews

SpaceViews is a monthly publication of the Boston chapter of the National Space Society. It focuses on recent developments in space policy, exploration, development, and cutting-edge research into new space-related technologies. There are midmonth updates and an archive of past issues.

URL:
http://www.seds.org/spaceviews/

Excerpt from the Net...

(from the Space Environment Laboratory)

Space Weather Outlook

SOLAR ACTIVITY SHOULD CONTINUE AT VERY LOW LEVELS FOR THE NEXT THREE DAYS.

THE GEOMAGNETIC FIELD IS EXPECTED TO BE QUIET TO UNSETTLED FOR THE NEXT THREE DAYS.

StarChild Project: Connecting NASA and K–12

More for K than 12, this great beginning science page for young children explores the mysteries of the universe and teaches fun facts in the process. Large type and graphics make it easy for young and old alike to visualize the wonders of space.

URL:
http://guinan.gsfc.nasa.gov/K12/StarChild.html

Stars and Galaxies Guide

Why buy the CD-ROM when you can interact with one online? This site offers a comprehensive multimedia guide to stars and galaxies including text, audio, and pictures taken from the Earth and Universe CD-ROM.

URL:
http://www.telescope.org/btl/

Sudbury Neutrino Observatory

The Sudbury Neutrino Observatory is being built below ground in the deepest section of the Creighton Mine in Ontario, Canada. More than two kilometers below the earth's surface, the detector will be housed within a man-made cavern the size of a ten-story apartment building. At the center of the cavern is a large acrylic vessel holding one thousand tons of radiation-free heavy water. The $300 million of heavy water necessary for conducting research at the Sudbury Neutrino Observatory is on loan from the Canadian government. Why? Turn to the SNO page and find out! (We wonder if they could have used Perrier instead.)

URL:
http://snodaq.phy.queensu.ca/SNO/sno.html

Sun's Internal Rotation and Magnetic Field

This site is maintained at The National Center for Atmospheric Research and contains a 17-minute animation showing the evolution of the Sun's internal rotation and magnetic field. Accompanying this animated model is an abstract of the research paper that details the scientific calculations used to produce the model.

URL:
http://www.ucar.edu/STAFFNOTES/
sunvideo.html

Swedish Institute of Space Physics

The Swedish Institute of Space Physics conducts basic research and education, and associated observatory activities in space physics. Here you'll find extensive information on the institute's projects, experiments, faculty, facilities, and equipment.

URL:
http://aurora.irf.se

Tour of the Universe

Why bother to boldly go where no one has gone before (and heaven help you if you're wearing a red shirt and your Captain says, "Scotty, beam *this one* down to the planet's surface *first*"), when you can point and click your way from here to Farpoint without a dangerous encounter? Take a graphical tour of the universe with cool pictures of stars, planets, and galaxies.

URL:
http://www.holli.com/~jshoup/space/

Views of the Solar System

Springtime is a lovely season to visit Jupiter and Mars. Blast off to this page that revolves around the Sun, the planets, and other heavenly bodies. In-depth information, great graphics from the Hubble and other major telescopes, and spectacular animation make this site out of this world.

URL:
http://www.c3.lanl.gov/~cjhamil/SolarSystem/
homepage.html

AUDIO

Acoustics Resources

An informative resource on acoustics including books, the *Sound and Vibration Electronic Digest*, *Acoustics Archives*, news about conferences and courses on acoustics, and links to related pages.

URL:
http://www.eng.auburn.edu/department/me/
research/acoustics/Acoustics.html

A B C D E F G H I J K L M N O P Q R S T U V W X Y Z

Cohen Acoustical

A professional consultation service, Cohen Acoustical offers technical papers available via email on such topics as "Virtual Environments Auralization and Architectural Considerations." The site also includes information on the company and clients they serve.

URL:
http://www.cohenacoustical.com/ca/

High Fidelity

For audiophiles only, these links lead to a wealth of high-end stereo resources such as hi-fi computer software, electronic parts and components, reviews, reports, A/B tests, an extensive list of Usenet groups, and even hi-fi humor.

URL:
http://www.unik.no/~robert/hifi

International Lung Sounds Association

Physicians, physiologists, physicists, and medical sound engineers will breathe easier knowing where to find respiration acoustics information. Learn about the history, organization, and membership of the ILSA. You'll also find abstracts, announcements, and a substantial reference section on books and papers about respiration acoustics.

URL:
http://www.umanitoba.ca/Medicine/Pediatrics/
ILSA/

Laboratory of Seismics and Acoustics

Delft University of Technology performs research on acoustic and elastodynamic wave theory, with applications in acoustic imaging and acoustic control.

URL:
http://wwwak.tn.tudelft.nl/index.html

OUCL Audio Archive

The Oxford University Computing Laboratory archive includes links to sound-related resources, newsgroups, Internet radio, music, software, and more.

URL:
http://www.comlab.ox.ac.uk/archive/comp/
audio.html

RealAudio

Now you can hear Internet audio streams instantly using Progressive Network's RealAudio player in conjunction with your web browser. Download your own copy of the RealAudio Player today.

URL:
http://www.realaudio.com

World Forum for Acoustic Ecology

Concerned with the state-of-the-world soundscape as an ecologically balanced entity? Of course you are. That's why WFAE's extensive menu of links to related sites focuses on research in acoustical ecology and communication, bioacoustics, psychoacoustics, sonic arts, electroacoustic music, and the natural and human-made soundscape.

URL:
http://interact.uoregon.edu/MediaLit/
WFAEHomePage

AUTOMOTIVE

Acoustic-n-Electronic

An online magazine for car audio enthusiasts and mobile electronics professionals, *Acoustic-n-Electronic* includes tutorials on highly technical subjects such as the fundamentals of electronic and acoustic theory, enclosure design, amplifiers, speakers, filters, etc.

URL:

http://www.ansouth.net/111/

Alternative Fuel Cars

If you're tired of having to dig deeper into your pocket every time a dictator in the Middle East threatens the world oil supply, support alternative fuel vehicles. On this page, you can learn about four snappy alternative fuel vehicles Chrysler is working on. The Patriot has a low-powered combustion engine that generates electricity to power the car. The Jeep ECCO is a runabout with a 1.5-liter, two-stroke engine; and Voyager and Concorde/Intrepid/Eagle Vision lines support flexible fuel needs—ranging from natural gas to methanol.

URL:

http://www.chryslercorp.com/environment/
alternative_fuels.html

Alternatives to Gas

Ford Motor Company is exploring methanol, ethanol, natural gas, and electricity as viable alternatives to gasoline-powered cars. Check out the Ecostar, an electric police vehicle being used in the U.K. today. This page also offers information about Ford's bi-fuel F-Series pickups; the Crown Victoria—the first full-size sedan to be mass-produced for natural gas; and other ventures into bio-gas and other environmentally friendlier fuels.

URL:

http://www.ford.com/corporate-info/
environment/GasAlt.html

**Need some great music?
Lend your ear to the Music:
MIDI section.**

The Web will really launch you.

Bowling's Automotive Programs

Are you plagued by curiosity about how Detroit calculates fuel injector sizes? Does not knowing the optimum gear-shift timing for your Beemer keep you awake at night? Provided to anyone interested in the technical aspects of automotive sciences, this collection of programs has something for everyone. From drive-shaft RPM computations to Holley jet environmental corrections, if you think it involves a computation, you'll find a computer program here to do that computation.

URL:

http://devserve.cebaf.gov/~bowling/auto.html

Chrysler History

If you're researching the history of the automotive industry or just looking for some interesting reading, try this page. Chrysler makes their internal newspaper—the *Chrysler Times*—available on the Web, and the issues go all the way back to the '20s. These pages are actually condensed versions of the paper. The stories and issues of the day make for some intriguing historical reading.

URL:

http://www.chryslercorp.com/library/
library.html

Back to the Future?

So you thought a Delorian running on beer cans and stale peanut butter sandwiches was only something that could happen in a Hollywood comedy. Well, think again. Rubbishgas is here now! Don't flip that garbage disposal switch again. Put those Tiger's Milk bars in your tank and listen to your wheels purr. Check the Alternative Fuel Cars and Alternatives to Gas pages to get the latest scoop on alternative-fuel vehicles.

A
B
C
D
E
F
G
H
I
J
K
L
M
N
O
P
Q
R
S
T
U
V
W
X
Y
Z

Chrysler Technology

Even though the automobile industry is a fiercely competitive marketplace, Chrysler makes a tremendous amount of their technology freely accessible to anyone on the Web. Read about Chrysler's 3/8-scale wind tunnel and how they use it with computational fluid dynamics analysis to measure lift, drag, yaw and pitch, and rolling moments on their body designs. Check out the environmental and electromagnetic test centers, the noise vibration and harshness lab, and the power train facility. You can even get a peek at Chrysler's concept cars here and read about their new technology.

URL:
> http://www.chryslercorp.com/technology/
> technology.html

Driving Green

This page is packed full of great tips from Ford on driving your car in an environmentally friendly way. Of course, there are the obvious suggestions—such as keeping your car's engine in proper tune—but there are also less obvious tips. For example, you'll find tips on oils to use, luggage racks, and driving with open windows.

URL:
> http://www.Ford.com/corporate-info/
> environment/DriveGreen.html

Driving Tips from Goodyear

This page is an excellent refresher for driving safety and care of your tires. What to do when you're stuck, highway hazards, and spare-tire care are among the topics of interest here. You'll also find links to other pages with information on tire storage, tire service, and cold-weather driving.

URL:
> http://www.goodyear.com/Home/HTML/
> Educational/Tips.html

Dynomation

No, this isn't the home page for a TV show from the '70s featuring a funny kid with a big bushy afro. Dynomation is a wave-action simulation program for optimizing the power output of an internal combustion engine. Your inputs to the program are things like the bore, stroke, rod length, compression ratio, and a slew of others. The outputs are horsepower, torque, volumetric efficiency, and other things that we mere computer guys don't understand.

URL:
> http://www.public.iastate.edu/~punk/
> dynomation/dma.html

Fil's Auto Corner: Oil Facts

Here's a helpful series of documents to help you choose the best oil for your vehicle's engine. Topics include viscosity; flash point; zinc; oil additives; sulfated ash; and many other important subjects that you need to understand before you pour oil into any engine.

URL:
> http://www.paranoia.com/~filipg/HTML/
> AUTO/F_oil_facts12.html

Ford News Briefs

Each week, Ford publishes a news digest for Ford employees and the automotive industry. These publications are available online for you to peruse. Topics are generally nontechnical, and the briefs have general business flavor focusing on subjects such as plant openings and closings, and consumer reports and ratings.

URL:
> http://www.Ford.com/corporate-info/news/
> NewsDigest.html

Gasoline FAQ

This collection of documents will answer every question you have ever had about gasoline and oil. Starting with the basics of oil refining, these documents become more complex as they discuss ways to improve the fuel economy of your automobile.

URL:
> http://www.paranoia.com/~filipg/HTML/
> AUTO/F_Gasoline.html

Goodyear Tire School

Did you ever wonder what all those cryptic hieroglyphics on your tires actually mean? Stop by the Goodyear Tire School for an interpretation. Here, you can find out about tread-wear patterns; traction; how temperature affects your tires; and what speed ratings mean and how to interpret them. There are also some excellent descriptions of alignment terms, an overview of wheel-alignment effects on tires, facts about tire balancing, and rotational tips.

URL:
> http://www.goodyear.com/Home/HTML/
> Educational/TireSchool/TireSchool.html

Semi-Synthetic Oils

Are semi-synthetic oils the wave of the future, or another fad that will fade away in a year or two? On this page you can read the answers to frequently asked questions about semi-synthetic oils from oil expert Norm Hudecki of the Valvoline Company.

URL:

> http://www.valvoline.com/mech_only/
> hudecki_archives/synth_semi_qa.html

Smoke Signals from Your Engine

If your car is puffing and belching clouds of smoke, it may be an indication that something's amiss under the hood. This page describes some of the possible causes based on the color, timing, and amount of smoke your car is emitting.

URL:

> http://www.valvoline.com/mech_only/
> hudecki_archives/engine_smoke.html

*Is your car trying to tell you something? If its voice sounds like your mother's, go directly to your therapy session. Otherwise, sign into the **Smoke Signals from Your Injun**—uh, make that **Engine** page.*

Synthetic Lubricants

When synthetic oils first hit the market, the oil companies claimed that frequent oil changes may have become a thing of the past! Now we know better, and the synthetic oil rage never did actually deliver. On this page, you can read the real story about synthetic lubricants—both their benefits and their drawbacks.

URL:

> http://www.valvoline.com/mech_only/
> hudecki_archives/synthetics_qa.html

Tips on Buying a New Car

A must-read guide before going to a car dealership, this one seems to have been written by someone in the industry (and in the know) and will help you defend yourself when you negotiate your new car price. Don't be ripped off by a slick salesperson—understand the process of selling new cars, "low balling" on trade-ins, and charging inflated prices for extended warranties and financing.

URL:

> http://mr2.com/TEXT/newcarfaq.txt

Tire Industry Safety Council

The TISC is a trade association supported by U.S. tire manufacturers to promote tire care and safety to consumers. There are two publications you can read from these pages: the "Motorist's Tire Care and Safety Guide," and the "RV Tire Care and Safety Guide." These hypertext articles provide valuable information on inflation pressures, tire inspection, good driving habits, conditions affecting tires, and other topics—such as selecting replacement tires and storage tips.

URL:

> http://www.tmn.com/tisc/index.html

AVIATION

Aeronautics and Airliners

Just in case you're building your own 747, you'll find airliner technology, design, construction, performance, human factors, operation, and histories of aircraft discussed here. Soar through an archive of past articles, or puddle-jump to other relevant Net sources. Many associations, companies, and individuals are linked to this important resource for aviation technology and innovation.

URL:

> gopher://wiretap.spies.com/00/Library/
> Document/geologic.tbl

Air Force Office of Scientific Research

AFOSR directs the Air Force's basic research program and creates new and advanced technology, enabling U.S. industries to produce world-class military and commercial products. Search this page's many databases for information on procurements, grants, program information, announcements, general agency information, and links to Air Force laboratories and related agencies.

URL:

> http://web.fie.com/web/fed/afr/

A B C D E F G H I J K L M N O P Q R S T U V W X Y Z

Aircraft Aerodynamics and Design Group

Navigate to this Stanford University page on applied aerodynamics and aircraft design. Research ranges from computational and experimental methods for aerodynamic analysis to studies of unconventional aircraft concepts. Test pilot their Wing Analysis Program for calculating the aerodynamic properties of aircraft wings.

URL:
    http://aero.stanford.edu/ADG.html

Aircraft Images

Glide your Web browser to this extensive archive with hundreds of top-flight images of aircraft. The pictures include various types of planes found throughout the history of aviation.

URL:
    http://lal.cs.byu.edu/planes/planes.html

Aviation Weather Reports

Before you jump into your Cessna for a short trip into the next county, or even if you plan on taking a Boeing 777 to Las Vegas for this year's Comdex, find out what the weather is like along the skyway. This gopher menu provides current weather, wind conditions, and forecasts for most of North America.

URL:
    gopher://geograf1.sbs.ohio-state.edu/1/
    wxascii/aviation/airways

AviationWeb

To become proficient at many of life's technical skills, you need to go to school. If you plan to be a pilot, then it's flight school for you. You'll find a database of them here on AviationWeb. Search by state, city, and/or ZIP code. Why simulate when you can experience real flight?

URL:
    http://www.aviationweb.com

Avion Online Newspaper

Avion, published by Embry Riddle Aeronautical University, is the first aerospace and aviation newspaper on the Internet. Devoted to a wide range of aeronautical topics, *Avion* reports on general aviation, commercial, and military subjects. You'll find timely news articles on shuttle launches, avionics, navigation, airshows, and pilots' associations. If you get lost in the clouds, search through past issues to get back on course.

URL:
    http://avion.db.erau.edu

Avro Arrow

During the late 1950s, the Avro Arrow, a twin-engine, supersonic military interceptor, was boosted on NIKE rockets to stratospheric heights. The history, technical specifications, and a number of images of the Canadian-built aircraft highlight the Arrow's career.

URL:
    http://calum.csclub.uwaterloo.ca/u/rkschmid/
    Arrow/AvroArrow.html

Excerpt from the Net...

(from *Highly Nonplanar Lifting Systems* at the Aircraft Aerodynamics and Design Group)

Nonplanar wings include configurations such as biplanes, box-planes, ring-wings, joined wings, and wings with winglets. Apart from configuration differences related to stability and trim, variations in nonplanar geometry represent one of the few major differences in aircraft conceptual design.

Such designs may be of interest because of their potential for lower vortex drag at a fixed span, a key constraint for many aircraft, including very large commercial transport concepts. However, several non-aerodynamic features are of interest as well including effects on stability and control, characteristics of wake vortices, and structural implications of the nonplanar design.

AVweb

AVweb may well be the premier starting point for aviation buffs on the Web. Sponsored by several organizations, including the Cessna Pilots Association and King Schools, AVweb will keep you up to date with the latest aviation-related news and events. AVweb also features editorial articles; safety briefs; information about noteworthy accidents; and links to follow for products, classifieds, brainteasers, and great places to fly.

URL:

http://www.avweb.com

AVweb

At one time or another, haven't we all dreamed of taking the controls of a real airplane and taking flight? Unfortunately, the house always seems to need a new roof or the car needs a new set of brake pads. But even if running down to the corner airdrome and forking over a down payment for a shiny new Cessna is out of the question, you can look at pictures and read about your favorite airplanes. Why experience the real thing, anyway, when you can experience virtual flight at a fraction of the thrill? But if you insist on reality, at least you can check into AVweb to read stories of, by, and about real aviators and aviation. And maybe when your ship . . . uh . . . plane comes in, maybe you, too, can write about flying.

Ballooning Online

Ever since the Montgolfier brothers made lighter-than-air ascent possible, people have been fascinated with balloons. Accomplishing the freedom to ride high over the earth with nothing more than the wind to guide you is both challenging and exhilarating. Learn more about balloon technology and what keeps balloonists up in the air on this high-flying page.

URL:

http://sunsite.unc.edu/ballooning/

Blimps-R-Us

A dirigible by any other name is still a blimp. These amazing airships have a long and storied history, but nowadays they're usually just relegated to flying over football stadiums. Discussion focuses on Airship Industries' Skyship 500HL, supported by excellent cutaway views of the gondola, controls, and envelope (the part that holds it up). Should you one day find yourself at the controls of one, link here for good descriptions of how blimps fly and how to control them.

URL:

http://www.iag.net/~zim/airship.html

Okay, we admit it. We like blimps. After all, they make great billboards and produce some pretty good aerial shots of ball games. But blimps are more than that — they lend an air of extravagance to any event. We've always wondered, though, how the people riding inside of blimps keep from talking like the chipmunks with all that helium around. Dock at the *Blimps-R-Us* page for the answer to this and any other question you have about these lighter-than-air ad campaigns.

College of Aeronautics

Located at LaGuardia Airport in Flushing, New York, this is the only institution in New York entirely devoted to programs preparing students for technology careers in the aviation and aerospace industries. A complete list of the programs and curricula, faculty, and student-related information is provided.

URL:

http://www.mordor.com/coa/coa.html

Want to be a doctor? Study up in Medicine.

A B C D E F G H I J K L M N O P Q R S T U V W X Y Z

FAA Information Services

The Federal Aviation Administration sets the rules for aviation standards in the U.S. Since these regulations are often subject to revision, this comprehensive site offers news releases and updates to keep you informed. Rules, safety tips, operations, air traffic, and other general aviation information are provided.

URLs:

gopher://gopher.faa.gov

http://www.faa.gov

Federal Aviation Regulations

Before you wing it, read the complete Federal Aviation Regulations. This document features a keyword search and contains definitions; airspace, air traffic, and procedural rules; certification procedures; airworthiness standards; navigational facilities; and airport regulations.

URL:

http://acro.harvard.edu/GA/fars.html

Hang Gliding and Para-Gliding

Ever since Daedalus and Icarus took wing, man has dreamed of soaring like the birds. Take wing with everything you need to know about hang gliding and para-gliding, including online digests, FAQs, photo Gallery, and movies.

URL:

http://cougar.stanford.edu:7878/
HGMPSHomePage.html

Helicopter Aviation

Don't get your rotors in a twist if you can't find helicopter information on the Net, because this is one page to hover over. Technical information and pictures of mechanical components, maneuvers, and aerodynamics help you operate and maintain helicopters. If that's not enough, there's a helicopter FAQ, and a separate section on the history of helicopters. It's enough to make your head spin.

URL:

http://world.std.com/~paulc

International Aerobatics Club

If you've ever dreamed of doing a barrel roll or an Immelmann, check out the International Aerobatics Club, a division of the Experimental Aircraft Association and the National Aeronautics Association. There is a great deal of competition and technical aerobatics-related information, and the page has an excellent FAQ on aerobatics. Take that, Red Baron!

URL:

http://niit1.harvard.edu/IAC/

Jane's Electronic Information System

For nearly a century, Jane's has published accurate and impartial books on defense, weaponry, civil aviation, transportation, and technology. Although this is a commercial page, you may perform limited database searches by aircraft term or model. Or use their online demo to retrieve text and images.

URL:

http://www.btg.com/janes/

Jet Fan

Read how jet fan technology works. Past, present, and futuristic applications regarding jet engines and applicable jet technology innovations are discussed in detail.

URL:

http://www.splitcycle.com.au

Military Aircraft Specifications Archive

Make a reconnaissance flight over this site to uncover information on warplanes from America, Britain, the former Soviet Union, Japan, Germany, France, and other countries. There are listings of acronyms and code names, military aircraft museums, the **rec.aviation.military** FAQ, and pictures of warplanes. Banzai!

URL:

ftp://byrd.mu.wvnet.edu/pub/history/military/
airforce

**Nobody sends you email?
Join a mailing list.**

**Having trouble reading
Usenet news? Type "news:"
into your Web browser.**

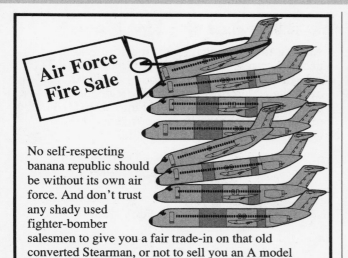

Air Force Fire Sale

No self-respecting banana republic should be without its own air force. And don't trust any shady used fighter-bomber salesmen to give you a fair trade-in on that old converted Stearman, or not to sell you an A model for a D model. Kick the tires and spin those turbines yourself. Be armed with a full complement of specs for every military aircraft when you pick out your next line of fighter or bomber aircraft. Wingover to the *Military Aircraft Specifications Archive*.

Official SR-71 Blackbird Site

If there's one bird that military aircraft buffs go cuckoo over, it's the Blackbird. Superseding the U-2 spy plane, and 20 years before Stealth technology, the sophisticated reconnaissance SR-71 flew at extremely high speeds and at great altitudes, and incorporated amazing radar-absorbing capabilities. Read about the incredible technology devised to produce this plane, track the status of each one, and view a collection of photos.

URL:
   http://www.primenet.com/~mikeq/sr71.html

Pilot Magazine

Pilot serves aviation-related concerns in Great Britain. Online are current and past online issues, plus sections on aviation instruction, jargon, reviews, pilots' associations, teaching and reference articles, flight tests, and touring. Pip, pip, tally ho, and all that sort of thing.

URL:
   http://www.hiway.co.uk/aviation/pilothom.html

Southern Aviator

If you live in the southern U.S. or plan to fly over it, don't miss this monthly online magazine. Technical, commercial, and hobbyist aviation topics are covered in detail. Excellent photographs accompany engaging articles.

URL:
   http://www.solopub.com/solopub/TSA.html

Ultralight Aircraft

To heck with United, Delta, and American Airlines with their high fares, *toe* room, recycled air-could-be-bug-spray, plastic food, and plastic attendants. Build your own plane! Ultralights are cheap, fun, and you can bring your own plastic food. You'll find a FAQ on ultralights; regulations regarding their use; and sections on safety and maintenance, clubs, manufacturers, newsgroups, mailing lists, classified ads, and events.

URL:
   http://www.cs.fredonia.edu/~stei0302/WWW/
   ULTRA/ultralight.html

University of Maryland Rotorcraft Center

If the technical and engineering aspects of helicopters and other rotary-powered flight vehicles make your world go 'round, bank over to this page of mechanical and aerodynamics reports and data. Links to other technical aerodynamics pages and a rotorcraft bibliography database server are also on this web landing pad.

URL:
   http://www.glue.umd.edu/~cjones/RC.html

A
B
C
D
E
F
G
H
I
J
K
L
M
N
O
P
Q
R
S
T
U
V
W
X
Y
Z

BIOLOGY

Animal Virus Information System

The AVIS database is maintained by the Bioinformatics Distributed Information Centre and is located at the University of Poona in India. At present, there is information on more than 1,000 animal viruses available. You'll find general information, descriptions of the database, and handy tips for searching.

URL:
http://bioinfo.ernet.in/www/avis/avis.html

Applied and Environmental Microbiology

Descriptions of basic and applied research into microbiology and microbial ecology are found here in journal citations. Foods, agriculture, industry, biotechnology, public health, and plants; plus references to biological properties of bacteria, fungi, protozoa, and much more await discovery.

URL:
http://www.asmusa.org/jnlsrc/aem1.htm

Artificial Life Forms

Biotopia is a computer program that simulates a Darwinistic eco system. Watch artificial life forms reproduce, eat, mutate, and die with this novel program that runs under DOS. Get a life! (Even if it's an artificial one.)

URL:
http://alife.santafe.edu/~liekens/

Artificial Life Online

Since it is quite unlikely that alien life-forms will present themselves to us for study in the near future, our only option is to try to create alternative life-forms ourselves—Artificial Life—literally "life made by Man rather than by Nature." These web pages include pointers to software, a bibliographic database on Artificial Life, and discussion forums.

URL:
http://alife.santafe.edu

BIO Online

Do the walk of life on BIO Online, one of the most comprehensive sites for biotechnology-related information and services on the Internet. Here is an extensive collection of news and information from companies, universities, and other biotechnology research institutions.

URL:
http://www.bio.com

Biodiversity and Biological Collections Gopher

This large collection of biology-related resources (fortunately with an index) includes images, directories of biologists, journal and newsletter info, items of interest to curators, and more.

URL:
gopher://muse.bio.cornell.edu

Biodiversity, Ecology, and the Environment

This navigational rest stop along the information highway contains a great deal of information and links to bioscience resources. A handy index of biological terms and a complete list of biomedical web sites make this Harvard site a great starting point.

URL:
http://golgi.harvard.edu/biopages/
biodiversity.html

Biologic Fluid Dynamics

While it's a captivating area of interest, biologic fluid dynamics is not about how babies drool and kindergartners wipe their noses on their sleeves. The Biologic Fluid Dynamics Group applies computational fluid dynamics to better understand the body's fluid flow and related complex systems. Take a box of tissue paper with you, just in case.

URL:

> http://giles.ualr.edu

Biological Sciences at UMN

People are people, no matter where you go. Even in Minnesota. So if you're going to check out biology information on the Internet, stop by the College of Biological Sciences at the University of Minnesota. We're fairly sure that their biology is pretty much like ours. You might want to check it out, though.

URL:

> http://biosci.cbs.umn.edu/cbs.html

Biotech Server

Biotech is an educational resource and research tool from Indiana University. This learning tool attracts students and enriches the public's knowledge of biology issues in the world today.

URL:

> http://biotech.chem.indiana.edu/pages/
> contents.html

Bipartite NLS Locator

Just yell, "AAAAAAKKKKKAAAAAAAAAAAKKKAAAAA" and when everyone stops to look at you quizzically, calmly explain that you're attempting to direct proteins into a nucleus from cytoplasm using a bipartite nuclear localization sequence (NLS). Not only will they be impressed, they're likely to genuflect. This page makes it easy to locate related NLS candidates. You can try this sequence: KRPAATKKAGQAKKKKL. It's pretty neat stuff. (But don't bother with KRAKATOA. It just blows up.)

URL:

> http://cy-mac.welc.cam.ac.uk/biploc_form.html

Bugs in the News!

Looking for a few good layperson's articles on microbiology, but were afraid to ask? Inoculate yourself with information on what allergies, antibiotics, genes, enzymes, the flu, *E. coli*, and microbiology itself are all about. You'll never be bugged wondering again.

URL:

> http://falcon.cc.ukans.edu/~jbrown/bugs.html

The Web will set you free.

CELLS Alive!

The marvels of microbiology burst upon your screen with this great page that describes how cells do their fabulous things. Watch breathlessly as an amoeboid human neutrophil hunts down and ingests an ovoid yeast. Thrill to videos of penicillin killing bacteria and gasp at apoptosis—when a cell commits suicide!

URL:

> http://www.comet.chv.va.us/quill/

Dictionary of Cell Biology

Published in 1989 and translated into many languages, the *Dictionary of Cell Biology* is now on the Net, with quick access to clearly written and cross-referenced definitions of terms used in modern biology literature.

URL:

> http://www.mblab.gla.ac.uk/~julian/Dict.html

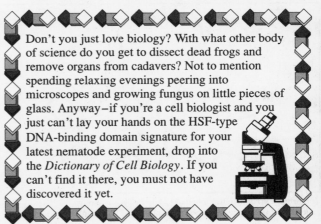

Don't you just love biology? With what other body of science do you get to dissect dead frogs and remove organs from cadavers? Not to mention spending relaxing evenings peering into microscopes and growing fungus on little pieces of glass. Anyway—if you're a cell biologist and you just can't lay your hands on the HSF-type DNA-binding domain signature for your latest nematode experiment, drop into the *Dictionary of Cell Biology*. If you can't find it there, you must not have discovered it yet.

Digital Anatomist

Here's your chance to slice and dice the human body in 3-D. Images are acquired as serial sections from frozen cadavers and clinical MR or CT scans. Photographs are taken of each frozen section; structures are manually traced, or segmented, on tracing paper; and the traces are digitized as input to the computer. And we thought Berkeley's Virtual Frog Dissection was neat.

URL:

> http://www1.biostr.washington.edu/
> DigitalAnatomist.html

A
B
C
D
E
F
G
H
I
J
K
L
M
N
O
P
Q
R
S
T
U
V
W
X
Y
Z

Essays in Virology

Virology always makes for great early morning reading—right up there with genetics and an extended session in **alt.religion.polka**. If you're still feeling groggy, try this first issue of essays covering viral evasion of host immune responses and virus-induced autoimmune disease. Coffee helps, as well.

URL:
   http://www.eps.lshtm.ac.uk/~npanjwan/eiv.html

Flavoprotein Research at Wake Forest University

Let Dr. Al Claiborne and Dr. Leslie Poole educate you on the research programs at the department of Biochemistry and Molecular Genetics. Included is information on school staff, facilities, labs, publications, and current research projects.

URL:
   http://invader.bgsm.wfu.edu

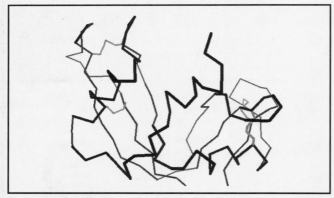

The protein structure Ribonuclease Calpha-SS. (You're right. It really does look like some creative Crayola work plastered by magnets to the fridge. See, there's a doggy, and that's daddy sitting at the computer...) Find out what it all really means at Wake Forest University's Flavoprotein Research pages.

Flow Cytometry at Aberystwyth

Flow cytometry is the technique where the physical and chemical characteristics of cells are measured as the cells pass in a fluid stream through a measuring point. This page offers extensive information on the scientific procedure of flow cytometry research, publications, and software.

URL:
   http://pcfcfh.dbs.aber.ac.uk/home.htm

INBio

Wouldn't it be wonderful if every country in the world set aside a quarter of its land as wilderness preserve? Costa Rica has done just that with its National Biodiversity Institute (INBio), responsible for studying and maintaining these rain forest parks. With text in English and Spanish, read scientific reports, rain forest information and species inventories, and browse the online collection of Costa Rican plants.

URL:
   http://www.inbio.ac.cr

Interactive Frog Dissection

Unlike Berkeley's Frog Dissection Kit, which simulates the task of dissecting a frog, this page prepares you for the real thing. Included are pictures, instructions, and QuickTime videos, illustrating the steps for preparing, incising, and exploring the organs and musculature of a frog.

URL:
   http://curry.edschool.virginia.edu/~insttech/frog/

IUBio Archive

Who wants to lug around a desktop Drosophila database when you can take a portable one with you? From Indiana University, the IUBio Archive includes many items to browse: molecular data, software, biology news and documents, and links to remote information sources in biology.

URL:
   http://iubio.bio.indiana.edu

Light Microscope History

Imagine where we'd be if microscopes had never been invented and all those little germs, bacteria, and diseases had remained undiscovered—we might not be writing this book and you might not be reading it! Pay tribute to the history of the microscope, and lenses in general, at this illuminating page.

URL:
   http://www.duke.edu/~tj/hist/hist_mic.html

Live Artificial Life Page

At long last, live computer simulations of artificial life using computational biological mathematics. Experience John Conway's Life, morphs, swarms, and forest fires. Quick, someone pinch me.

URL:
   http://www.fusebox.com/cb/alife.html

Excerpt from the Net...

(from the Live Artificial Life Page)

Warning: The following links lead to interactive programs and live animated simulations. Do not attempt to view these pages while eating a bowl of Fruity Pebbles.

Matthew's Nature Page

Matthew (we never do learn the chap's last name) provides a substantial offering of animal and plant information, nature organizations, and archives. The collection is well organized and focuses primarily on resources in the U.K.

URL:
http://rfhsun1.rfhsm.ac.uk:81/golly/
naturpag.html

Microbial Underground

Forget those crowded classrooms and registration fees. Take what may be the first online course on *Medical Bacteriology* on the Internet at this U.K. site. There's other material as well on medical and molecular biology, with pertinent related links.

URL:
http://www.ch.ic.ac.uk/medbact/

Microbiology at UCT

The scientists at University of Cape Town, South Africa, are on the cutting edge of microbiology as evinced from this up-to-date page. Read news updates about the latest viruses such as ebola, haemorrhagic fever, and hantavirus; try out their tutorial on molecular virology; and check out their virus database.

URL:
http://www.uct.ac.za/microbiology/

Microbiology BBS

If you're a biologist, you'll want to circulate over to this Internet BBS for exchanging mail, files, and information among others in your profession. Check out different online forums that allow for discussion of specific topics within the field of biology—industry, academics, and private practice.

URL:
http://www.microbiol.org/microbbs/

MOOSE

And here we thought the primary source of protein was found in Ovaltine, but it turns out it comes from MOOSE—the Macromolecular Object Oriented Search Engine at the Department of Biochemistry and Molecular Biophysics, Columbia University. MOOSE provides quick searching of a number of features of macromolecular structure as found or derived from data in the Protein Data Bank.

URL:
http://cuhhca.hhmi.columbia.edu/moose/

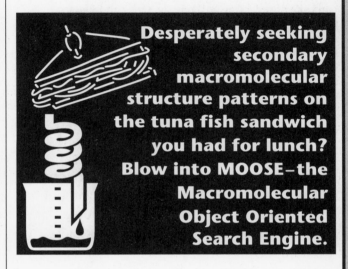

Desperately seeking secondary macromolecular structure patterns on the tuna fish sandwich you had for lunch? Blow into MOOSE—the Macromolecular Object Oriented Search Engine.

Swiss 2-D Page

This two-dimensional polyacrylamide gel electrophoresis database contains data on proteins identified on various reference maps. Highly interactive, you'll find these proteins on the 2-D Page maps—or in a region of a 2-D Page map where one might expect to find a protein from SWISS-PROT.

URL:
http://expasy.hcuge.ch/ch2d/ch2d-top.html

USGS Listing of Biology Resources

The U.S. Geological Survey has organized links to resources of many branches of biology. This page includes everything from bio-computing to agricultural genetic technology.

URL:
http://www.usgs.gov/network/science/biology/

Virology at Australian National University

Use the databases at ANU to search for classifications and descriptions of viruses. The Index Virum presents lists of virus taxa nomenclature, and a second lists their capacities. The database is built on DELTA, the DEscription Language for TAxonomy.

URL:

   http://life.anu.edu.au/./viruses/virus.html

Virtual Frog Dissection Kit

Put away those scalpels and let the frogs live! Now you can take apart a frog without the mess and nausea normally induced by those poor formaldehyde-soaked critters. The Virtual Frog Dissection Kit, available in many languages, allows interactive dissection of a frog—including the capability to make on-the-fly movies. Kids of all ages will leap at the chance to jump in and start hacking away using the keyboard and mouse. This is one place where the frogs don't croak.

URL:

   http://www-itg.lbl.gov/ITG.hm.pg.docs/dissect/

All we can say is, it's about time someone developed a way to dissect a frog without all the nasty blood and guts. Of course . . . virtual frogs . . . why didn't we think of it before?! Sort of takes the fun out of biology class, though. Now how are pubescent boys supposed to terrorize the girls in algebra class?

World Wide Web Server for Virology

Bugs beware! There is probably no more thorough source for virus information on the Web than here at this site at the University of Wisconsin. Among other resources, it is the home of the American Society for Virology. You'll find virology-related news and journal articles, computer visualizations and topographical maps of viruses, virus sequences, digitized images of viruses by electron microscope, and much, much more.

URL:

   http://www.bocklabs.wisc.edu/welcome.html

BIOTECHNOLOGY

Biological Data Transport

Here's your link to thousands of biotech companies on the Web. Open the Biotech Registry and if they're on the Web, you'll find them here.

URL:

   http://www.data-transport.com

The Biophysical Society

Biophysics fields include photobiology, bioenergetics, molecular biology, information theory, and biological control systems, among others. For information about the boundless beauties of biophysics, begin with this home page and work your way through its links to journals, employment, meetings, newsgroups, and many other related sites.

URL:

   http://molbio.cbs.umn.edu/biophys/biophys.html

Biospace: Where Biotechnology Meets Cyberspace

An excellent link to the biotechnology industry, with Biospace you can find the people, products, companies, and resources you need to research a biotech project. There are links to biotech companies, industry resources, a gene pool, and a catalog. Biospace is searchable by keyword.

URL:

   http://www.biospace.com

Biotech Web Sites and Documents

This comprehensive list of useful biotechnology sites on the Web includes government agencies, company listings, educational resources, databases, and many other resources.

URL:

   http://inform.umd.edu:86/EdRes/Topic/
   AgrEnv/Biotech/.www.html

Biotechnology at the Virtual Library

This great list of biotechnology links at the Web Virtual Library is maintained by Cato Creative Systems. It's categorized by subjects, including education, sources of information, research, publications, pharmacology and toxicology, genetics research, and many others.

URL:

   http://www.cato.com/interweb/cato/biotech/

The Net is humanity's greatest achievement.

Biotechnology Dictionary

When you're looking for a protein in the catabolic repression system, you'll find "cyclic adenosine monophosphate receptor protein" in the BioTech Dictionary. Simply type in your favorite biotech babble and out pops a cogent definition, understandable by anyone with a Ph.D. in biochemistry (or a relative of Jonas Salk). If you have a fast connection, or a lot of patience, you can view the entire BioTech dictionary.

URL:
http://biotech.chem.indiana.edu/pages/
dictionary.html

Biotechnology Information Center

The National Agriculture Library created this useful page to help you find and navigate the biotech resources on the Web. Links are provided to publications, patents, educational resources, bibliographies, and many other sources of biotech information.

URL:
http://www.inform.umd.edu:8080/EdRes/Topic/
AgrEnv/Biotech/

Biotechnology Permits

Importing and moving genetically engineered plants and microorganisms is highly regulated. The U.S. Department of Agriculture and Iowa State University have set up this page to explain regulations and how to obtain a permit to move biotechnological products.

URL:
http://www.aphis.usda.gov/bbep/bp/

DNA Vaccine Web

The DNA Vaccine Web is an interesting site focusing on vaccine construction and provides information and protocols for fighting disease using genetics. There's information about plasmids and diseases, and articles of biotechnology information.

URL:
http://www.genweb.com/Dnavax/dnavax.html

Federally-Funded Research in the U.S.

Connections to the major government agencies that distribute federal research money, including the U.S. Department of Agriculture, National Institute of Health, National Science Foundation, Advanced Technology Program, Small Business Innovation Research, and the Community of Science-Expertise, Facilities and Inventions.

URL:
http://medoc.gdb.org/best/stc/us-rd-im.html

National Center for Biotechnology Information GenBank

Have you discovered a gene sequence and wondered if you were the first? Check the NCBI GenBank—a genetic sequence database created by the NIH that collects all known DNA sequences. If you like, you can enter your sequences into the GenBank for other scientists to use.

URL:
http://www.ncbi.nlm.nih.gov

Excerpt from the Net...

(from the NCBI GenBank)

GenBank is the NIH genetic sequence database, a collection of all known DNA sequences. A five-page description is available. There are approximately 385,000,000 bases and 556,000 sequences as of October 1995. As an example, you may view the record for the neurofibromatosis gene. The complete release notes for the current version of

NIEHS Laboratory of Molecular Biophysics

The National Institute of Environmental Health Services offers a molecular biophysics laboratory that gives you access to biophysics-related databases with state-of-the-art search engines and accurate and thorough reporting.

URL:
http://lmb.niehs.nih.gov/LMB/home.html

A
B
C
D
E
F
G
H
I
J
K
L
M
N
O
P
Q
R
S
T
U
V
W
X
Y
Z

WIS Plasma Laboratory

Weizeman Institute of Science does research on many things involving plasma, including its interaction with certain magnetic and electrical fields. This program provides names of involved scientists, lists by country of plasma servers, and helpful software resources.

URL:

http://plasma-gate.weizmann.ac.il

BOTANY

Abstracts from UCSD

Understanding flower development on a molecular level is the goal of this lab at the University of California at San Diego. The studies are conducted on a weed called *Arabidopsis thaliana*. Here, you'll find a variety of abstracts from recent publications.

URL:

http://www-biology.ucsd.edu/others/
yanofsky/abstracts.html

Algy's Herb Page

Whether you pronounce it "herb" or "erb," Algy's got it all! There are links to hundreds of specific herb-related sites. Topics are broken down into growing herbs, cooking with herbs, medicinal use, herb catalogs, and a herbal discussion group. Take your pick of text or color and graphics pages.

URL:

http://frank.mtsu.edu/~sward/herb/herbpic.html

Arabidopsis Thaliana Database

Do you have a hankering for small flowering plants as model organisms for the plant kingdom? If so, come and commune with information about this member of the *Brassica* family that enjoys no major agronomic importance, yet conveys a world of information pertaining to its haploids. If this hasn't intrigued you enough so far, you still haven't dialed into its DNA map yet. Wonders abound!

URL:

http://weeds.mgh.harvard.edu

Arboricultural Stuff

A nice graphic of how to prune plants and trees accompanies this page that features guidelines for correct cutting and tree treatments. There are examples of some common tree myths, do's and don'ts of pruning, and ways to treat your tree's or plant's wounds if you wind up blowing it.

URL:

http://www.sccs.swarthmore.edu/~justin/
Docs/arbor.html

Australian National Botanic Gardens

Discover all the botanical Web resources that Australia has to offer at this server. There are links to items for students and teachers, for serious botanists and for botanical gardens managers; plus other environmental and biodiversity services and Australian regional information. The page contains a searchable index based on key words.

URL:

http://osprey.anbg.gov.au/anbg/
anbg-introduction.html

Bonsai and the Tea Ceremony

Bonsai is more than miniature gardening—it's an art form. Pruning, fertilizing, and watering are only a part of bonsai. Along with an understanding of these beautiful creations comes an understanding of the Oriental mind. This essay attempts to translate the art and philosophies of bonsai.

URL:

http://www.pass.wayne.edu/~dan/tea.html

Bonsai Dictionary

Are you trying to get into bonsai, but find yourself struggling with terms like "Eda-nuki" and "Kuro-tuschi"? Look them up in the Bonsai Dictionary.

URL:

http://www.pass.wayne.edu/~dan/
Bonsai_Dictionary.html

Kids, check out Kids and Amateurs.

Botanists' Catalog

Here is an enormous page containing hundreds of pointers and descriptions to botanical resources around the Internet. Categories include: Algae, Arboreta, Botanical Gardens, Bryophytes, Carnivorous Plants, Chromosome Numbers, Floras, Fungi, Images, Paleobotany, Plant Genetics, Poisons, Specimen databases, Threatened Plants, and more.

URL:
http://www.helsinki.fi/~rlampine/botany.html

Botany Resource

Cornell University has cultivated an excellent bunch of botanical resources from across the country. A great place to start for budding botanists on the Internet.

URL:
http://muse.bio.cornell.edu/cgi-bin/hl?botany

Cactus Mall

Browse through the Cactus Mall and find all you need to know about cacti and succulent plants. A picture gallery with descriptions of the plant and links to other prickly resources are just part of what you will find.

URL:
http://www.demon.co.uk/mace/cacmall.html

Carnivorous Plant Database

This database includes over 3,000 entries with exhaustive nomenclatural synopses and photos of all carnivorous plants, including the famous *Dionaea Muscipula*. Not sure what that is? Plug it into their search engine and find out!

URL:
http://www.hpl.hp.com/bot/cp_home

Looking for a science project? Look in Kids and Amateurs.

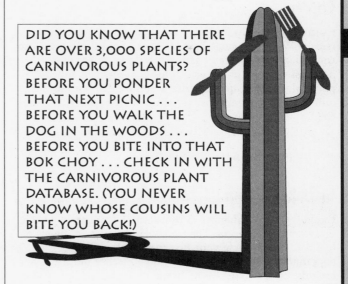

DID YOU KNOW THAT THERE ARE OVER 3,000 SPECIES OF CARNIVOROUS PLANTS? BEFORE YOU PONDER THAT NEXT PICNIC . . . BEFORE YOU WALK THE DOG IN THE WOODS . . . BEFORE YOU BITE INTO THAT BOK CHOY . . . CHECK IN WITH THE CARNIVOROUS PLANT DATABASE. (YOU NEVER KNOW WHOSE COUSINS WILL BITE YOU BACK!)

Center for Plant Conservation

Plant yourself here at the CPC if you're an aficionado of rare arboreta. CPC maintains 486 of America's rarest plant species in live, protective custody in the National Collection of Endangered Plants. See how you can help by visiting its search-engine database sometime soon.

URL:
http://www.mobot.org/CPC/welcome.html

Environmental Biology Program

BYOBB (bring you own boda bag) when you visit a unique assemblage of plants from the world's arid and semiarid regions. The Environmental Biology Program at Boyce Thompson Institute in central Arizona displays, studies, and preserves this diverse collection of dandy desert delights. Take a virtual tour of boojums, cardons, and saguaros, and learn more about the program from the comfort of your own oasis.

URL:
http://birch.cit.cornell.edu

GardenNet

GardenNet is a great resource center for garden enthusiasts. Read the online *GardenNet* magazine and consult the Ardent Gardener to learn great planting tips and tricks. Visit some online gardens and check out garden associations, upcoming events, books, and multimedia gardening resources.

URL:
http://www.olympus.net/gardens/welcome.html

A B C D E F G H I J K L M N O P Q R S T U V W X Y Z

GardenWeb

Cultivate your passion for posies, plants, and people with GardenWeb. This page hosts forums that allow you to tap the collective wisdom of experienced gardeners and botanists. Post a query or initiate a discussion on a particular plant topic. There's a mailing list, contests, online botanical gardens to visit, and tips and tricks from the Cyber-Plantsman.

URL:
http://www.gardenweb.com

Greenhouse Tour

Explore the greenhouses of the Botany Department of the University of Georgia. The tour includes several pages of beautiful graphics of your favorite plants with explanations of their classifications. Pitcher plants, orchids, bougainvillea, and bromeliads are just a few of the many colorful plants you can find at this web site.

URL:
http://dogwood.botany.uga.edu

Guelph Greenhouse

The Botany Department at the University of Guelph has included a greenhouse of plant graphics worth browsing. You can view images of pineapples, Venus flytraps, orchids, cacti, and even the ancient plants once used to make paper. If you are more interested in the microscopic world of plants, take a peak at the Confocal Facility to see graphics of pollen grains, microtubules, and chloroplasts.

URL:
http://www.uoguelph.ca/CBS/Botany/
botany.htm

Herbal Hall

According to Algy's Herb Page, this is the Web's largest medicinal herb resource. There is a discussion list for professional herbalists, topical articles on medicinal herb usage, book reviews, frequent herb news updates, FDA information, links to many related sources, and great graphics to boot!

URL:
http://www.crl.com/~robbee/herbal.html

Anyone can browse the Web.

Internet Bonsai Club

Enthusiasts of purposely pruned miniature trees and plants should cut over to this growing page of Bonsai-related resources. You'll find a Bonsai FAQ, species care guide, the USDA Growing Bonsai pamphlet, commercial suppliers, and bonsai book lists and reviews. Be sure to check out the icons submitted for the Internet Bonsai Club Icon contest! Other pages are simply dwarfed by comparison.

URL:
http://www.pass.wayne.edu/~dan/bonsai.html

Know Your Fruits and Berries

Is your favorite fruit a pepo or a pome? This colorful page teaches you the finer points of fruit botany. Just point to a fruit and learn its classification. Check out the "Fruit Key" for more images of delicious fruits. This site is highly interactive and encourages licking your monitor for best results.

URL:
http://arnica.csustan.edu/maps/fruit.html

Is a kiwi fruit a fruit or a berry? Find the answer on the fruit image map at Cal State's Biological Sciences Server. (We'll bet you get it wrong.)

Mr. Banks' Trees of the World

Compiled by a fourth grade class and other students from around the world, this resource is usable and valuable to anyone who wants to identify a tree. Trees of the World is an informative collection that focuses on trees by the states or countries in which they're found. Excellent photographs of each tree and information accompany each tree entry.

URL:
http://www.hipark.austin.isd.tenet.edu/home/
trees/main.html

Orchids, Orchids, and More Orchids

Coming in a variety of colors and shapes, these popular plants seem to run rampant in Wisconsin. Photos of the species, descriptive information, and a range map are included in Orchids of Wisconsin. Just to be fair, there are a few images of out-of-state orchids.

URL:
http://www.library.wisc.edu/Biotech/demo/orchid/Orchids_of_Wisconsin/Orchids_of_Wisconsin.html

Plant Fossil Record

The Plant Fossil Record database includes descriptions of thousands of extinct plants, plus modern genera with fossil species.

URL:
http://sunrae.uel.ac.uk/palaeo/pfr2/pfr.htm

Plant Pathology at Rothamsted Experimental Station

Diagnose and cure your photosynthetic pal with help from RES's sections on plant disease symptoms. Solve your plant's problem using their handy electron micrographs of viruses and viroids. If that doesn't work, you can always sing to your flowering friend while checking out all their links to other resources.

URL:
http://www.res.bbsrc.ac.uk/plantpath/

Red Hot Chili Peppers

These spicy plants that originated in South America are a delightful (or painful) experience for the taste buds. There are about 26 species of peppers. Everyone knows about bell peppers and jalapenos, but how about those hairy rocotos?

URL:
http://chile.ucdmc.ucdavis.edu:8000/www/intro.html

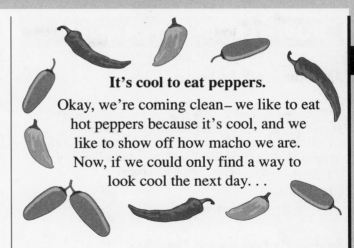

It's cool to eat peppers.
Okay, we're coming clean– we like to eat hot peppers because it's cool, and we like to show off how macho we are. Now, if we could only find a way to look cool the next day. . .

South Carolina Botanical Gardens

Research, relax, and renovate your botanical preclusions with this site that focuses on a holistic perception of botanical information and resources. Stroll through the online South Carolina garden, and keep up with botanical science, news, and well-being.

URL:
http://agweb.clemson.edu/hort/scbg/intro.html

Viruses of Plants in Australia

It's true, plants get sick, too. This database displays Australian-specific data on plant viruses, their distribution, host plants, character lists, vectors, and agricultural aspects. But, have you ever heard of a flower with a bad case of hay fever?

URL:
http://life.anu.edu.au/viruses/Aussi/aussi.html

World Wide Botany Resources

With links to the mosses of China and mushrooms of Slovenia, virtually everything that grows can be found on this beautifully groomed and well-maintained page.

URL:
http://meena.cc.uregina.ca/~liushus/bio/botany.html

A B C D E F G H I J K L M N O P Q R S T U V W X Y Z

BUSINESS AND FINANCE

American Stock Exchange

As one of the largest stock markets in the world, the AMEX is compelled to offer this useful collection of pages providing you with market summaries, news, and information on options and derivatives. There's also a complete listing of AMEX-listed companies that you can peruse by symbol, name, and home page.

URL:
   http://www.amex.com

Beginner's Investment Checklist

Interesting documents about beginner and small time investing. Included are a list of books, advice for asset allocation, and links to other documents and investment newsletters.

URL:
   http://www.investorweb.com/begin.htm

Benefits of Incorporating

Company Corporation provides information on the advantages of transforming your business into a corporation, even if you're a one-person enterprise. Find out more information about incorporating, including the two main reasons that people do it—tax sheltering and liability protection.

URL:
   http://virtumall.com/CC/inc.html

Business and Commerce

If the business of America is business, you'd be wise to consult this directory of business resource categories. You'll find a wealth of information on business administration, policy, strategies, finance, law, regulatory issues, commodity and consumer prices, domestic and foreign currency statistics, trends, news, and other general reading.

URL:
   http://galaxy.einet.net/galaxy/
      Business-and-Commerce.html

Everyone should be on the Web.

Business Critical Issues in Document Automation

Receive a free guide to document automation by ordering online from this Standard Register page. Learn why this subject should be treated as a business strategy, and how it can benefit your company through cost reduction, faster response time, and better decision making. The guide is both an overview and a practical road map for migrating a company from paper to digital information.

URL:
   http://www.stdreg.com/highlite.htm

Excerpt from the Net...

(from the web of the Standard Register Company)

While U.S. businesses spend $8-10 billion each year on paper forms, it has been estimated that by 1996 over half of all office information will be digital in form. The migration from paper to electronic forms is rapidly becoming a stampede, and many companies will be left in the dust...

Business Services

This page is sponsored by the All Business Network and offers a free one-stop online access point for many types of business and consumer services. U.S. Postal Service information includes address and ZIP code lookup, U.S. postage rates, stamp-related information, and a postal information locator. You can track your FedEx and UPS package from links to either company's databases. There is a resource directory full of banking, accounting, legal, investment, and business supply information and links; and last, but not least, you can search for info on any U.S. city's Chamber of Commerce.

URL:
   http://www.all-biz.com/

captive.com

Captive insurance organizations include insurance companies that are owned and controlled by their insured. **captive.com** provides location data and information on captives, insurance companies, self-insureds, risk-retention groups, purchasing groups, associations, reciprocals, and public-entity pools. You'll find a news section, a conference and seminar calendar, and an ask-the-expert forum.

URL:
   http://www.captive.com

CIB Supersite

The Chartered Institute of Bankers (CIB) is one of the world's foremost professional bodies in banking and finance. The CIB Supersite offers links to many web sites that relate to banking, finance, investment, and electronic payments. These include a list of banks online; credit/debit/charge/smart card companies; ATM networks; virtual banks, such as First Virtual; banking and finance libraries; newsgroups; and online magazines about banking and finance.

URL:
http://www.qmw.ac.uk/~cib/supersite.html

CNN Financial Network

Get the latest news on financial markets and business on CNN's Financial Network. This page offers current market levels for the DJIA and NASDAQ indices, links to breaking stories, stock quotes, and a resource center with economic calendars, search tools, and links to other financial resources on the Net.

URL:
http://cnnfn.com

CommerceNet

CommerceNet is developing secure commerce on the Internet based on the secure HTTP protocol. The CommerceNet site contains requests for proposals for supporting Internet-based electronic commerce, news, and papers on electronic commerce. Visual examples and demonstrations are also available to help you understand CommerceNet and its services. Some of the information here is available to CommerceNet members only; membership information is provided.

URL:
http://www.commerce.net/

Company Listings

Provided by InvestorWEB, this is a very useful list with investor information about hundreds of public companies. The list items are links to each company's home page on the Web, and each link is tagged with a symbol indicating whether or not you can find investment information there.

URL:
http://www.investorweb.com/company.htm

DHL Information

Chances are that you live in one of the 217 countries that DHL serves, and you can access online shipping information to and from your country and any of the other 216 from DHL's web page. You can view a list or a world map of countries served and obtain a service center address near you.

URL:
http://www.dhl.com/dhlinfo/1.html

Domestic or International Costs of Incorporating

Find out how much it will cost your domestic or international company to incorporate in any of the 50 states. This guide lists each state's fee to create "For Profit" and "Non-Profit" corporations, and the annual registered agent fee.

URL:
http://www.incorporate.com/tcc/select1.html

Electronic Commerce Australia

ECA is a leader in Australia of promoting and coordinating electronic commerce. ECA promotes the use of email, EDI, fax, mobile technology, EFT, computer to fax, and electronic forms for conducting electronic commerce business activities. Their site offers details of topics discussed at conferences, news, a list of electronic commerce publications and standards for sale, and training information.

URL:
http://www.sofcom.com.au/ECA/

Electronic Commerce Resources

This page features a large collection of articles, papers, and links that pertain to electronic commerce. Included are sections on network payment mechanisms, pricing information goods, protocols for Internet commerce, commercial sites, marketing, and legal issues.

URL:
http://gopher.econ.lsa.umich.edu/EconInternet/
Commerce.html

Are you on the Web?

A
B
C
D
E
F
G
H
I
J
K
L
M
N
O
P
Q
R
S
T
U
V
W
X
Y
Z

Anyone can browse the Net.

FedEx Airbill Tracking

The next time you use Federal Express, you can track the status of exactly where your package is by accessing this page on the Internet. Type in the FedEx package tracking number in the field provided, and the delivery information will be displayed. There is also a software download option for package tracking.

URL:
    http://www.fedex.com/track_it.html

Finance Links at the Virtual Library

An enormous list of sites relating to finance available on the Web. This is a great place to start looking for any type of financial information.

URL:
    http://www.cob.ohio-state.edu/dept/fin/cern/
    cernnew.htm

FinanCenter

Are you thinking of refinancing your home loan but unsure of how to calculate the benefits? Welcome to FinanCenter! This great free service provides personal-finance calculation programs, current interest rate quotes, advisory reports, and convenient access to professionals and their services. Sections allow you to calculate the key elements of vehicle loans and leases; compare purchase and finance options; compare the costs and benefits of competing credit cards; and consolidate your debt to lower your payments and save taxes.

URL:
    http://www.financenter.com/resources/

FMA Guide to Finance on the Internet

A categorized list of financial information on the Internet. Read news about stocks, mutual funds, associations, conferences, and investments, and follow many other interesting links.

URL:
    http://www.webspace.com/~fma/link.htm

FundLink

If you're shopping around for a good mutual fund, don't miss this site. Follow links to information on specific funds, review research, compare performance, or check out the certain funds in the spotlight. There's also a great collection of financial software including tax calculators, fund quotes, databases of fund transactions, and other commercial and freeware packages.

URL:
    http://www.webcom.com/~fundlink/

Galaxy Electronic Commerce Directory

This is a large directory of information resources about the practice and theory of electronic commerce. It covers electronic payment protocols, markets, standards, strategies, banking on the Web, digital money, and market trends. The site also offers upcoming event guides; payment software; product and service descriptions; and pointers to related collections, discussion groups, companies, and organizations.

URL:
    http://galaxy.einet.net/galaxy/
    Business-and-Commerce/
    Electronic-Commerce.html

Thinking of doing business on the Internet? Don't even think of plying your trade without plowing through the **Galaxy Electronic Commerce Directory**.

Guide to Going Public

Thinking about going public, but leery of paying accountants and lawyers to tell you how to do it? Here's a great little guide to get you started. Read it and know for certain that you'll need an army of lawyers and accountants.

URL:
    http://www.winternet.com/~grbarron/sec.html

Hermes

Hermes is a research project on the commercial uses of the World Wide Web. The aims of the Hermes project are to understand why people and businesses decide to provide and use commercial resources on the Web (isn't this obvious?), to develop reliable methods for tracking and predicting trends, and to provide commerce-related information to the Web. The site offers analysis of several Web surveys, details and results of current activities in the project, and links to other research projects on Internet commerce.

URL:
> http://www.umich.edu/~sgupta/hermes/

The Insider

Here's an omnibus packed with personal investment resources, price quotes, charts, and stock ticker information. Search for information on mutual funds, bonds, interest rates, commodities, and derivatives. The Insider also provides financial newsletters, periodicals, lists of public companies on the Web, security exchange information, portfolio tracking resources, and currency data.

URL:
> http://networth.galt.com/www/home/insider/
> insider.htm

Internet Banking and Security

This white paper about banking on the Internet begins by outlining trends in retail banking, and introduces the Internet as an insecure, open network. Then it goes on to discuss Internet banking, followed by an introduction to Internet security that covers cryptography, firewalls, and trusted operating systems. Finally, it concludes by suggesting that banks need to have a presence on the Internet now, before other types of companies move in and take over financial relationships with bank customers.

URL:
> http://www.sfnb.com/fivepaces/wpaper.html

Internet Commerce Essays and Links

Here's yet another collection of essays and links relating to electronic payments on the Internet. These essays are in HTML format and cover smart contracts, multinational small business, and off-line digital cash. Links to Internet payment systems—such as Digicash, Mondex, and First Virtual—are provided. There are also links to many other sites that pertain to Internet commerce, payment systems, business, and network security.

URL:
> http://www.digicash.com/~nick/commerce.html

Internet Marketing Discussion List

This mailing list for discussing the marketing of services, ideas, and items on the Internet is a good source of information for anyone who wants to create Internet sites, market products, and develop payment systems. Topics on the list include secure ordering systems, digital cash, trademarks, copyrights, and advertising. There are also extensive archives that you can search by keyword or chronologically. One word of caution: this is a very active list.

URL:
> http://www.popco.com/hyper/
> internet-marketing/

Mortgage Calculator

How much house or car can you afford? Fill in the blanks with an interest rate, principal loan balance, amortization period, and prepayment information. Click "Calculate Mortgage" and this useful program will tell you exactly what each payment will be for the entire length of the mortgage.

URL:
> http://ibc.wustl.edu/mort.html

Newspapers and Periodicals

Get the scoop on business in America and the world through this list of online journals and magazines. Some of the major players are *Inc.*, *Worth*, *Wall Street News*, and *TRADE*—a daily international trade publication.

URL:
> http://galaxy.einet.net/galaxy/
> Business-and-Commerce/
> Business-General-Resources/
> Newspapers-and-Periodicals.html

Online Market Research Internet Survey

For business to flourish on the Internet, we need solid demographic data about Internet users and online service subscribers. Some of the results of perhaps the first statistically defensible survey of Internet and online service users can be found here. Highlights of the study are the true number of U.S. Internet users, features that attract web surfers to a web site, and factors that will transform a web surfer into a web buyer.

URL:
> http://www.ora.com/survey/

A
B
C
D
E
F
G
H
I
J
K
L
M
N
O
P
Q
R
S
T
U
V
W
X
Y
Z

Online Small Business Workshop

Eighty percent of new businesses fail within the first three years; but if the owners had been able to profit from this wonderful business page, perhaps more enterprises would have survived! Designed to provide techniques for developing your ideas and putting them to work, this great online resource makes starting a new venture seem like a piece of cake. Topics covered include evaluating and protecting your ideas; concepts for small businesses; patents; trademarks; copyright; marketing basics and research; sales forecasting; and financing.

URL:

> http://www.sb.gov.bc.ca/smallbus/workshop/
> workshop.html

Professional Development Associates

If you're in charge of hiring, you can analyze your prospective salespeople to see whether they'll be productive or not by sending for a free email copy of *Twelve Types of Sales Call Reluctance*. Then, to be fair, don't forget to read Arthur Miller's *Death of a Salesman*.

URL:

> http://www.infoanalytic.com/salespower/

Sample Business Plan

Start your new business off on the right foot by consulting this sample business plan from the Canada/British Columbia Business Service Centre. Use the plan as a reference guide when you prepare your own. It will help you answer your lenders' questions and organize your strategy in a logical fashion.

URL:

> http://www.sb.gov.bc.ca/smallbus/workshop/
> market/sample/sample.html

SEC EDGAR Database

One of the classic resources that validates the existence of the Net, the Securities and Exchange Commission's EDGAR Database of Corporate Information used to be something companies paid big commissions to get access to. Find out how a company is performing by perusing its latest corporate report; check out what its stock is trading at; scan its balance sheets; or simply find out where the business is located.

URL:

> http://town.hall.org/edgar/edgar.html

Security APL Quote Server

Get continuously updated Dow Jones Thirty Industrial and NASDAQ composite stock market quotes, plus other news and company profiles online. This service is free for 60 days from the time you register.

URL:

> http://www.secapl.com/cgi-bin/qs

Selling Information on the Internet

The **www-buyinfo** mailing list is for discussing topics on information commerce over the Internet. Other relevant topics include the privacy of financial transactions, authentication of payers, and the efficiency of payment schemes. There is also a large hypermail and FTP archive of all the messages ever sent to the list, pointers to payment schemes, and other relevant web pages.

URL:

> http://www.research.att.com/www-buyinfo/

Excerpt from the Net...

(from the Online Small Business Workshop)

Many people think that all you need to start a business is one good idea. While that can sometimes be true, the best foundation for business success is to have several possibilities for creating profits. If the main idea does not test out well, an alternative is available to be tried.

It is a worthwhile exercise to explore additional products, services, procedures and processes that may enhance your initial idea. This section provides you with forty concepts that may help you with this process.

Some business people are skilled at carrying out this process on their own, while others may want to assemble a "team" - people with different talents and skills who can interact to produce useful ideas and evaluate their potential.

Ongoing business success is often linked to the ability to produce ideas on a continuing basis, testing each to determine its suitability. This process ensures that a business stays in touch with the real opportunities of the marketplace.

It's already begun...
It's taken multiple millennia to develop mankind's greatest medium–the Internet–and less than a decade for us to start using it for junk mail. Now you, too, can consume vast quantities of high-speed digital bandwidth to promote your product. Buy your way into the **www-buyinfo Mailing List Home Page** and find out how you can help make the Internet emulate the U.S. Postal Service.

Southern California New Business Association

It's great to live in Southern California; and if you do and you're actually making a good living, so much the better! Network with other business professionals by consulting SCNBA's calendar of upcoming meeting and lecture events for San Diego and Orange Counties.

URL:
    http://www.catalog.com/rmg/scnba.htm

StockMaster

Get the latest stock market and mutual fund information from this experimental server at MIT. Visual charts and graphs make it easy to spot hot stocks and chart the market with historical data files.

URL:
    http://www.ai.mit.edu/stocks.html

U.S. Census Bureau

As the "Factfinder for the Nation," the Census Bureau collects data about the people and economy of the U. S. and produces a wide variety of statistical data products. Demographic information is an important tool in business, and you can take advantage of this data online. These include printed reports, statistical briefs, and computer files on tape and CD-ROM media. Enter the name of your city and find out tons of interesting statistics.

URL:
    http://www.census.gov

UPS Package Tracking

United Parcel Service packages can be tracked online via this web site. Simply enter the package number into the appropriate field, and find out its status and expected delivery time. There is also a software download option called MaxiTrac for package tracking.

URL:
    http://www.ups.com/tracking/tracking.html

Yahoo Business Summary

Yahoo, the famous web directory service, maintains a concise and up-to-date business news section. Select hypertext headlines to read breaking business news articles from Reuters NewMedia, or peruse the past week's stories.

URL:
    http://www.yahoo.com/headlines/current/
    business/

Your Life

If you are just beginning to learn about personal finances and what you should be doing with your money, this is a simple and easy-to-use collection of documents about investment and financing for laymen. There are simple explanations about the advantages of buying a house instead of renting one, and simple answers to other questions of all sorts.

URL:
    http://www.yourlife.com

BUSINESS PAYMENT SYSTEMS

Electronic Commerce, Payment Systems, and Security

IBM Zurich provides this page with links to information on payment projects and systems, payment services via the Internet, intellectual property rights, mailing lists, payment protocols, security, cryptography, EDI, and links to financial institutes on the Internet.

URL:
    http://www.zurich.ibm.ch/Technology/Security/
    sirene/outsideworld/ecommerce.html

A
B
C
D
E
F
G
H
I
J
K
L
M
N
O
P
Q
R
S
T
U
V
W
X
Y
Z

Electronic Payment Schemes

This address will lead you to a page on the W3C organization's Web server that collects a commented list of proposals for Internet electronic payments. It offers a roadmap to electronic payments, working group information, speeches and slide presentations, specifications for the Visa STT and Mastercard SEPP protocols, and draft papers. The payments roadmap presents a layered protocol model, a general payment protocol model, a long list of Internet payment schemes and proposals, conference news, and related articles.

URL:
   http://www.w3.org/hypertext/WWW/Payments/

Internet Money Survey

As the number of businesses on the Internet is growing exponentially, the need for secure payment mechanisms has become critical. This questionnaire was designed to identify people's attitudes toward using money on the Internet. An analysis of the responses and aggregated raw results are presented here; for example, 81.9 percent of the respondents expressed a willingness to use money on the Internet.

URL:
   http://graph.ms.ic.ac.uk/money

Millicent

This article describes Millicent—an examination of the feasibility of small-scale commercial transactions over electronic networks. Millicent's developers hope to reduce transaction costs, increase transaction rates, and provide a higher level of confidence to vendors. From the ground up, Millicent is designed to support purchases of less than one cent. The Millicent protocol specifications are available here in the PDF and PostScript formats, and slides are also available in PowerPoint, PDF, and PostScript formats.

URL:
   http://www.research.digital.com/SRC/millicent/

Want to see some cool resources? Check out Internet Resources.

Netscape Commerce Products

Netscape has developed a range of software and protocols for performing electronic commerce on the Internet. The Netscape Commerce Server is a high-performance system for conducting secure electronic commerce on the Internet, providing data encryption, data integrity, and user authorization. Data communications are based on open standards—including HTML, HTTP, the Common Gateway Interface, and the Secure Sockets Layer (SSL) protocol. The Merchant System can securely handle transactions from Internet users. Technical details and specifications of the security standards on which these products are based are also available here.

URL:
   http://home.netscape.com/comprod/
   netscape_products.html

Network Commerce and Intelligent Agents

This directory of research papers on digital currency and related topics offers pointers to commercial sites that promote or design network commerce systems. You can also read about the details of payment protocols, digital currencies and providers, marketing resources and data, and intelligent agent resources.

URL:
   http://www.spp.umich.edu/telecom/
   net-commerce.html

Network Money

This page brings together a collection of articles on the future of banking and commerce on the Internet. A feature of the GNN Personal Finance Center, there's an introduction to electronic commerce, a reader survey on Internet banking, an interview discussing the importance of transaction security, and pointers to network money sites.

URL:
   http://nearnet.gnn.com/gnn/meta/finance/feat/
   emoney.home.html

Network Payment Mechanisms and Digital Cash

If you're researching network payment mechanisms and digital cash, this is the place to begin your work—a collection of papers, articles, reports, press releases, discussions, implementation tools, links to payment development and related sites, and more. This page has a particularly good section on Internet payment methods with links and details about almost every payment system designed for use on the Internet.

URL:
   http://ganges.cs.tcd.ie/mepeirce/project.html

We like money. Admit it . . . you like money too. Without it, how would we buy all this cool stuff we need to plug into the Internet?

Deposit yourself into the *Network Money* pages, and find out how you can convert your real money into a loose collection of electrons that flow freely through cyberspace.

OpenMarket

OpenMarket offers a collection of products for Internet commerce. They've designed a transaction management system that can host an electronic marketplace, enable business-to-business transactions, and offer home banking or bill-payment services. Available here are product documentation, technical papers on their payment mechanisms, company information, and articles on security.

URL:

http://www.openmarket.com

Payment Mechanisms Designed for the Internet

This is a collection of links and pointers to existing payment schemes that were designed for, or are being used on, the Internet. It includes links and articles on such systems as Bitbux, CyberCash, First Virtual, IKP, NetBank, NetBill, NetCheque, NetMarket, and Security First Network Bank, as well as electronic cash systems like DigiCash's Ecash, NetCash, Mondex, and Magic Money.

URL:

http://ganges.cs.tcd.ie/mepeirce/Project/
oninternet.html

Payment Protocol Discussions

The E-PAYMENT mailing list is a discussion forum for payment protocols based on cryptography and online authorization.

Mailing List:

Address: **majordomo@cc.bellcore.com**
Body of Message: **subscribe e-payment**

The Web is too cool.

RediCheck

RediCheck offers a system for online payment and ordering using customer's regular checking accounts. The RediCheck system doesn't require any additional hardware or software and is based on account identifiers, passwords, and email verification, much like the First Virtual system. Their web page offers press releases, a FAQ, an application form for a free customer account, pointers to places to shop, and a description of how the system works.

URL:

http://www.redi-check.com

Secure Electronic Payment Protocol

To secure bankcard transactions over open networks like the Internet, IBM, Netscape, GTE, CyberCash, and MasterCard have teamed up to develop a Secure Electronic Payment Protocol (SEPP). The spec they came up with is an open, vendor-neutral, non-proprietary, license-free specification for securing online transactions and is divided into four parts: business requirements, a functional specification, a payment system specification, and a certificate management specification. The full SEPP specification is available in PostScript or Word 6.0 formats. An acronym list and glossary are also provided.

Mailing List:

Address: **majordomo@cc.bellcore.com**
Body of Message: **subscribe ietf-payments**

URL:

http://www.mastercard.com/Sepp/sepptoc.htm

Work of Stefan Brands

Stefan Brands is a Ph.D. student who has been conducting research for several years in the area of untraceable off-line electronic cash and privacy-protecting mechanisms for digital credentials. Here, you can find Stefan's work on Internet cash systems and his publications in PostScript form.

URL:

http://www.cwi.nl/~brands/

A B C D E F G H I J K L M N O P Q R S T U V W X Y Z

Yahoo Electronic Commerce Directory

The Yahoo directory will come up with enough links on electronic commerce and digital currency resources to drown any researcher. It has sections and details on conferences, mailing lists, Electronic Data Interchange (EDI), digital money, transaction clearing, shopping centers, newsletters, software, Internet businesses, and much more. There's also an academic research survey on Internet payment systems and articles on agent-assisted consumer shopping, the law of electronic commerce, smart cards, and various payment protocols.

URL:

  http://www.yahoo.com/Business_and_Economy/
  Electronic_Commerce/

BUSINESS PAYMENT SYSTEMS: CASH

CAFÉ

Based on recent research in public key cryptography, the CAFÉ Project is a European undertaking to develop an off-line electronic cash system for the public using smart card technology. CAFÉ will be anonymous and highly secure, protecting the issuer against fraud and the holder against loss. On this page, you can get a software simulation package for Windows that shows how CAFÉ will work and what you will be able to do with it.

URL:

  http://www.digicash.com/products/projects/
  cafe.html

Why bother converting only some of your money into digital dollars? Buy into CAFÉ, and convert your whole wallet! CAFÉ . . . the electronic wallet for the information age.

Consumer Software

Here's the place to come for software for any popular computer platform to use with electronic payment systems. Software is available for a variety of payment systems, including CheckFree, CyberCash, DigiCash, First Virtual, NetBank, and VentureTech Online.

URL:

  http://www.yahoo.com/Business_and_Economy/
  Companies/Financial_Services/Digital_Money/

DigiCash

DigiCash develops payment technology products that use smart cards, software only, and hybrid systems that use a combination of hardware and software. DigiCash's technology is based on advances in public key cryptography developed by the company's founder, David Chaum. On their home page, DigiCash offers details about the company; technical papers and reports; press releases; and product descriptions, including Ecash, smart cards, road toll technology, CAFÉ, and encryption tools.

URL:

  http://www.digicash.com

DigiCash Smart Cards

This page contains introductory descriptions of some of the smart card mask technologies available from DigiCash. Some of these technologies use public key cryptography and digital signatures. Also available are pictures of specific smart card products, and details of the hardware and software development tools for use with these cards.

URL:

  http://www.digicash.com/products/
  smartcard.html

Digital Cash Articles

This Gopher server houses a collection of articles on digital cash and payment systems—mostly from the Cypherpunks mailing list. Included are a bibliography; discussions of the protocols designed by David Chaum; legality issues; problems and ideas for solving them; payment protocol proposals; technical opinions; explanations of untraceable and anonymous electronic cash; and other related topics.

URL:

  gopher://idea.sec.dsi.unimi.it:70/11/docs/
  dig_cash

Digital Cash Mini-FAQ

This document, which is actually a short FAQ on electronic cash, is an excellent introduction to the topic of digital cash. It explains how digital cash is possible; the different kinds of digital cash; the double-spending problem; and the types of digital cash systems, such as online and off-line.

URL:

> http://world.std.com/~franl/crypto/
> digicash-minifaq.html

Digital Money

Assembled by Brad Cox, this page brings together a large collection of articles, press releases, and links to sites that relate to electronic currency. Included are papers and articles on many electronic cash protocols, articles about the future of electronic commerce, privacy issues, and links to cybershops using electronic cash.

URL:

> http://web.gmu.edu/bcox/ElectronicProperty/
> 00ElectronicMoney.html

Ecash

Based on public key cryptography, Ecash is an anonymous, secure, electronic cash payment system for the Internet. One of the most important features of Ecash is that it has the same level of privacy as paper cash. A payer can choose to be known or remain anonymous. On these pages, the technical concepts and mechanisms used in Ecash are presented, along with press releases, news, and publications. There's also a trial system running where you can get 100 cyberbucks for free to use at a large variety of experimental Ecash shops that are up and running.

URL:

> http://www.digicash.com/ecash/ecash-home.html

Don't forget to swing by the virtual ATM and pick up some ecash.

Planning an outing? Check the weather in Meteorology.

Electronic Money and Money in History

This page offers a wealth of information about money. Pointers here cover many aspects of money, including its evolution from the dawn of society up through the emerging electronic commerce of today. With essays on the history of money and money transfers, pointers to interesting resources on money around the world, links to electronic currency resources and systems, resources on non-conventional financial systems, and articles about Internet economics, this page is must-see Internet.

URL:

> http://www.ex.ac.uk/~RDavies/arian/money.html

Electronic Payment over Open Networks

This five-part article is a tremendously useful introduction to the problem of securing electronic payments over the Internet and other open networks. Written by two employees of IBM Zurich, the article delves into the problems and some solutions for electronic commerce. The article is illustrated, and covered topics include online versus off-line systems, tamper-resistant hardware, cryptography, and buyer anonymity.

URL:

> http://www.zurich.ibm.ch/Technology/Security/
> publications/1995/JaWa95.dir/JaWa95e.html

Imagine the possibilities! No more waiting in line while you wonder if you'll get the good-looking teller or the ugly one. No more condescending loan officers asking what you intend to use for collateral. No more hoping you'll get the ugly teller so that you don't look like a fool to the good-looking one when you explain that you need a cashier's check because your rent check bounced. Experience virtual banking – the newest way to avoid those embarrassing real-life situations.

A B C D E F G H I J K L M N O P Q R S T U V W X Y Z

First Bank of Internet

FBOI is an electronic payment system that uses email to transfer funds from a prepaid, PIN-protected Visa Automated Teller Machine (ATM) card. Forgeries and impersonations are prevented by using PGP software for every transaction. This mail server will send you the details on how to sign up for the FBOI system; how to obtain and use PGP; and the technical concepts, features, and security of FBOI.

Mail:
>   Address: **fboi@netcom.com**
>   Subject: **information**

Hack Ecash

Truly secure digital cash has the potential to revolutionize electronic commerce as we know it. It is vital that it be made as secure as possible. The Ecash system from DigiCash is now being used to transmit real money over the Internet. In order to find and patch any security flaws, Community ConneXion is offering a T-shirt to anyone who can find holes in the Ecash system. This page offers some ideas for possible hacks and additional information about the contest.

URL:
>   **http://www.c2.org/hackecash/**

LETSystems

A LETSystem (Local Exchange Trading System) is a community trading network that uses a local currency to facilitate the indirect exchange of goods and services between users. This web page provides a large amount of detail on the theory, practice, and development of LETSystems, and how they can be implemented in an electronic community or on the Internet.

URL:
>   **http://www.u-net.com/gmlets/**

Mark Twain Bank

The Mark Twain Bank was the first bank connected to the Internet to use real money in the Ecash system designed by DigiCash. Account holders can buy Ecash worth real money and use it to pay for objects and services on the Internet. On their web pages, the bank makes available their press releases, the basics of Ecash, an application form for opening an account, and pointers to merchant shops that accept Ecash.

URL:
>   **http://www.marktwain.com**

Mondex

The Mondex system uses a smart card to store electronic cash that you can use to pay for goods and services in much the same way as you use cash. Mondex value can be sent and received instantly across phone lines, networks, and the Internet. The Mondex Home page gives a history of the Mondex scheme, FAQs on the system, pictures of the card and other Mondex devices, balance readers, an electronic wallet, a list of manufacturers of Mondex devices, news, and contact addresses. There are currently several pilot tests of Mondex underway throughout the world.

URL:
>   **http://www.mondex.com/mondex/home.htm**

NetBank Payment System

NetBank is an electronic cash system using email, aimed primarily at email order businesses. NetBank's web pages offer an introduction to the system, example transactions, a FAQ, and details of how to use PGP to encrypt the email messages the system uses.

URL:
>   **http://www.teleport.com/~netcash/**

NetCash

NetCash is a framework for electronic currency under development at the University of Southern California. This system offers security, anonymity, scalability, acceptability, and interoperability. It was designed to allow payments over unsecure networks such as the Internet. Articles, papers, licensing information, details of software distribution, a slide presentation, and pointers to other network payment systems can be found here.

URL:
>   **http://nii-server.isi.edu/info/netcash/**

> ## Going shopping on the Web? Remember to pick up some ecash.

NetCheque

Using NetCheque, you can send electronic checks to other NetCheque users in the form of an encrypted email message. Check signatures are authenticated using Kerberos. The strengths of the NetCheque system are its security, reliability, scalability, and efficiency. On this page, you can find technical and user documentation, papers, account application forms, software demonstrations, pointers to the NetCheque software for popular Unix systems, a list of NetCheque merchants, and pointers to other payment systems.

URL:

http://nii-server.isi.edu/info/NetCheque/

NetChex

NetChex is a system that uses proprietary security software to allow users to write electronic checks. This page offers a description of the NetChex system, its security features, corporate information, links to merchants that accept NetChex checks, member and merchant application forms, and a free NetChex client software package.

URL:

http://www.netchex.com

Semesterwork on Electronic Cash

Here's a collection of articles and descriptions of electronic payment systems presented by two Swiss students. On this page you can find a list of shops on the Internet where you can spend your money; discussions of payment systems, including online credit cards, Ecash, First Virtual, Downtown Anywhere, and CyberCash; definitions and descriptions of memory cards, smart cards, and electronic fund transfers; and FAQs about network money.

URL:

http://www.itr.ch/~pklomp/electroniccash.html

Smart Card Cyber Show

This page features an excellent presentation on smart card technology in both English and French. It gives a detailed history of the development of smart cards from the 70s—covering the advances and products available each year since 1974. The site also offers product specifications from corporate leaders, details of trade shows and conferences, international news on smart cards and their applications, close examinations of specific smart cards, and articles and reports on major card issuers. If you are interested in smart card technology and products, this site is definitely worth checking out.

URL:

http://www.cardshow.com

Thoughts on Digital Cash

This pointer will transport you to a short discussion on the enabling technology of the underground economy and digital cash. It includes a good essay on digital cash and monetary freedom, and pointers to anonymous digital cash resources like the Magic Money system of Product Cypher. There are also many links and views about the electronic frontier in general, tools for privacy, cryptography issues and protocols, and much more.

URL:

http://www.c2.org

A
B
C
D
E
F
G
H
I
J
K
L
M
N
O
P
Q
R
S
T
U
V
W
X
Y
Z

USA Check

With USA Check, merchants get paid for their merchandise by getting the customer's checking account information. The account details are sent by phone, fax, or email to the USA Check center where actual paper check drafts are prepared and sent to the merchant for deposit. A short description of the system, fees, and contact information is given here.

URL:
http://www.valleynet.net/~usaweb/usachek.htm

BUSINESS PAYMENT SYSTEMS: CREDIT

Anonymous Payment, Publishing, Privacy

This page details research at AT&T's Bell Labs involving electronic commerce. Read papers about AT&T's anonymous credit card system, and the anonymous Internet Mercantile Protocol. Also available are papers about how to transfer confidential information over a network in a way that makes it extremely difficult for an attacker to intercept. Finally, there are papers on privacy and details of how to protect the identity of information buyers.

URL:
http://www.research.att.com/lateinfo/projects/ecom.html

CARI

CARI (Collect All Relevant Information) is an Internet Voice Robot system that uses virtual credit cards to provide for secure transactions on the Web. The CARI home page offers a FAQ list explaining how the system works; visual diagrams showing the important concepts; an online demo where you're assigned a temporary virtual credit card; and a description of features, specifications, and requirements.

URL:
http://www.netresource.com/itp/cari.html

CheckFree

CheckFree is a simple payment system that eliminates paper checks and allows you to issue bills or make payments online. With the CheckFree Wallet software and a major credit card, you can buy goods, services, or information from online merchants. The wallet is a small, stand-alone application that works with all major Web browsers. The software is available for Windows and the Mac, and you can get it here. Also available are FAQs, press releases, and details about the company that produces CheckFree.

URL:
http://www.checkfree.com

CyberCash

The CyberCash Secure Internet Payment Service offers to bridge the security gap between the Internet and the world's banks. CyberCash gives Internet merchants the ability to process credit card transactions safely over the Internet, and the Money Payments Service provides a way for merchants to be paid in cash. You can download the software you need here for Windows, Macintosh, and Unix machines. There is also a users' guide, a trouble-shooting FAQ, and a list of merchants that accept CyberCash-protected payments.

URL:
http://www.cybercash.com

Downtown Anywhere

Downtown Anywhere is a virtual city with a real economy based on existing credit cards, and offering a commercial environment for shoppers and merchants. This system uses account numbers and personal payment passwords. A FAQ list covering the system and the available services is provided, and you can visit the actual storefronts to spend your money.

URL:
http://www.awa.com

So now you've got all this ecash, and you don't know where to spend it . . . How about a virtual shopping spree at **Downtown Anywhere?**

First Virtual

First Virtual is a simple credit card system that uses email for buying and selling information. FV uses no encryption because no sensitive financial information, such as credit card numbers, travel across the Internet. Here, you can read about how the system works, FAQs, descriptions of how to obtain an account, and the addresses of online information shops where you can purchase documents.

URL:
http://www.fv.com

Internet Keyed Payment Protocols

IKP is a family of secure payment protocols for making payments over open networks. Proposed by the security group of IBM Research, IKP implements credit card transactions between customers and merchants using existing financial networks for approval. These protocols can also apply to other payment models, such as debit cards and electronic checks. An overview is given here, and the full paper is available in PostScript format. There is also a publication list on IKP and pointers to other related protocols and mailing lists.

URL:
http://www.zurich.ibm.com/Technology/
Security/extern/ecommerce/iKP.html

Nobody sends you email?
Join a mailing list.

NetBill

NetBill allows goods and services to be purchased over the Internet. The protocol NetBill uses is specially designed to handle low-cost items—for example, journal articles at 10 cents per page. The consumer is guaranteed the certified delivery of goods before payment is processed, and the merchant is guaranteed that the consumer cannot access the goods until payment has been received. NetBill's pages provide a good overview of the system, news, their research goals, several publications and papers, a FAQ, pointers to other electronic commerce resources, and a description of tools used to develop NetBill.

URL:
http://www.ini.cmu.edu/netbill/

NetMarket

NetMarket provides accounts based on credit cards that you can use to purchase goods and services from participating merchants. The system uses account identifiers and passwords. Their site offers a history of the company, press releases, instructions on setting up a free account, and places to shop.

URL:
http://netmarket.com/nm/pages/home

Secure Transactions Technology

STT was developed jointly by Microsoft and Visa to allow credit card holders to make secure purchases across open networks. SST uses both secret-key and public-key cryptography, and provides information confidentiality. The full STT spec is available in PostScript format here.

URL:
http://www.w3.org/hypertext/WWW/Payments/
STT.html

A B C D E F G H I J K L M N O P Q R S T U V W X Y Z

CHEMISTRY

Acid/Base Calculator

Use this feature-packed Windows program to apply activity corrections and to plot and manipulate speciation diagrams, titration curves, buffer-capacity plots, and bound-proton diagrams.

URL:

http://evanslab.chem.umn.edu/acidbase.htm

If your mind is wasting away on acid, just say no . . . to those tedious pH calculations. Download the Acid/Base Calculator from the University of Minnesota—before you burn out any more brain cells on your latest acid/base

American Chemical Society

The American Chemical Society Gopher contains supplementary material pages from the *Journal of the American Chemical Society*, a Copyright Transfer Form for authors publishing in the ACS journals, and instructions for authors for each of the 24 peer-reviewed journals.

URL:

gopher://acsinfo.acs.org

The Net is humanity's greatest achievement.

American Institute of Chemical Engineers

AIChE provides leadership in advancing the chemical engineering profession to meet the needs of society. It also promotes excellence in the development and practice of chemical engineering.

URLs:

http://www.che.ufl.edu/WWW-CHE/aiche/
http://www.et.byu.edu/student-chapters/aiche/aiche.html

Boil, boil, toil, and trouble. Trouble, indeed, unless you know how to reach AIChE on the Web.

Amino Acids

What would life be without amino acids, those tiny building blocks that make up DNA? Why, they're the basis for all organic material on Earth. Including you! Check out descriptions and models of alanine, isoleucine, guanine, cysteine, and other cool aminos in this page of primordial soup.

URL:

http://www.chemie.fu-berlin.de/chemistry/bio/amino-acids.html

Analytical Chemistry

The combination of analytical chemistry and academia generates a reaction like no other. The results include precipitous meetings, symposia, notices, calls for papers, grants, fellowship opportunities, publications, legislation, public policy, databases, and additional links to resources. Wear gloves and eye protection.

URL:

http://nexus.chemistry.duq.edu/analytical/analytical.html

CAST Mailing List

The CAST (for Computing and Systems Technology) mailing list promotes the activities of the CAST division of the American Institute of Chemical Engineers. This list provides a means for distributing calls for papers, meeting announcements, professional opportunities, and related information. This page describes the list, how to join it, and also stores the archives of the list.

URL:

http://control.cheg.nd.edu/cast10/

Do you like computers and chemistry? Join the CAST mailing list.

Chemical Mailing Lists

This page features mailing lists on virtually every topic in chemistry—from amalgam-of-mercury health issues to the Young Scientists' Network.

URL:

http://bionmr1.rug.ac.be/chemistry/
overview.html

Chemist's Art Gallery

This site in Finland contains many chemistry-related animations. These movies were produced by Finland's Center for Scientific Computing. Among the selections here are animations of small polymer molecule diffusion, proteins, chromosomes, and viruses. There are also links to visualizations and animations at other locations.

URL:

http://www.csc.fi/lul/chem/graphics.html

Chemistry Courses

Take a general chemistry, organic chemistry, or biochemistry course on the Internet. While not a substitute for actual college courses, these online lecture presentations, study guides, and summaries can broaden your understanding of chemistry. You can even take an online exam to test your newly acquired knowledge.

URL:

http://odin.chemistry.uakron.edu/genobc/

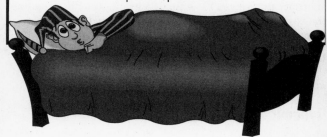

From time to time, we all experience anxiety because we just didn't get enough of chemistry class in school. If your lack of chemistry knowledge keeps you awake at night, your problems are over. Take a chemistry class on the Web and never again feel inferior because you don't know your specific gravity from your specific problem!

Chemistry Hypermedia at Virginia Tech

Usenet and IRC aside, there are many ways to get a decent education on the Internet. One of the best ways to learn about analytical, organic, and physical chemistry is through this web page. Immerse yourself in volumes of information, tons of computer graphics, movies, and Virtual Reality Modeling Language (VRML) models.

URL:

http://www.chem.vt.edu/chem-ed/
vt-chem-ed.html

Chemistry Index

More resources for chemistry exist here than you can shake a test tube at, and they're listed both in English and German. General science resources on biochemistry and molecular modeling, databases, software, chemist's address books, and even selected biographies of chemists are just a mouse click away.

URL:

http://www.chemie.fu-berlin.de/chemistry/

A B C D E F G H I J K L M N O P Q R S T U V W X Y Z

> ## Kids, check out Kids and Amateurs.

Chemistry Newsgroups

Develop some human catalytic interaction through these Usenet groups on different facets of chemistry. All the elements necessary for chemical bonding are here.

URLs:
    news:sci.chem.electrochem
    news:sci.chem.labware
    news:sci.chem.organomet

Chemistry Software Reviews

Thinking about getting that new chromatography software, but wondering what the experts are saying about it? Before you blow that 25 grand, consider reading the reviews posted here. Then you'll know if the whole thing's a real gas or just a lot of hot air.

URL:
    http://www.liv.ac.uk/ctichem/swrev.html

Dynamic Publications

Several authors shed lumens on the creation and absorption of photons on this multimedia page using Excel spreadsheets and QuickTime videos. Watch simulations and view excellent graphics of oscillating dipoles, harmonic oscillators, and dynamic quantum rigid-rotor trajectories. Quite illuminating!

URL:
    http://jchemed.chem.wisc.edu/DynaPub/
    DynaPub.html

EPA Toxic Substance Fact Sheets

This is a directory of nasty stuff. Whether you ingest, inject, inhale, or apply these chemicals to your skin, you're likely to be in a world of hurt. For example, breathe or touch tetrachloroethane and you can become unconscious, cause damage to your liver or kidneys, or die! These fact sheets document hundreds of dangerous chemicals, including identification, symptoms from exposure, precautions, and much more.

URL:
    gopher://ecosys.drdr.virginia.edu/11/library/
    gen/toxics

Gateway to Physical Chemistry

Discover why quantum mechanics is the foundation of chemistry. Explore related subjects such as statistical mechanics and spectroscopy with graphics and animation at this informative site. Tip: Some of it has to do with wave-particle duality and the Schrödinger equation.

URL:
    http://www-wilson.ucsd.edu/education/
    samplegateway.html

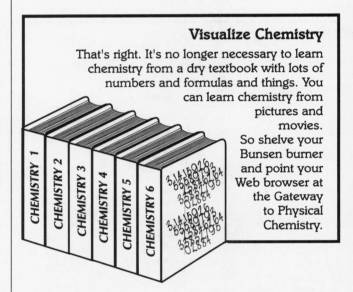

Visualize Chemistry

That's right. It's no longer necessary to learn chemistry from a dry textbook with lots of numbers and formulas and things. You can learn chemistry from pictures and movies. So shelve your Bunsen burner and point your Web browser at the Gateway to Physical Chemistry.

Journal of Chemical Education Software

JCE produces chemistry software for the classroom and laboratory. With a library of more than 50 instructional titles, programs range from introductory courses to advanced research lab software. Many of the programs have lengthy browsable abstracts and graphics online.

URL:
    http://jchemed.chem.wisc.edu

Journal of Computer-Aided Molecular Design

Published bimonthly, this extensive online journal is a forum for the dissemination of molecule analysis and computer-based design applications. Topics include theoretical and quantum chemistry, modeling, protein engineering, automated molecule generation, and drug design.

URL:
    http://wucmd.wustl.edu/jcamd/jcamd.html

Material Safety Data Sheets

Oh no, OSHA's coming for a visit! Thousands of chemicals, compounds, and materials are listed by product name in this database, including the formula for each and its physical properties, appearance, toxicity level, and hazard warnings. Information on storage and transportation of each substance is also provided.

URL:

    gopher://atlas.chem.utah.edu/11/MSDS

Mathematics and Molecules

Mathematics and Molecules (MathMol) is a K-12 online resource for chemistry students and teachers. Using VRML and innovative features, students can navigate through hypermedia textbooks, a library of molecular images, and a series of online interactive quizzes and challenges related to chemistry topics.

URL:

    http://www.nyu.edu/pages/mathmol/

Moviemol Molecular Animation Program

Moviemol is an easy-to-use molecular display and animation program for visualization and animation of molecular structures. The program may be obtained free of charge for academic researchers. It runs on PCs, IBM RISC/6000, and SGI workstations. There are videos and demos produced with Moviemol, plus links to other visualization programs.

URLs:

    ftp://chem-ftp.mps.ohio-state.edu

    http://chem-www.mps.ohio-state.edu/~lars/
    moviemol.html

NIH Molecular Modeling

Scrutinize your favorite molecules in the Brookhaven Protein Databank "PDB at a Glance." Use the "Molecules R Us" form to choose from eight different categories of molecule structures. Then pick an image and select your desired type of output. Our favorite is "space filling." Click on "Submit Request" and voila! There's your perfect little world.

URL:

    http://www.nih.gov/molecular_modeling/
    pdb_at_a_glance.html

Looking for a science project? Look in Kids and Amateurs.

Ozone Depletion

There's no zone like the ozone, and when it goes away, so do we. This site contains a mammoth FAQ on this important oxygen isotope, with relevant subtopics on the Antarctic ozone hole, how ozone breaks down in the stratosphere, and the effects of UV radiation on chemical structures.

URL:

    http://www.cis.ohio-state.edu/hypertext/faq/
    usenet/ozone-depletion/top.html

Paracelsus

The famous chemist, alchemist, and Renaissance doctor, Paracelsus, has his own web page. Pictures of medieval woodcuts, some from Paracelsus's own books, enhance the page's informative text.

URL:

    http://www.nlm.nih.gov/hmd.dir/paracelsus.dir/
    paracelsus_1.html

Periodic Table of the Elements

What happens when Aunt Beatrice calls and asks you—since you're the computer genius—for the valence shell orbital Rmax of ytterbium? Stammer not. Just put dear Aunt Bea on hold and link to Gav's Periodic Table of the Elements. A clickable interface awaits with lots of handy enthalpies for all your favorite atoms. Impress her with your boundless molecular knowledge on the history and characteristics of these universal tinkertoys. Then, just maybe, she'll keep you in her will.

URLs:

    http://ripple.bu.edu/Gavin/PeriodicTable/
    web-elements-home.html
    http://www-c8.lanl.gov/infosys/html/periodic/
    periodic-main.html
    http://www.cs.ubc.ca/elements/tab/periodic-table

A
B
C
D
E
F
G
H
I
J
K
L
M
N
O
P
Q
R
S
T
U
V
W
X
Y
Z

Quick, what's Ga? The abbreviation for a southeastern state? Nope. It's Gallium, the 31st element in the periodic table. It has an atomic weight of 69.72, boils at 2,237 degrees C, and melts at 29.8 degrees C. But since you're looking for something to boil eggs in, it will take a long time for your stove to reach 2,237 degrees C. Instead, try Hg (you know, Mercury). It boils at just over 350 degrees C and makes for healthy and great-tasting egg salad sandwiches. Really! Find out more on the Periodic Table of Elements at the University of British Columbia. Bon appétit!

Poly-Links—Plastics and Polymers

Caution. This page isn't real. It's made of plastic. This index of Internet polymer and plastics resources features an online plastics and polymers industry trade magazine, a database of industry terminology, a chat service, a job bank, a list of plastics engineering resources, and sections on polymer toxicity and safety.

URL:

http://www.polymers.com

Polymer Liquid Crystals Textbook

Case Western University sponsors this electronic textbook covering the fields of polymers and liquid crystals. Aimed at the college freshman level, this site lets you study polymer growth and liquid crystal changes through a "virtual laboratory." There are movies, animations, and Macintosh and Windows simulations.

URL:

http://abalone.phys.cwru.edu/files/
textbook.html

PrepNet

Study trace analysis, microwave chemistry, atmospheric pressure, and closed vessel digestions in the Analytical Sample Preparation and Microwave Chemistry Center at Duquesne University. You will find an article on sample preparation and automation, and a searchable database of research articles is to be implemented shortly.

URL:

http://nexus.chemistry.duq.edu/sampleprep/
prepnet.html

Presentations, Talks, and Workshops

Chemists are invited to submit papers and workshop presentations to this page. Contributions include articles on hyperactive molecules, simulations, and the future of chemistry on the World Wide Web.

URL:

http://www.ch.ic.ac.uk/talks/

Protein Images

Everyone loves proteins. They're yummy and good for you. They're kind of small, though. But thanks to the wonders of electron microscopes, now you can see what they look like in this seminal collection of over 500 images. You'll find everything from aconitase to zinc fingers.

URL:

http://expasy.hcuge.ch/pub/Graphics/IMAGES/
GIF/

Representation of Molecular Model Rendering Techniques

If thallium superconductor animations turn you into a quivering globule of ectoplasm, consult this compendium of techniques to shed the light of day on your favorite molecules. A sample of the animated or illustrated topics discussed here includes wire frame, ball and stick, space-filling models, surfaces, spin densities, vibrations, and minimum-energy reaction paths.

URL:

http://scsg9.unige.ch/eng/toc.html

Understanding Our Planet Through Chemistry

Unfortunately, many people cope with the complexity of our world by resorting to chemical abuse. Well, here's a safe way to understand the world through chemistry. This poster page, sponsored by the U.S. Geological Survey, illustrates how chemical processes affect the Earth. Topics from volcanic activity to natural and man-made pollution are discussed as influences on global chemistry. Discussions include articles on air bubbles, amber, dinosaurs, and acid rain. A selection of photos accompanies the text of each article.

URL:

http://helios.cr.usgs.gov/gips/aii-home.htm

COMPRESSION

Compression Discussion

You would expect people interested in compression to be of few words, but the folks in **comp.compression** expand on all topics related to data compression technology with pictures, voice, sound, video, data, and more.

URL:

news:comp.compression

Compression FAQ

Find out all about crunching files and more in this one-stop guide to all types of compression technology. Covered are such formats as zip, tar, lzw, TIFF, JPEG, MPEG, JBIG, and MHEG. Part 1 is oriented toward the practical uses of compression programs, Part 2 explains how compression works, and Part 3 deals with image compression.

URL:

http://www.cis.ohio-state.edu/hypertext/faq/
usenet/compression-faq/top.html

Data Compression White Paper

This detailed report from Motorola Information Systems Group discusses the importance of using data compression between branch office networks. Informative illustrations and diagrams highlight this paper that explains the time and cost savings that can be achieved with data compression.

URL:

http://www.mot.com/MIMS/ISG/Papers/
Data_Compress/

File Compression Techniques

Simple and straightforward instructions on compressing and decompressing files using popular tools such as PKZIP, gzip, and StuffIt.

URL:

http://www.vanderbilt.edu/VUCC/Docs/UN/
NETWORKS/NETWORKS_002/
networks_002.html

Anyone can browse the Web.

Gzip Info

A hypertext version of the gzip information manual documents the features and uses of this popular Unix file compression utility. Unlike archivers, gzip only compresses single files, but it does so with tremendous efficiency. Used in conjunction with tar, gzip can effectively compress multiple files.

URL:

http://www.ai.mit.edu/!info/gzip.info/!!first

StuffIt

StuffIt is a popular file-archiving tool from Aladdin Systems. The company offers shareware versions of StuffIt Lite, DropStuff, and StuffIt Expander for downloading. Files compressed in StuffIt format (.sit), as well as other archival or encoded formats, can be simply dragged onto StuffIt Expander, which will dutifully expand or convert them. Versions of StuffIt Expander are available for both Macintosh and Windows.

URL:

http://www.aladdinsys.com

That's right. Don't leave that bloated Mac file lying around on your desktop, StuffIt! Yes, this amazing software utility will take your Macintosh files and put them where the Finder won't find them. How much is this valuable offer worth? Act now and get DropStuff and StuffIt Expander free of charge—a value at half the price. So remember, DropStuff and StuffIt!

WinZip

One of the best compression utilities for Windows, WinZip lets you effortlessly manage ZIP files without requiring fumbling around at the DOS prompt with PKZIP and PKUNZIP. It features built-in support for popular Internet file formats, including tar, gzip, and Unix compress. The latest shareware version is available for downloading.

URL:

http://www.winzip.com

A
B
C
D
E
F
G
H
I
J
K
L
M
N
O
P
Q
R
S
T
U
V
W
X
Y
Z

COMPUTER PROCESSORS AND CPUs

Advanced Micro Devices

Advanced Micro Devices, Inc. is the fourth-largest U.S. merchant-supplier of integrated circuits. Focusing on the personal and networked computing and communications markets, AMD produces microprocessors; embedded processors and related peripherals; memories; and programmable logic devices and circuits for communications and networking applications. Specifications, data sheets, related information, and product support are available.

URL:
   http://www.amd.com

Chip Directory

Here's an easy-to-use index of computer chips and manufacturers. Just click on the first few numbers of a chip you're interested in (say, 65 for 6502), and then find your chip from an exhaustive listing. The site is extensively indexed with a handy hypertext glossary.

URL:
   http://www.xs4all.nl/~ganswijk/chipdir/

Intel Pentium Processor

News and technical information on the Intel Pentium are listed here. Articles on each Pentium variant are provided in multiple languages. Among the technical information is an explanation of the well-known, floating-point division flaw. Don't forget to download your Pentium Processor Wallpaper and Screen Saver.

URL:
   http://www.intel.com/procs/pentium/

Intel Secrets Inside

Sure, it's "Intel Inside" all right, but what exactly is lurking in the dark, dusty case of your computer? According to this page, rife with warnings and disclaimers, a number of undocumented features and other scary secrets about Intel's processors have been withheld from you. Learn about previously undocumented opcodes like AAM, AAD, SALC, ICEBP, UMOV, LOADALL, and others. Beware of Intel illuminati.

URL:
   http://www.x86.org

Macintosh Clock Chipping

Learn how to speed up your Mac by swapping its crystal oscillator for one that spins a little faster—and a little hotter. By shaving cycles off your Mac's lifespan, you can gain a few runtime cycles in the short term. Benchmarking software and operating system patches are available for download.

URL:
   http://bambam.cchem.berkeley.edu/~schrier/mhz.html

MIPS Technologies

MIPS—which stands for millions of instructions per second—is a fitting name for this company that creates high-performance microprocessors, such as the R4400. At 200MHz, the R4400 has a rating of 141 SPECint and 143 SPECfp. That means it's damn fast. The MIPS web site, titled "The Architectural Wonder," uses an interesting Egyptian motif. But the extensive graphics nearly necessitate having an R4400 in your PC. Fortunately, there's a text index.

URL:
   http://www.mips.com

Motorola Microprocessor and Memories Technology Group

The Microprocessor and Memories Technology Group manufactures and markets many different types of high-performance CPUs for computer and embedded systems. Here you'll find information on High Performance Embedded Systems, which include the popular M68000 family; RISC processors such as the PowerPC; and Dynamic Memories Products.

URL:
   http://pirs.aus.sps.mot.com

Need help getting connected? See Rick's book The World Wide Web Complete Reference.

Pentium Jokes

Take a break for a little levity. In early 1995, a flaw in Intel's Pentium processor microcode that could cause incorrect results from the FDIV opcode was discovered—an instruction for performing floating-point division. News of the flaw spread through the Internet, only to be downplayed for months by Intel and large companies like IBM with reputations riding on the Intel chip. This situation presented an opportunity for computer humorists everywhere to take jabs at one of the world's largest microprocessor manufacturers. While Pentiums were having problems dividing, the jokes were multiplying with exacting precision. Here is a collection of some of the best.

URL:

http://vinny.csd.mu.edu/pentium.html

Pentium Pro

Just what is it that makes the Pentium Pro (or P6) tick? Find all the answers to your questions here. Read all about the Pro's dynamic execution technology, which includes multiple-branch prediction, dataflow analysis, and speculative execution. Check out the die photos and view the slide show describing the technology that went into this amazing piece of silicon.

URL:

http://www.intel.com/procs/p6/

PowerPC News

Published twice a month, *PowerPC News* is a free electronic magazine for those interested in keeping up with the news on Motorola's PowerPC microprocessors. Subscribers can elect to have articles emailed for faster *PowerPC News* access.

URL:

http://power.globalnews.com/ppchome.htm

Processor Count vs. Problem Size

This technical report discusses the correlation of problem size and microprocessor performance when using multiple processors on problems of fixed size. A model is presented that can be used to predict quantitatively how the problem size must increase in order to maintain a given level of efficiency as the number of processors increases.

URL:

http://www.nas.nasa.gov/NAS/TechReports/ RNRreports/otherpeople/RNR-90-010.html

Excerpt from the Net...

(from Pentium Jokes)

Intel stock was down 3.749999932 points today in heavy trading. — Brad Templeton

Rockwell Semiconductor Systems

Rockwell is the leading manufacturer of chipsets used in modems. The modem you're using to surf the Web probably has a Rockwell chip in it. Semiconductor Systems also specializes in remote access networking products; Global Positioning System (GPS)-based products; and packet wireless data technology that transmits data in small bursts over standard radio waves.

URL:

http://www.rockwell.com/rockwell/ bus_units/telecomm.html

COMPUTER SCIENCE

Acronym Reference List

This large list provides expansions for the many acronyms you're likely to run into in networking, telecommunications, and many other areas of computer science. It's divided into alphabetical sections so that you can quickly find the acronym you're looking for.

URL:

http://www.hill.com/acrolist.html

The Ada Project

The Ada Project (TAP) is a clearinghouse for information and resources about women in computing. TAP includes information on conferences; projects; discussion groups and organizations; fellowships and grants; notable women in the computer science field; and other electronically accessible information sites. TAP also maintains a substantive bibliography of references.

URL:

http://www.cs.yale.edu/HTML/YALE/CS/ HyPlans/tap/tap.html

A B C D E F G H I J K L M N O P Q R S T U V W X Y Z

Advanced Computer Architectures

Do you ever wonder what is actually on the bleeding edge of computer technology today? Wonder not. Just check out the Advanced Computer Architecture page at U.C. Irvine and read all the latest on superscalar microprocessors, interconnection networks, and multithreading.

URL:
   http://www.eng.uci.edu/comp.arch/

BABEL: A Glossary of Computer Abbreviations and Acronyms

Here's a towering glossary containing thousands of computer-related abbreviations and acronyms. This list is presented alphabetically, and some of the words are accompanied by historical information. The list is revised three times a year—in January, May and September.

URL:
   http://www.access.digex.net/~ikind/
   babel95c.html

Berkeley NOW Project

No, this isn't a program for women. Nor is it a group demanding use of the Berkeley variant of the Unix operating system. The Berkeley NOW project is a program underway at UC Berkeley to create a network of workstations (NOW) to show that such a network can act as a distributed supercomputer offering price/performance ratios far superior to traditional massively parallel processing (MPP) architectures. (But don't we already know this?)

URL:
   http://now.cs.berkeley.edu

Having trouble sleeping? See Business Payment Systems.

Coherent Structure in Turbulent Fluid Flow

Get behind the controls of the Cray Y-MP C90 and other supercomputers in this National Center for Atmospheric Research (NCAR) demonstration of astrophysical fluid dynamics. Simulations of turbulent fluid flow are processed on a Cray, then transferred back via the national network to NCAR where they are analyzed both visually and statistically.

URL:
   http://www.ucar.edu/Restemp.html

Crazy About Constraints!

Here's a different kind of web site—dedicated to providing resources and information about the theory of constraints, the thinking process, synchronous manufacturing, and a whole host of other seemingly non sequitur topics.

URL:
   http://www.lm.com/~dshu/toc/cac.html

Data Diffusion Machine

If you're plagued by the lack of scalability beyond the dozens of processors of shared memory machines, you need to find out about the Data Diffusion Machine (DDM). This machine leapfrogs over the shortcomings of shared memory machines by providing a virtual memory abstraction on top of a distributed memory machine. But, who doesn't know that?

URL:
   http://www.pact.srf.ac.uk/DDM/

Excerpt from the Net...

(from the pages of the Data Diffusion Machine)

Shared memory machines are convenient for programming but do not scale beyond tens of processors. The Data Diffusion Machine (DDM) overcomes this problem by providing a virtual memory abstraction on top of a distributed memory machine. A DDM appears to the user as a conventional shared memory machine but is implemented using a distributed memory architecture. This approach is generally known as Virtual Shared Memory, or VSM.

ENIAC

Fifty years ago, a group of scientists and engineers at the University of Pennsylvania's Moore School of Electrical Engineering quietly inaugurated a revolutionary way of managing information. They called it the ENIAC (Electronic Numerical Integrator and Computer). It gave rise to the modern computer industry and would eventually transform people's lives to a degree that even its inventors could not have imagined. Come join the big 50-year birthday bash. And, yes, ferrite core memory retrofits are appropriate gifts for the venerable machine.

URL:

http://www.seas.upenn.edu/~museum/

Formal Methods Archive

This archive offers many resources on formal methods, and Z notation in particular. It provides references to introductory articles, individual notations, methods and tools, papers, information on meetings, pointers to ongoing projects, relevant newsgroups, links to many formal methods repositories containing articles and software, and much more.

URL:

http://www.comlab.ox.ac.uk/archive/
formal-methods.html

Free On-line Dictionary of Computing

The FOLDOC is a searchable dictionary of acronyms, jargon, programming languages, tools, architecture, operating systems, networking, theory, conventions, standards, mathematics, telecomms, electronics, institutions, companies, projects, products, history, in fact—anything to do with computing. Entries are cross-referenced to each other and to related resources elsewhere on the Net. You can also just browse the entries randomly—learning as you go.

URL:

http://wombat.doc.ic.ac.uk/

Help for Undergrads in Computer Science

Richard Suchoza offers help to computer science undergraduates who are having problems with their assignments. Richard's intent is not to give step-by-step instructions on how to complete an assignment but, rather, to give a guideline as to what your solution should entail and where, if applicable, to look for further information. This web form requires that you enter your name, email address, course title, the time assignment is due, and a description of the problem.

URL:

http://www.cs.pitt.edu/~suchoza/tutor.html

Hopper, Grace, Admiral

Who was Admiral Grace Hopper? She invented the charming and elegant programming language, COBOL. The Navy was so pleased with her work that they considered embossing her profile on every punch card. Well, not exactly. Discover the true stories about the esteemed Admiral on the Web.

URLs:

http://www.cs.yale.EDU/HTML/YALE/CS/
HyPlans/tap/Files/hopper-medal.html
http://www.cs.yale.EDU/HTML/YALE/CS/
HyPlans/tap/Files/hopper-obit.html
http://www.cs.yale.EDU/HTML/YALE/CS/
HyPlans/tap/Files/hopper-story.html
http://www.cs.yale.EDU/HTML/YALE/CS/
HyPlans/tap/Files/hopper-wit.html

Multigrid Computing Algorithms

Multigrid (MG) or multilevel algorithms are very fast solvers used in numerical analysis, physics, and computing. For example, they may be used to solve partial differential equations, or in fluid dynamics and graphics programming with texture mapping. The page provides WAIS searchable databases.

URL:

http://src.doc.ic.ac.uk/bySubject/Computing/
Algorithms.html

National Center for Supercomputing Applications

Until the mid-1980s, many of today's most prominent U.S. scientists had to study supercomputing abroad since access to high-performance computers was primarily restricted to the military. So with federal funding, NCSA has fostered desktop computing and scientific visualization with the leading edge of supercomputing. On their page, you'll find more information on news, job opportunities, multimedia exhibits, software, tools, computational resources, and publications.

URL:

http://www.ncsa.uiuc.edu/General/
NCSAHome.html

NCSA Publications

The National Center for Supercomputing Applications regularly publishes online magazines. The first is *access online*, NCSA's general-interest magazine. Then browse the *Technical Resources Catalog* listing NCSA software, user guides, technical reports, preprints and reprints, and publications by NCSA staff researchers.

URL:

http://www.ncsa.uiuc.edu/Pubs/PubsIntro.html

A B C D E F G H I J K L M N O P Q R S T U V W X Y Z

Neural Net Tutor

NN-Tutor is a highly graphical, neural net engine you can use to adapt to your own pattern-matching and other neural network applications. But, in addition to a full-fledged engine, NN-Tutor is a full-fledged tutorial that will start you at the very beginning and teach all about neural networks.

URL:

http://mmink.com/mmink/dossiers/attg/
nntutor.hmtl

NeuroLab

NeuroLab is a neural network library for a graphical simulation program called Extend. With Extend and NeuroLab, you construct a neural net visually— arranging special blocks representing your input and output layers in much the same way you construct an electronic circuit with circuit-simulation software.

URL:

http://mustang.mikuni.com/neurolab/

NeuroSolutions

Get a free Windows demo of NeuroSolutions—a robust and powerful neural network design and simulation tool. The actual program runs both under Windows and on Unix workstations. Here on the NeuroSolutions home page, you can read about this product, step through an informative primer on neural networks, and check some other related products.

URL:

http://www.nd.com

Newsbytes Pacifica

Keep abreast of the computer industry with daily news coverage in text and pictures from Newsbytes Pacifica. More than 180 media outlets are licensed to publish Newsbytes wire material. The wire service reports about 30 stories each day, filed by 19 correspondents worldwide. All reporting is firsthand, original, and objective. News is gathered from independent sources, trade shows, and interviews with top industry professionals.

URL:

http://www.islandtel.com/newsbytes/

Principia Cybernetica

The aim of the Principia Cybernetica project is the computer-supported collaborative development of an evolutionary-systemic philosophy. Simply stated, it tackles classical philosophical questions with the help of the most recent cybernetic theories and technologies. This extensive site includes background material, glossaries, and papers.

URL:

http://pespmc1.vub.ac.be/

Excerpt from the Net...

(from the pages of Principia Cybernetica)

Cybernetics and Systems Science (also: "(General) Systems Theory" or "Systems Research") constitute a somewhat fuzzily defined academic domain, that touches virtually all traditional disciplines, from mathematics, technology and biology to philosophy and the social sciences. It is more specifically related to the recently developing "sciences of complexity", including AI, neural networks, dynamical systems, chaos, and complex adaptive systems.

Propagator Neural Network Software

Propagator is yet another neural network design tool. This one, though, runs on Macs, as well as Windows and Sun workstations. Propagator offers features that allow you to adjust a number of parameters, and highly flexible, dynamic graphing options like scatter plots and bar graphs.

URL:

http://www.he.net/~gator/features.html

Reconfigurable Logic

Everyone needs their logic reconfigured once in a while. BYU's Reconfigurable Logic Lab is a great place to start. This lab explores designing and building computers that can adapt to situations. Many technology papers are available here, including some with topics such as dynamic instruction set computers, nano processors, neural networks, and run-time reconfigurable artificial neural networks. If that won't reconfigure your logic, nothing will.

URL:

http://splish.ee.byu.edu

Technical Reports Archive

Here's a large list of over 300 sites that contain technical computer science reports and papers. Given for each site are the FTP address, comments, and a human contact email address. The comments give some idea of the area or discipline covered by the papers at that site.

URL:

http://www.rdt.monash.edu.au/tr/siteslist.html

Unified Computer Science Technical Reports

Search this database with over 6,000 references to computer science reports and resources, complete with descriptions and hypertext links that make them easy for you to download.

URL:

http://www.cs.indiana.edu/cstr/search

Don't Reinvent the Wheel
Looking for data on a computer science topic? Query the Unified Computer Science Technical Reports database.

Upside Magazine

Upside, an online magazine, provides technology executives with provocative, insightful analyses of the individuals and companies leading the digital revolution. *Upside* tracks the thoughts and activities of technology visionaries while examining the strategies companies use to gain competitive advantages.

URL:

http://www.upside.com

Virus Descriptions

Search for hundreds of computer viruses and learn how to identify them with this handy database. The names alone are worth the look. For example, "AntiPascal" sounds like a Franco-Italian salad, but this Bulgarian virus hunts down and deletes any .PAS and .BAK files on your drive.

URL:

http://www.datafellows.fi/vir-desc.htm

The Internet is too cool.

VME FAQ

Virtual Memory Environment? Vacuum Missing Earrings? Very Messy Elephants? Just what does VME stand for? It's a question that's been nagging even the most nerdy hackers between compiles. Now you can find out the answer, and more about Motorola's VMEbus on the VME FAQ page.

URL:

http://www.ee.ualberta.ca/archive/vmefaq.html

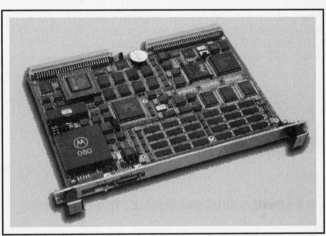

The Heurikon Nitro60 VME board with 64-bit data path. Find out what it does with the VME FAQ.

CRIMINOLOGY

Alcatraz

A great document about one of the most famous and expensive maximum security prisons—take a virtual tour of the island and read about some of the great attempts to escape it. You can even listen to some pretty eerie sound clips from prisoners incarcerated on Alcatraz. Read, too, about the interesting history of Alcatraz even before it became the most famous federal prison.

URL:

http://woodstock.rmro.nps.gov/alcatraz/

Escape to Alcatraz

Buckle yourself in for a breathtaking tour of one of the most famous prisons in the world. Just hope that your Internet connection doesn't go down while you're there . . . or you'll be just another statistic on the Alcatraz page.

Assault Prevention Information Network

Your brain is your best weapon in an assault situation, so feed your brain some valuable information about how to best protect yourself. This site will help you develop safety strategies for yourself and your children and will locate self-defense courses in your area. There are also links to other self-defense and firearms sites.

URL:
    http://galaxy.einet.net/galaxy/Community/
    Safety/Assault-Prevention/apin/
    APINindex.html

Cecil Greek's Criminal Justice Page

Cecil Greek has put together an amazing assortment of pages about criminal justice, law enforcement, prisons, and the death penalty. Some of these pages are Cecil's own, other links lead to resources elsewhere on the Net. Topics include criminal justice resources, a list of searchable law sites, federal criminal justice agencies, international criminal justice resources, police agencies, and criminal information agencies.

URL:
    http://www.stpt.usf.edu/~greek/cj.html

Central Intelligence Agency

Believe it or not, you can connect yourself to the CIA home page. Find out what's new in the CIA, read CIA publications, learn more about the agency, or look for other intelligence links on the Internet.

URL:
    http://www.odci.gov/cia/

Death Row U.S.A.

What percentage of the death row population is white? Black? Latino? What race were the victims of most of these murderers? The answers to these and other questions assaulting racial harmony are right here in black and white. There are also statistics on states and other entities that have death penalties, how many they've carried out, and how many convicts are waiting for someone to flip the switch.

URL:
    http://pathfinder.com/vibe/v3/1august95/
    docs/drowinfo.html

Federal Bureau of Investigation

Anyone interested in criminology should visit the home page of the FBI where you can learn about "the Bureau" and watch the progress of on-going investigations. Included are the FBI's "Ten Most Wanted Fugitives" list (including pictures and descriptions); information on investigations, such as bombings and the Ruby Ridge incident; and general information about the FBI.

URL:
    http://www.fbi.gov

Internet Crime Archives

Visit this disturbing site to learn more about serial killers caught, and those still on the loose. Included are photos of killers and their sordid acts. There are sections on famous mass murderers and killer cults.

URL:
    http://www.mayhem.net/Crime/archives.html

Law Enforcement Sites on the Web

This gigantic list of law enforcement sites on the Web will provide you with links to city police departments, terrorist sites, missing children sites, legal and court pages, and everything else you could possibly think of relating to law enforcement. This list is so big that they had to split it into two!

URL:
    http://www.geopages.com/CapitolHill/1814/
    ira.html

Planning an outing? Check the weather in Meteorology.

Missing Children and Adults

This is an excellent list of missing persons sites on the Internet. The list includes missing persons organizations in other countries.

URL:

http://www.geopages.com/CapitolHill/
1814/ira.html#missing

Missing a Loved One?

Get the word out on the Web. A great many law enforcement agencies and others check missing persons lists on the Web. You may be able to make a difference too. Check the Missing Children and Adults section of the Law Enforcement Sites on the Web page.

National Center for Missing and Exploited Children

The NCMEC helps parents locate their missing children by providing database services and other resources on the Internet. There's also information about other organizations and publications that reach out to anyone in this especially difficult trouble.

URL:

http://www.missingkids.org

New York Police Department

Check out the latest from NYPD—including information about missing persons, the NYC most-wanted list, precinct maps, community affairs, and how best not to get a ticket in the big apple. How rough is New York? Find out in the up-to-date police blotter.

URL:

http://www.ci.nyc.ny.us/nyclink/html/nypd/
finest.html

Going shopping on the Web? Remember to pick up some ecash.

Prison-Related Resources

A collection of prison-related links on the Internet including information about control units, the death penalty, police abuse, political prisoners, prison labor, uprisings, and women behind bars.

URL:

http://www.cs.oberlin.edu/students/
pjaques/prison/home.html

Public Area of Cop Net

Providing you with numerous links to forensic resources—many of which are crime labs and forensic institutes—Cop Net is a great place to browse for law enforcement information. Topics here include electronic crime, firearms, traffic, and wanted and missing persons.

URL:

http://police.sas.ab.ca/

U.S. Department of State Heroes

The Department of State offers reward money for information leading to the capture of wanted terrorists. Here on their web page, they provide information about wanted terrorists and the cowardly crimes they commit. There's also audio clips from Charlton Heston and Charles Bronson talking about the fight against terrorism.

URL:

http://www.clark.net/pub/heroes/

United States Department of Justice

This is a connection to the largest law firm in the United States: the Department of Justice. Find out information about current investigations and activities of the DOJ. Use this connection to search crime databases or to connect yourself to other government sites on the Web.

URL:

http://www.usdoj.gov

A B C D E F G H I J K L M N O P Q R S T U V W X Y Z

CRYPTOGRAPHY

Amiga PGP Page

The Amiga PGP has everything an Amiga owner needs to know about PGP. A good overview of PGP is given here, along with patent and export issues. The importance of encrypting email and using digital signatures is also discussed. Finally, this page also offers pointers for PGP software and lists utilities specifically for the Amiga.

URL:

http://www.cco.caltech.edu/
~rknop/amiga_pgp26.html

Ciphers and Their Weaknesses

Before you rely on your ciphers, check out these descriptions of some well-known ciphers, followed by examinations of how to go about attacking them. These pages include simple ciphers like the Caesar and Augustus ciphers and the one time pad, as well as polygram ciphers such as the Playfair cipher and Hill's Matrix cipher. Modern cryptosystems such as PGP and the RSA public key algorithm are examined. There is also a good description of how PGP works.

URL:

http://rschp2.anu.edu.au:8080/cipher.html

Cryptographic Software

This page offers a comprehensive list of various cryptography software packages—from email encryption to electronic cash. Some of the names sound like they come from a spy novel. Check out Stealth, CrypDisk, SPX, and many others.

URL:

http://www.cs.hut.fi/ssh/crypto/software.html

Cryptography Archive

This archive of cryptographic materials includes implementations of the Data Encryption Standard (DES); tools for cryptanalysis for several popular platforms; PGP utilities and interfaces; brute-force cracking programs; text documents on IDEA implementation; DigiCash, Enigma, and Euroclipper; and many other useful utilities and algorithm implementations for the cryptographer.

URL:

ftp://ftp.ox.ac.uk/pub/crypto/

Cryptography, PGP, and Your Privacy

Cryptography is the study of the transformation of data into a form unreadable by anyone without a secret decryption key. This page offers FAQs, references, papers, service and organization details, encryption software for many platforms, and links to other sites that pertain to cryptography. It also has large sections on the PGP software for performing public key encryption, and discussions and writings about the hot privacy issues facing modern society.

URL:

http://draco.centerline.com:8080/
~franl/crypto.html

Cryptography: An Introduction

Secret languages, codes, and ciphers aren't just children's play, or things controlled by only the military. In today's society, cryptography plays a widespread role. This page provides an introduction to cryptology, the terminology of cryptography, algorithms, and other information in "cleartext."

URL:

http://www.cs.hut.fi/ssh/crypto/intro.html

Cryptography got you pulling your hair out by the roots?

Are acronyms like PGP, SSL, and DES driving you CRAZY?

Key into *Cryptography: An Introduction*, and decipher the secrets of digital security.

EFFector Online Newsletter

Published online by the Electronic Frontier Foundation, an organization dedicated to maintaining the rights of an individual in the emerging electronic infrastructure, this newsletter contains current news articles on the politics, decisions, and events that pertain to privacy, cryptography, and computer security.

URL:
> http://www.eff.org/pub/EFF/Newsletters/
> EFFector/

International PGP Home Page

PGP is subject to U.S. export restrictions due to the cryptographic algorithms it employs. This page is provided as a service to non-U.S. PGP users. It contains an overview of PGP, links to the latest international members of the family of PGP versions, documentation, foreign language modules, FAQs, and PGP-related products and services. In true international spirit, the page is available in English, French, Norwegian, and Swedish.

URL:
> http://www.ifi.uio.no/~staalesc/PGP/

Introduction to Encryption

Cryptology, cryptography, cryptanalysis, or encryption. After reading you will be able to decipher these and other terms used in the art of secret writing. This article reveals processes you can use to hide what you are really trying to say.

URL:
> http://www.quadralay.com/www/Crypt/Dolphin/
> intro.html

Lawries Cryptography Bibliography

This searchable bibliography contains the publication details of over 800 articles on various aspects of cryptography and computer security. Entries in this database are in Bib/Refer format, and the search results are returned in this format as well.

URL:
> http://mnementh.cs.adfa.oz.au/htbin/bib_lpb

Like lizards? Check out Herpetology.

Need some great music? Lend your ear to the Music: MIDI section.

Macintosh Cryptography

The web page is dedicated to using cryptography with the Macintosh family of computers and is home to the Macintosh Cryptography Interface Project—a group working to improve cryptographic interfaces for Macintosh software. The page also offers accessories for MacPGP, related FTP sites, FAQ lists on RSA and PGP, and services for PGP public key searches.

URL:
> http://uts.cc.utexas.edu/~grgcombs/htmls/
> crypto.html

MD2, MD4, and MD5 Message Digest Algorithms

MD2, MD4, and MD5 are a collection of one-way hash functions designed by Ron Rivest that produce a 128-bit hash. MD2 and MD5 are used in Privacy Enhanced Mail (PEM). MD5 is an improvement on MD4, and MD2 is the slowest and least secure of the three. These RFCs give an executive summary, a description, references, and security considerations for each MD algorithm.

URLs:
> ftp://ds.internic.net/rfc/rfc1319.txt
> ftp://ds.internic.net/rfc/rfc1320.txt
> ftp://ds.internic.net/rfc/rfc1321.txt

Netsurfer's Focus on Cryptography and Privacy

As a good introduction to cryptography and its useful applications today for the layman, this page has articles about cracking encrypted code, key certification, secure email, export issues, steganography, electronic cash, anonymous remailers, databases, clipper and digital telephony, trusting software, and more. Throughout each article there are links to other web resources where you can get more detailed information, and there is also a set of additional resources and printed references.

URL:
> http://www.netsurf.com/nsf/v01/03/nsf.01.03.html

A
B
C
D
E
F
G
H
I
J
K
L
M
N
O
P
Q
R
S
T
U
V
W
X
Y
Z

PGP Discussion and Help

This Usenet newsgroup is the place for discussions, questions, book reviews, applications, problems, articles, and arguments about PGP, the freely available public key encryption utility written by Phil Zimmermann.

URL:
> news:alt.security.pgp

Pretty Good Privacy

PGP is an electronic privacy program created by Phil Zimmermann that implements public key encryption using the RSA algorithm and the IDEA symmetric key algorithm. It is freely available for non-commercial users on Unix, DOS, Macintosh, VMS, OS/2, and other platforms. This excellent page on PGP describes where to download the appropriate version of the program, descriptions of books about PGP written by Phil Zimmermann, PGP documentation and related utilities, links and advice on PGP public keys, and links to other PGP information sources.

URL:
> http://world.std.com/~franl/pgp/

Pretty Good Privacy

We like our privacy. Without it, how could we surf the Net in our underwear? Take a peek at Pretty Good Privacy for more security than you can shake your public key at.

Quadralay's Cryptography Archive

Quadralay's Cryptography Archive is a large collection of articles, FAQs, reviews, and links to sources of information on cryptography. It contains sections on the Clipper chip, Digital Encryption Standard (DES), Digital Telephony, Kerberos, laws and restrictions that relate to cryptography, RIPEM, RSA, Tempest, and other subjects. This server is also searchable for specific information.

URL:
> http://www.quadralay.com/www/Crypt/
> Crypt.html

Ron Rivest's Cryptography and Security Page

Rivest's cryptography pages offer many pointers to web pages that pertain to cryptography and security. It's organized into sections, including bibliographies, government sources, nonprofit organizations, commercial enterprises, newsgroups, newsletters, FAQs, alert sites, computer security people, algorithms, protocols, and other topics.

URL:
> http://theory.lcs.mit.edu/~rivest/
> crypto-security.html

RSA Cryptography FAQs

These FAQs have explanations and information about many areas of today's cryptography topics, including public key cryptography; digital signatures; MD2, MD4, and MD5; patents; RSA; DES; key management; certificates; digital time stamping; one-way functions; factoring; Clipper; Digital Signature Standard (DSS), and others.

URL:
> http://www.rsa.com/rsalabs/faq/

RSA Data Security

RSA Data Security is a recognized world leader in cryptography, with millions of copies of its encryption and authentication software installed and in use worldwide. RSA's technologies form the global standards for public-key encryption and digital signatures, and are part of existing and proposed standards for the Internet, CCITT, ISO, ANSI, PKCS, IEEE, and business and financial networks. RSA's site offers FAQs on general cryptography, the RSA algorithm, key management, factoring and discrete logs, DES, Capstone, Clipper, DSS, NIST, and NSA. There are also press releases, free evaluation copies of RSA software for Windows, conference information, and links to other crypto- and security-related sites.

URL:
> http://www.rsa.com

> ## The Web will really launch you.

sci.crypt Cryptography FAQ

This very comprehensive and detailed overview of many aspects of cryptography starts by defining basic cryptography and providing references for those who want to learn about this subject. It then explains cryptosystems and attacks in mathematical terms. Also explained and examined here are product ciphers, public key cryptography, and digital signatures, including the common algorithms used by such systems. Finally, a good list of references divided into subject areas is provided.

URL:
> http://www.cis.ohio-state.edu/hypertext/faq/
> usenet/cryptography-faq/part01/faq.html

Secret Key Algorithms

There are a variety of algorithms used in encrypting data. Public key algorithms and secret key algorithms are compared here, and a variety of them are described. Don't let your attacker guess your hash function or block cipher. Find out which are most secure and how they work in this very extensive, advanced page.

URL:
> http://www.cs.hut.fi/ssh/crypto/
> algorithms.html#symmetric

Software Archive

This page contains links to cryptography software packages for popular computer platforms. It contains sections for email encryption; voice encryption; secure communications and authentication; file encryption; steganography; electronic cash; cryptographic libraries; and cryptanalysis.

URL:
> http://www.cs.hut.fi/crypto/software.html

Steganography Archive

The Steganography Archive houses a collection of steganography software, much of which is available in C source code. The archive includes Texto, Stegola, Hide-and-seek, and Wnstorm, all of which allow you to hide your data in various formats that include pictures and English sentences.

URL:
> ftp://ftp.csua.berkeley.edu/pub/cypherpunks/
> steganography/

Survey of Cryptographic Products

This survey of products that pertain to cryptography around the world includes information on over a thousand products that have been identified, including 455 products coming from countries outside of the U.S. Summary statistics of the survey are presented, along with the methods used to identify products and information sources.

URL:
> http://www.tis.com/crypto/crypto-survey.html

Texto

Texto is a steganography program written in C that transforms uuencoded or PGP ASCII-armored data into English sentences and back again. The program's output should be close enough to English text so that it will slip by any kind of automated mail scanning. Texto works just like a simple substitution cipher and replaces each of the 64 ASCII symbols used by PGP ASCII armor or uuencode with an English word.

URL:
> ftp://ftp.csua.berkeley.edu/pub/cypherpunks/
> steganography/texto.tar.z

Your Privacy

Privacy of the individual and the prevention of a Big Brother Society like that described by George Orwell in the novel *1984* are issues presented by the material at this site. It contains press releases on current privacy issues, details of how to protect your privacy and anonymity, and links to privacy organizations. There is a good collection of writings and articles on privacy, including "A Parable of Privacy," "Cyberwire Dispatches," and "Tools for Privacy." A list of privacy-related Usenet newsgroups is also provided.

URL:
> http://world.std.com/~franl/privacy/

A
B
C
D
E
F
G
H
I
J
K
L
M
N
O
P
Q
R
S
T
U
V
W
X
Y
Z

DATA COMMUNICATION

Black Box Reference Center

The nuts, bolts, and terminators of networking and data communications are explained in the Black Box Reference Center. First, check out the Glossary of Data Communications Terms to familiarize yourself with the lingo. Next, explore detailed tutorials on local and wide area networks, different types of Ethernet networks, data transmission methods, high-speed digital services, protocols, wireless communication, cables, video, and more.

URL:
    http://www.blackbox.com/bb/refer.html/tigf012

d.Comm Magazine

d.Comm covers current information technology issues, including issues about desktop computing, networking, communications, and so on. This online magazine also maintains an archive of past articles.

URL:
    http://www.d-comm.com

Dan Kegel's ISDN Page

This definitive page on ISDN is nothing short of an Internet treasure. In the confusing, complicated, and fast-moving world of ISDN, this page brings together news and announcements on FCC tariffs, standards, compatibility, hardware, and software. At this site, a substantial effort has been made to explain what ISDN is, how it works, and how you can learn more about it. If you haven't made the move from modems to ISDN yet, bond with this page first.

URL:
    http://www.alumni.caltech.edu/~dank/isdn/

Versit

Read about the joint venture of Apple, AT&T, IBM, and Siemens to create interoperability specifications between communications and computing. Versit specs are based on existing standards and technologies, and are available from the consortium. Versit does not develop commercial products. You'll need the freely available Adobe Acrobat program to read several of the specifications.

URL:
    http://www.versit.com

DATABASES

Access Discussion Group

Got questions that the manuals just don't address? Does paying for technical support have you checking the couches for loose change? Turn to the forum where Access experts and novices alike come for the quickest and most diverse opinions on Access design and programming problems—Usenet.

URL:
    news:comp.databases.ms-access

Access Shareware Programs

This page lists oodles of valuable shareware software packages written for Microsoft Access. Among the goodies here are developer's tools, contact managers, utilities and information to make the transition to Access easier, and dozens of other useful programs and documents. If you use Access, you must see this site before you begin your next project.

URL:
    http://www.jumbo.com/bus/win/access/

Excerpt from the Net

(from Dan Kegel's ISDN Page)

The FBI recently revealed that it plans to spend $500 million to equip phone companies to allow tapping of 0.25% of all phone lines in the country on short notice. They claim they won't tap any more than they do now, about 1000 a year. That's about $50,000 per wiretap, if the equipment lasts 10 years. Seems like an incredibly expensive way to catch crooks to me, and a serious temptation for whoever's in power to tap 250,000 or so phones. That would be enough to listen in on all the politicians in the country. . . . or all the journalists. . . .

Chicago Access User Group

You don't have to live in Chicago to benefit from the Chicago Access User's group. This is an independent group of developers and users of Microsoft Access and related products. On their web pages, the CAUG publishes articles about using and developing applications in Access, and provides an easy form to fill out to join the CAUG mailing list.

URL:

http://www.imginfo.com/caug.htm

Colin's FoxPro Page

The lengths to which some people will go to provide others with easy access to valuable resources is truly amazing. This site is an excellent example of an enthusiast who has brought together all of the most important resources on the Internet that relate to the FoxPro database management system. Colin's FoxPro Page offers links to FoxPro talk on IRC, FoxPro bulletin boards, web sites, mailing lists, FAQs, user groups, news, and so much more.

URL:

http://www.state.sd.us/people/colink/
foxpage.htm

FoxPage

For all the latest news and rumors about FoxPro—one of Microsoft's database products—burrow into the FoxPage. Check the tip of the week, enter into contests to guess the ship dates for upcoming products, read and join FoxPro user groups, and investigate a collection of useful links to other FoxPro resources on the Web.

URL:

http://turnpike.net/emporium/B/bmc/
foxpage.htm

Informix Software Resources

Informix is a long-time developer of high-end, relational database management systems for a variety of system and network platforms. Popular Informix products include INFORMIX-SQL, INFORMIX-4GL, C-ISAM, NewEra, and a variety of database servers and connectivity products. Find the technical specs for all of these products on Informix's pages and join in user and developer discussions on the listed newsgroup.

URLs:

news:comp.databases.informix

http://www.informix.com

Surf the net with a cup of coffee.

XBase Applications and Programming

If you're into any aspect of XBase programming, this page is a must-see. It offers links to the (also listed) XBase-related Usenet newsgroups; FoxPro product information and books; dBASE resources, including pages for both DOS and Windows versions; Clipper; SBT; Cykic Software; and links to a collection of XBase third-party providers and developers.

URLs:

news:comp.databases.xbase.fox
news:comp.databases.xbase.misc

DOMESTIC SCIENCE

Care and Cleaning of Your Oriental Rug

Master some tips and tricks for keeping your Oriental rug clean and healthy. Learn how to vacuum it, pad it, and remove those messy spots and stains. If you find any pointers on how to turn your rug into a flying carpet, let us know!

URL:

http://www.unilk.com/~yervant/care.html

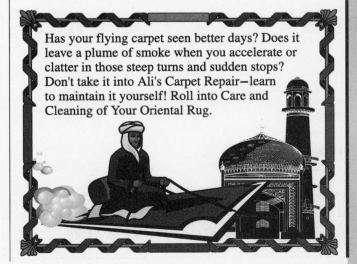

Has your flying carpet seen better days? Does it leave a plume of smoke when you accelerate or clatter in those steep turns and sudden stops? Don't take it into Ali's Carpet Repair—learn to maintain it yourself! Roll into Care and Cleaning of Your Oriental Rug.

A
B
C
D
E
F
G
H
I
J
K
L
M
N
O
P
Q
R
S
T
U
V
W
X
Y
Z

Chocolate Recipes

Ah, chocolate, that marvelous concoction produced from the cacao bean that adds taste to our lives and pounds to our waists. Find a collection of pointers (including **rec.food.chocolate** and Godiva recipes) to scrumptious cocoa treats for any occasion and excuse!

URL:

http://www.qrc.com/~sholubek/choco/
recipes.htm

Food Safety and Handling

Ah, dinner time—and you're sitting down to a nice hot plate of . . . what the heck is *that*? Not sure if it's safe to eat those leftovers that turn more colors than a chameleon? Use this guide offered by the University of Florida to determine whether something is edible. Topics include feeding children, meat and fish safety, and food preparation in times of natural disaster.

URL:

http://hammock.ifas.ufl.edu/text/he/31521.html

Household Pests

Has your house been invaded by insects? Maybe you should read this pamphlet produced by the University of Florida about pest and insect control. Included are environmentally safe solutions to problems with ants, carpet beetles, termites, book lice, moths, and even those horrible cockroaches.

URL:

http://hammock.ifas.ufl.edu/text/mg/19794.html

Laundry

You, too, can get a master's degree in laundry. Sure, there's science going on every time you wash those rank gym socks of yours. To get them springtime fresh requires a delicate balance of chemical, mechanical, and thermal energy. This web site covers the very basics of clothing care, such as sorting your undies from your jeans, pretreating stains, water hardness, ironing, fabric types, and energy-conservation measures.

URL:

http://hammock.ifas.ufl.edu/text/he/7812.html

The Web will set you free.

Piano Finish Care

Enhance your home's decor and preserve the value of your piano by learning the proper techniques of dusting and polishing. Avoid that waxy buildup before you tinkle the ivories by following these time-tested guidelines brought to you by none other than the Piano Technicians Guild.

URL:

http://www.prairienet.org/arts/ptg/ptgtb5.htm

Pool Maintenance

Dive in to the do's and don'ts of keeping your pool clean with this handy guide. Learn the ins and outs of leaf skimming, brushing, vacuuming, and cleaning the skimmers and filters. The idea is to remove debris before it sinks to the bottom of your pool, where it's far more difficult to remove and may cause a stain. After all that work, you'll hopefully still have time and energy enough to go for a swim!

URL:

http://waterworks.olin.com/docs/
maintenance.html

Pork Recipes

Trying to think of a new way to prepare that pork roast in the freezer? Wonder what the difference is between Canadian bacon and traditional bacon, or spareribs and back ribs? This site, brought to you by the National Pork Producers Council, has the answers to all of your pork questions, including nutrition facts, food-safety hints, descriptions of the different cuts, and preparation tips and recipes.

URL:

http://www.nppc.org/food&cooking.html

Random Research Questions

Ever pondered the effects of Mandated Lead-abatement Legislation on Indigestion? How about the effects of Random Research Questions on Civilization? Now scientists needing grants, professors up for tenure, and eighth grade science fair contestants can rejoice. Your days of worry are over! This wacky page is capable of generating countless valuable research questions to aid you in your career.

URL:

http://www.coedu.usf.edu/behavior/research/
research.html

Seafood Recipes

Even if this entry doesn't fall into the category of science, research, or technology, and if you love seafood, we guarantee that you'll still enjoy making good use of this page often. Discover kilobytes galore of seafood recipes for your next soiree at this delectable page of succulent, oceanic gourmet suggestions.

URL:

ftp://ftp.neosoft.com/usenet/
rec.food.recipes/shellfish/

Turkey and all the Trimmings

It's a shame that the Indians and the pilgrims couldn't access this web page from *Good Housekeeping* magazine since they might have learned some new tricks and cooking hints for the first Thanksgiving. Plan your next holiday dinner by reading up on the secrets for a tender turkey, four smooth ways to mash potatoes, and the ins and outs of disposable roasting pans. (Of course, roasting pans wouldn't have been disposable for anyone way back at the first Thanksgiving.)

URL:

http://homearts.com/gh/food/11turkf1.htm

Waste Management

Thinking about cleaning up your ways? The University of Florida has a series of handbooks that will teach you how to reuse and recycle your refuse. Topics include composting, reusing packaging materials, and "enviroshopping."

URL:

http://hammock.ifas.ufl.edu/text/he/19819.html

Need help getting connected? See Rick's book The World Wide Web Complete Reference.

One person's trash is another person's treasure. If you want to learn how to use your refuse rather than lose it, check out Waste Management.

Waxing Your Car

It's Saturday afternoon and you've got that big date at the drive-in with little Susie tonight. Time to get the T-Bird in shape, and polish that baby so well that Susie can see herself in the shine. Wait a minute, bub, how're you gonna go about doing it without getting your goose cooked? By calling up this page, that's how! Get the whole skinny from a master waxer, including tips on application, technique, waxes and polishes, and that all-important paint job.

URL:

http://www.realtime.net/~drl/
mercedes/meguiars

Want to see some cool resources? Check out Internet Resources.

A B C D E F G H I J K L M N O P Q R S T U V W X Y Z

EARTH SCIENCE

Auroral Tomography

The Auroral Large Imaging System (ALIS) is located in Sweden, and its purpose is to determine the 3D spatial structure of the aurora. This requires several stations strategically placed. A basic introduction to tomography is given, as well as a simulation of ALIS tomography. There is also a map showing the position of each of the ALIS stations.

URL:
http://snake.irf.se/~bjorn/tomography/tomografi.html

CIESIN

The Consortium International Earth Science Information Network (CIESIN) is dedicated to furthering the interdisciplinary study of global environmental changes. CIESIN specializes in the access and integration of physical, natural, and socioeconomic information across scientific disciplines.

URL:
http://www.ciesin.org

Comprehensive Oil and Gas Information System

Compared to the World Wide Web, using a gopher might be considered a crude tool—but when you're looking for crude, here's the gopher you want digging for you. COGIS gives you easy access to articles and documents that contain information on crude oil, coal, jet fuel, industrial energy consumption, and much more. Developed by the Energy Information Administration's Office of Oil and Gas.

URL:
gopher://una.hh.lib.umich.edu/11/ebb/energy

Earth Observing System

One of the most important discoveries of our missions into space has been the rediscovery of the planet on which we live. Link to the EOS web site to access hypertext documents and data from the *Mission to Planet Earth* project.

URL:
http://spso2.gsfc.nasa.gov/spso_homepage.html

It always helps to see things from a different perspective, so it's a good thing that we have satellites flying overhead to report back on how messed up things are here on Earth. Or are they? Find out what the satellites are telling the scientists, and what the scientists are telling us at the Earth Observing System page.

Explorer Outline for Natural Science

This extensive hypertext database of K-12 natural science resources includes lessons, curriculum notes, and software. The Explorer is a network database system for contributing, organizing, and delivering educational resources.

URL:
http://unite.ukans.edu/Browser/UNITEResource/Layer_Natural_Science.html

Hydrologic Cycle Archive Center

The Hydrologic Data Search, Retrieval, and Order system (HyDRO) provides access to an extensive amount of earth-science information. A commercial service from the kind folks at the Hydrologic Cycle DAAC at Marshall Space Flight Center, this site offers information that may be helpful to earth-science researchers and educators.

URL:
http://wwwdaac.msfc.nasa.gov

The Web will set you free.

National Geophysical Data Center

If you live in southern California, the National Snow and Ice Data Center or hurricane maps may not be of immediate interest to you, but you'll enjoy the National Geophysical Data Center page. This site has information on solid earth geophysics, solar-terrestrial physics, marine geology and geophysics, paleoclimate research, and a satellite data archive.

URLS:

ftp://ftp.ngdc.noaa.gov

gopher://gopher.ngdc.noaa.gov

http://www.ngdc.noaa.gov/ngdc.html

Ocean Drilling Program

What's at the bottom of the ocean? The U.S. National Science Foundation and 18 other countries know. They're investigating the history of the ocean floor and the entire nature of the crust beneath it using a scientific drilling ship. Coveted by dentists everywhere, this massive drill is poking holes at the bottom of the sea to learn more about whether it's possible to drain the whole thing. Don't believe me? Check it out.

URL:

http://www-odp.tamu.edu

Online Resources for Earth Scientists

If you're a scientist of the Earth, this page contains helpful pointers to digital documents, news sources, software, data sets, and additional Net gems.

URL:

http://www.bris.ac.uk/Depts/Geol/gig/ores/ores.html

Rainbows

This page has everything you wanted to know about rainbows and some more information about rainbows you didn't know you wanted to know! It explains why rainbows are shaped like bows, what causes the colors, and what causes double rainbows. The only secret not revealed here is where to find the pot of gold.

URL:

http://www.unidata.ucar.edu/staff/blynds/rnbw.html

What's Shakin' in California?

What was that jolt? Just a sonic boom, or is there a new land bridge to Catalina? Or maybe San Francisco's now an island. Find out fast in this newsgroup.

URL:

news:ca.earthquakes

ECOLOGY

40 Tips to Go Green

This flyer, distributed by the Jalan Hijau (which means "Go Green" in Malay) offers 40 wise tips for maintaining a green world while you are at home, at work, on the road, and shopping.

URL:

http://www.ncb.gov.sg/jkj/env/greentips.html

The Aleutians—Chain of Life

The Aleutian Islands encompass a pristine biosphere reserve consisting of 200 islands. Over a thousand miles off the Alaskan mainland, the islands offer a uniqueness difficult to surpass. This breathtaking wildlife environment is home to over 250 species of birds, and other rich and diverse fauna and flora. Visit this page to learn that the Aleutians are a nesting habitat for over 10 million seabirds and contain the world's largest colony of tufted puffins.

URL:

http://bluegoose.arw.r9.fws.gov/nwrsfiles/RefugeSystemLeaflets/R7/AlaskaMaritimeNWR/AleutianIslands.html

Canada's Biospheres

Visit six diverse Canadian biospheres without donning your parka. Each is described in detail, with data provided on the history of the preserve and its geographical location. You'll find a summary of the reserve's physical features, comprehensive information on fauna and vegetation, and each reserve's conservation management plan.

URL:

http://www.iatech.com/unesco/map.htm

Psst . . . hey, buddy! Wanna check out a tufted puffin? For a peep at puffins and other breathtaking Alaskan wildlife, point your browser toward the Aleutians—Chain of Life page.

Center for Clean Technology

There's nothing worse than dirty technology, unless you're one of those techno-abhorrent types. But they're not like that in Los Angeles, which is why the fine folks at UCLA provide information on pollution prevention, combustion, air toxics, water and wastewater treatment, intermedia transport, risks and systems analyses for the control of toxics, and other really clean things.

URL:
http://cct.seas.ucla.edu

Centre for Landscape Research Network

It's tough to improve on the original landscaper's design—but when the situation calls for it, check out this page offered by the University of Toronto. There is a Landscape Virtual Library filled with software for interactive and integrating CAD, GIS, remote sensing, multimedia and virtual worlds, teaching, publications, jobs, newsgroups, upcoming conferences and events, and collaborative projects. The page even includes a search engine for landscape-related topics.

URL:
http://www.clr.toronto.edu:1080/clr.html

Cooperative Research Center for Freshwater Ecology

Australia really takes its ecology quite seriously, and for good reason. With a plethora of rare indigenous plants and animals, there's nothing else like it on Earth. Learn more about the Wonder Down-Under's freshwater resources, hydrology, aquatic plants, fish, invertebrates, microbes, algae, and zooplankton, plus what scientists are doing to preserve the continent's unique resources.

URL:
http://lake.canberra.edu.au/crcfe/crchome.html

Ecological Modeling

Of course, everyone loves to build models, especially when it comes to ecology. This Web server provides easy access to ecological modeling resources, such as simulation software and databases.

URL:
http://dino.wiz.uni-kassel.de/ecobas.html

Ecological Society of America

This venerable society, founded in 1915, seeks to promote understanding of the relationships between organisms and their past, present, and future environments. Featured is the ESA Newsletter, environmental policy updates, reports, an assessment of the Endangered Species Act, and sections on statistical and theoretical ecology.

URL:
http://www.sdsc.edu/CGI/
cgi-bin/mfs.cgi/01/var/spool/gn/SDSC/
Research/Comp_Bio/ESA/ESA.html

Ecology and Environmental Physiology

McMaster University in Canada, adjacent to the Royal Botanical Gardens, provides an excellent atmosphere in which to study ecology. Its faculty's interests range from macro communities to molecular levels of organization.

URL:
http://www.science.mcmaster.ca/Biology/
faculty/Ecology.html

The Net is humanity's greatest achievement.

Ecology and Evolution

The fruit-fly-minded can check out "Drosophila developmental theory," while budding fossil hunters can read up on paleontology, paleobotany, and paleoecology. Explore a guide to biological field stations, and browse the archives. You'll also find *A Biologist's Guide to Internet Resources* in English and French.

URL:

gopher://sunsite.unc.edu:70/11/../.pub/academic/biology/ecology%2bevolution

Ecology and Evolutionary Biology at the University of Connecticut

Search for your favorite carniverous plant, take a campus tree walking tour, or catch up on detailed angiosperm family descriptions at Uconn's informative ecology page. Particularly interesting are the mammoth online EEB Greenhouse collections database and the gardens of the Bartlett Arboretum. See links to other eco resources plus the local weather forecast.

URL:

http://florawww.eeb.uconn.edu

EcoNet

Weed killer in your tap water, nuclear testing, and other cheery concerns are the topics of the day on EcoNet, a community resource for the preservation and sustainability of the environment.

URLs:

gopher://gopher.econet.apc.org/11/environment

http://www.igc.apc.org/econet/

EcoWeb

Gather your cans and bottles together, and see how fun, easy, and rewarding recycling can be on EcoWeb. Spend a virtual week with the University's recycling program learning the tips, tricks, and factoids of recycling. Reserve the EcoChat Conference Area for conferences and group discussions. While you're at it, browse their archive and perform keyword searches in the ecosystem database. You may never throw anything away again!

URL:

http://ecosys.drdr.virginia.edu/EcoWeb.html

Environmental and Energy Study Institute

EESI is a part of the U.S. Agency for International Development (USAID)—an agency of the U.S. federal government that conducts foreign assistance and humanitarian aid to advance political and economic interests abroad. The EESI promotes environmentally sustainable societies through social and economic transitions.

URL:

http://www.info.usaid.gov

Environmental Data from Ford

On this web page, Ford Motor Company outlines federal, and the more stringent California emission laws. Ford explains, from their point of view, the Clean Air Act, California's low-emission vehicle act, the Corporate Average Fuel Economy (CAFE) standards, the Superfund Amendments Reauthorization Act (SARA), and Ford's progress toward meeting these standards.

URL:

http://www.ford.com/corporate-info/environment/Envidata.html

Excerpt from the Net...

(from A Biologist's guide to Internet Resources —Una R. Smith)

The professionally-oriented newsgroups and mailing lists follow certain conventions of etiquette. These are none other than those used by most people at public events such as academic conferences. In fact, most of the science-related newsgroups (and mailing lists) are very much like mid-sized meetings of any professional society, except that they never end. The participants come and go as they please, but the discussion and exchange of ideas and information continues.

Environmental Education Programs

This page outlines the environmental education programs created and sponsored by the Ford Motor Company. Programs include *Designing the Environment, Earthquest: The Challenge Begins, Chicago Children's Museum, Zoo Atlanta,* and *Ocean Planet.*

URL:

http://www.ford.com/corporate-info/
environment/Design.html

Environmental Measures at Chrysler

OK—this is a car company, but Chrysler is doing a lot to clean up their manufacturing processes, their products, and to build more environmentally friendly vehicles that disassemble and recycle easily. Read about their efforts in recycling and conservation. Kudos to Chrysler, and keep up the good work.

URL:

http://www.chryslercorp.com/environment/
recycling.html

Environmental Resource Center

ERC sponsors an Online Query System with a large database of Earth observation images and photos. Perform custom searches through the NOAA National Environmental Data Referral Service.

URL:

http://www.clearlake.ibm.com/ERC/

Fish Ecology

There's something fishy on this Swedish gopher, and it's not just the pickled herring. Join academic personnel involved in empirical and theoretical research and assessment issues relating to fish, fisheries, and ecology. This substantial archive includes discussion lists, group postings, bibliographic services, a researcher address book, a software repository, and much more.

URL:

gopher://searn.sunet.se

"Say, Gunnar, what's that smell?"
"Well, Johann, it's the Fish Ecology page I've just pulled up."
"*Ufta*, Gunnar, when you go Net-surfing, you really reel them in!"

Friends of the Earth

It's not easy being green, but it's an important attribute for conducting environmentally safe scientific experiments today. Friends of the Earth offers alternative methods of scientific experimentation that protect the Earth and atmosphere, including ideas on pest control, rechargeable batteries, and inexhaustible electrical sources.

URL:

http://www.foe.co.uk/CAT/

Global Action and Information Network

GAIN collects environmental information to encourage behaviors that help us maintain a sustainable world. Think of GAIN as your mother ("Don't forget to turn off the water and the light when you've finished brushing your teeth"), but on a larger scale. Read up on current legislative actions, articles, membership information, and, of course, links to other sources.

URL:

http://www.nceet.snre.umich.edu/
GAIN/GAIN.html

Global Recycling Network

GRN is a clearinghouse and information service for recycled resources, surplus manufactured goods, and outdated or used machinery throughout the world. GRN's web pages offer an online reference library to track recycling companies and organizations, materials, and specific items, plus a glossary of recycling terms and links to other recycling resources.

URL:

http://grn.com/grn

InSites

The Waste Management Research and Education Institute of the University of Tennessee publishes this quarterly newsletter containing essays by top environmental researchers. The Institute focuses on biotechnology and solid, hazardous, and nuclear waste management.

URL:

http://www.ra.utk.edu/eerc/insites/

Institute of Terrestrial Ecology

Based in Great Britain, the ITE is responsible for research into all aspects of the environment and its resources with particular emphasis placed on terrestrial ecosystems. Research material includes the characteristics of individual plant and animal species, dynamic interaction among atmospheric processes, soil properties, and surface water quality.

URL:

http://www.nmw.ac.uk/ite/

International Center for Tropical Ecology

Missouri may not be Maui. Nevertheless, the University of Missouri's tropical ecology program let's you explore their graduate studies program, read the latest newsletter on tropical topics, and peruse other ecologically related links.

URL:

http://ecology.umsl.edu

Lawrence Livermore National Laboratory

LLNL focuses on global security, ecology, and bioscience. This page provides organization, historical, and general information about LLNL; lists published papers, reports, periodicals, research results, and projects; and explains the benefits of research from LLNL.

URL:

http://www.llnl.gov

Linkages

If you're vitally concerned about the environment, you'll want to stay informed through Linkages, an online clearinghouse for upcoming international meetings and conferences related to the health of the planet. These pages have details and summaries of environmental conferences, abstracts, and guidelines for future environmental conventions.

URL:

http://www.mbnet.mb.ca/linkages/

Man and the Biosphere

Man and the Biosphere Programme (MAB) is playing a major ecological role in Canada. Biosphere reserves are created with the long-range goal of documenting ecological patterns.

URL:

http://www.iatech.com/unesco/biofaq.htm

National Materials Exchange Network

The exchange is a nationally recognized marketplace for trading and recycling used and surplus materials and goods. You can sign up immediately—there is no cost—and advertise your surplus materials or your need for used or surplus goods. The NMEN database is fully automated and searchable through these web pages.

URL:

http://www.earthcycle.com/g/p/earthcycle

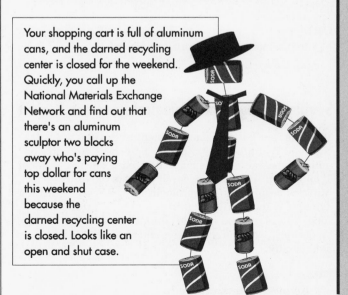

Your shopping cart is full of aluminum cans, and the darned recycling center is closed for the weekend. Quickly, you call up the National Materials Exchange Network and find out that there's an aluminum sculptor two blocks away who's paying top dollar for cans this weekend because the darned recycling center is closed. Looks like an open and shut case.

A B C D E F G H I J K L M N O P Q R S T U V W X Y Z

National Wetlands Inventory

God Bless America, wetlands that I love! Plant a bookmark here and see all the many species of flora that grow, mature, and thrive while completely swamped. The inventory is categorized by region—and once you find your favorite area, you can download files to your own pad.

URL:

http://www.nwi.fws.gov/Ecology.html

Project Water

Learn about planet Earth's most precious (and plentiful) substance—water—in this multimedia presentation offered by EnviroMedia. Explore geologic, chemical, and mythological aspects of the ocean planet.

URL:

http://www.aloha.com/~enviro/pwpage.html

Rangeland Ecology and Management

Home, home page on the range . . . where you can visit the S.M. Tracy Herbarium to study your favorite native Texas plants. Texas A&M's rangeland studies figure prominently in the department's curriculum, as seen by many of the links to related resources. You'll learn bushels at the Ranching Systems Group, the Rangeland Watershed Laboratory, the Grazingland Animal Nutrition Laboratory, and the Rangeland Rehabilitation and Restoration Ecology Laboratory.

URL:

http://rasc-sparc.tamu.edu/rlem/

Sevilleta Long-Term Ecological Research Project

Visit the real Land of Enchantment at Sevilleta, a National Wildlife Refuge that encompasses 400 square miles of central New Mexico. The area is one of a number of long-term ecological research projects in North America designed to test ecological hypotheses regarding dynamic responses of organisms in a biome transition zone. Topics covered include climate, meteorology, topography, soils, archaeology, and the region's biodiversity.

URL:

http://sevilleta.unm.edu

Sierra Club

Founded by the legendary John Muir himself, the Sierra Club is dedicated to Muir's spirit. Take an online field trip to the Critical Ecoregions Program, and read *The Planet*, a Sierra Club news publication. There are many links to information on outings, local Sierra Club chapters, books, mailing lists, and current Sierra Club SC-Action alerts.

URL:

http://www.sierraclub.org

Solar Cooking Archive

One day we may see solar cooking instructions on packaged food. But will it taste better? Offering creative and useful methods of solar cooking, this electronic journal documents construction plans for a solar oven, and tips and tricks for better cooking. Good news for those with leftovers: you can cook with aluminum foil in the sun without causing a nuclear meltdown in your kitchen.

URL:

http://www.accessone.com/~sbcn/search.htm

Tellus Institute

The Tellus Institute is a nonprofit organization that brings together scientists, planners, and policy analysts to address resource management and environmental issues. The Tellus web pages offer extensive information in many fields of science, including resource management, energy, solid waste, and the environment.

URL:

http://www.tellus.com/

U.S. Long-Term Ecological Research Program

A biome by any other name would still be an ecosystem, and this network of North American, Alaskan, and Antarctican sites encourages collaborative research and data sharing with respect to different ecosystems. See the impact of long-term research on you and your favorite biome.

URLs:

gopher://lternet.washington.edu:70/00/
    AboutLTER/g-hist.txt

http://lternet.edu

University of Maine's Wildlife Ecology

The rain in Maine falls mainly on the moose. (Er, something like that.) Maine is a beautiful state with lots of great wildlife, but not at night. The Department of Wildlife Ecology at Orono maintains descriptions of ongoing projects, research interests, available positions, and links to other ecology and biology web sites.

URL:

http://wlm13.umenfa.maine.edu/w4v1.html

EDUCATION

4000 Years of Women in Science

There are still many who believe that people are categorically different, thus the need for pages like this one. However, relative to all other species, there are minor differences between the human sexes. This page helps us remember this obvious fact by documenting the contributions women have made in science.

URL:

http://www.astr.ua.edu/4000WS/4000WS.html

African Americans in the Sciences

In 1887, Granville T. Woods patented the Synchronous Multiplex Railway Telegraph, which enabled communications between stations and moving trains. During his lifetime, Woods would be awarded over 60 patents. His is one of the engaging and inspirational biographies of African American scientists, complete with pictures and bibliographies.

URL:

http://www.lib.lsu.edu/lib/chem/display/
faces.html

AskERIC

The Educational Resources Information Center (ERIC) is a federally funded educational information system providing extensive education-related literature and resources to teachers, administrators, and learning specialists.

URL:

http://ericir.syr.edu

Civnet

Civnet is a collection of civics-teaching resources and discourses on civil society. Teachers will want to explore the Civics Teaching Resource Library, which includes online textbooks and lesson plans. There are also sections on civil society and civic journalism, and recent writings on civil liberty.

URL:

http://www.infomall.org/Showcase/civnet/

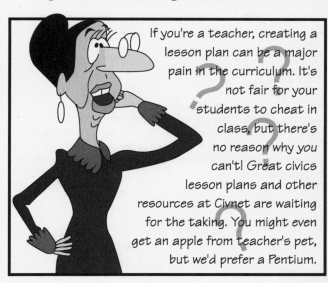

If you're a teacher, creating a lesson plan can be a major pain in the curriculum. It's not fair for your students to cheat in class, but there's no reason why you can't! Great civics lesson plans and other resources at Civnet are waiting for the taking. You might even get an apple from teacher's pet, but we'd prefer a Pentium.

CNN Center

Focusing on a career in TV news coverage? Take a comprehensive guided tour behind the scenes at CNN that covers every aspect of information gathering and reporting. You'll learn about satellite operations, assignment desks, control rooms, editing suites, feature production, international bureaus, Washington coverage, and the daily editorial meetings of the network.

URL:

http://cee.indiana.edu/rd/cnncenter.html

College and University Home Pages

Just passed your SAT test with flying colors, but not exactly sure where to go next? Figure out which school to attend with this enormous list of over 2000 college and university home pages. You can view the full list of institutions, or condense the search to a more specific region. International universities are included too.

URL:

http://www.mit.edu:8001/people/cdemello/
univ.html

A B C D E F G H I J K L M N O P Q R S T U V W X Y Z

Cornell Theory Center Math and Science Gateway

The Gateway provides links to resources in mathematics and science for educators and students in grades 9-12. This page features standard math and science subjects, plus online museum field trips, a list of schools on the Web, and other links.

URL:
http://www.tc.cornell.edu/Edu/MathSciGateway/

DeweyWeb

DeweyWeb is an experimental program that promotes using the Web to provide information and facilitate communications between students throughout the world. Students can spend part of every school day in a "global classroom" with WebICS, an ongoing series of innovative educational exercises.

URL:
http://ics.soe.umich.edu

DIALOG

A substantial scientific information resource from Knight-Ridder Information, Inc., DIALOG offers access to more than 450 online databases and 45 CD-ROM titles. Bluesheet guides give you a look into DIALOG's vast content, available by subscription.

URL:
http://www.dialog.com

Discovery Channel Online

Explore your world online with Discovery Channel Online. This handsome web site complements the equally fascinating 10-year-old TV channel with original articles on science, history, and new technology. Of particular note is the Knapsack feature. The Knapsack collects the results of an Internet-wide search query—even when you're not online—and stores it until you open your "pack."

URL:
http://www.discovery.com

You may have a hard time deciding what to watch tonight—the Discovery Channel or the Discovery Channel Online. Just remember: you can't control your TV with a mouse, and you can't control your computer with a remote. At least, not yet. . .

Education-related Newsgroups

"Education is what survives when what has been learnt has been forgotten." —B.F. Skinner. Keep track of what you've learnt with these educational newsgroups. Read and correspond with teachers and educators about new technologies, trends, methodology, and employment opportunities.

URL:
news:comp.edu

EdWeb

The EducationWeb Project explores the world of educational computing and networking in an easy-to-use, hypertext guide. You can hunt down online educational resources around the world, learn about trends in education policy and information infrastructure development, and examine success stories of students using computers in the classroom.

URL:
http://k12.cnidr.org:90

From Now On

Here's an excellent educational technology journal for teachers and educators who want to familiarize themselves and their students with the Internet. Articles range from "Libraries of the Future" to "Grazing the Net—Skills Required to Make Insight while Researching with the Net" and "Cutting to the Chase: Leading Teachers and Students to the 'Right Stuff' with WWW Curriculum Pages." Find sections with information on virtual museums and workshops, and strategies for student learning on the Net.

URL:
http://www.pacificrim.net/~mckenzie/

Idea Futures

Welcome to Idea Futures, Robin Hanson's answer to the dubious grant system perpetuated by those who should be pastured instead of tenured. In the world of Idea Futures, projects involve real payoffs based on, well, real results and benefits. Writes Hanson, "This would be an 'idea futures' market, which I offer as an alternative to existing academic social institutions. Somewhat like a corn futures market, where one can bet on the future price of corn, here one bets on the future settlement of a present scientific controversy." Here you can test out the concept, develop consensus, and have some fun.

URL:
    http://zero.arc.ab.ca/IF.shtml

International Centre for Distance Learning

The Open University's ICDL is a centralized repository of information on distance education efforts throughout the world. Services here include a bibliography of abstracts on distance education, access to a searchable database, details of conferences, entire chapters from the book *Mindweave: Communication, Computers and Distance Education*, and the ICDL newsletter. The database contains details of over 28,000 courses and programs; more than 800 institutions; and information on thousands of books, journal articles, research reports, and papers relating to the theory and practice of distance education.

URL:
    http://acacia.open.ac.uk/

K-12+ Servers

One of the best things about being a kid (and even about being a teacher) these days is that you can communicate with other children all over the world via the Internet. This page shows you how to begin doing just that, with lists of K-12 servers categorized by state and country. There are schools, school districts, departments of education, and miscellaneous educational sites.

URL:
    http://www.tenet.edu/education/main.html

LEGO Engineering

So you thought LEGOs were nothing but child's play. Fostering creativity and imagination, LEGOs make designing fun. LEGO Dacta blocks are new, and have a computer interface. Your prototypes will come alive with this new generation of LEGOs.

URL:
    http://amy.me.tufts.edu/LDAPS/Docs/lego.html

Quality Education Data

Chart the implementation of technology in U.S. schools through this innovative education page. There are 18 graphs that delineate the schools' progress (or lack, thereof) in the use of computers and the Internet in American classrooms. Users are free to download these graphics and use them in presentations to school boards, parents' groups, or teachers' conferences.

URL:
    http://www.infomall.org/Showcase/QED/

School Computing

Mainly for K-12 educators, this Gopher contains articles like "K12 Network: Global Education Through Telecommunications," "Electronic Tools in Actual Classroom Settings," "E-Mail Discussion Groups of Interest to Educators," "Telnet Interactive Access and FTP Sites of Interest to the Educator," and "Worldwide Teacher/Student Network."

URL:
    gopher://rain.psg.com/11/schools

University Park Elementary School

This Fairbanks, Alaska elementary school publishes *The Falcons' Nest,* an online journal for K-12 Internauts. You can read about life in Alaska, take a trip to Mount McKinley, view pictures of our largest state, and correspond with the students at University Park. There are links to other Web resources that the school uses most frequently.

URL:
    http://www2.northstar.k12.ak.us/schools/
    upk/upk.home.html

It may be freezing in the far north, but that doesn't mean that you have to put your mukluks on to enjoy the benefits of an Alaskan vacation. Mush your mouse on over to the **University Park Elementary School** web site, and brighten up the igloo of your mind with their heartwarming page.

A
B
C
D
E
F
G
H
I
J
K
L
M
N
O
P
Q
R
S
T
U
V
W
X
Y
Z

EGYPTOLOGY

Computer-Aided Egyptological Research

Computers are adept at organizing and accessing information. CCER at Utrecht University in The Netherlands specializes in the application of computers in Egyptology. CCER facilitates the sharing and exchange of data on an international level. Standards and conventions for the input of Egyptological data have been and are being developed. These include multilingual thesauri, field lists, and directions for standardizing the data input.

URL:
http://www.ccer.ggl.ruu.nl/ccer/ccer_newton.html

Djoser Complex

There are more pyramids in Egypt than the three familiar ones near the Sphinx. The oldest, perhaps, is at Saqqara, built for the Pharoah Djoser (or Zoser) by the world's first famous architect, Imhotep. Subsequent generations thought Imhotep's work to be so important that they made him a god. Sort of like I. M. Pei, these days. View Imhotep's greatest accomplishments via a clickable map on this page.

URL:
http://ccat.sas.upenn.edu/arth/zoser/zoser.html

Eastern Desert of Egypt

In *Casablanca*, "Everyone comes to Rick's." But in Africa, "Everyone goes to Egypt." The Archaeological Survey of the Eastern Desert of Egypt answers questions about the trade routes, particularly the transdesert routes, between the Nile Valley and the Red Sea, which linked the civilizations of the Mediterranean with those of the Indian Ocean between 300 B.C. and A.D. 400. Reload it again, Sam.

URL:
http://rome.classics.lsa.umich.edu/projects/coptos/desert.html

Egypt Pictures

If a picture is worth a thousand words, this site has about 50,000 things to say about Egypt. Happily, you get to see the pictures instead of reading more words. But if your service provider charges you by the hour for Internet access, they're going to be even happier.

URL:
http://www.sas.upenn.edu/African_Studies/Egypt_GIFS/menu_Egypt.html

Egyptian Information Highway

Not a superhighway yet, by any stretch of the imagination, but a highway, albeit, under construction. The world-famous Egyptian Museum home page is here with some good .gifs of jewelry, sculpture, statuary, tomb furnishings, and a photo of the mummy of Rameses the Great. There is additional info on Ancient Egypt, a list of 140 museums, Egyptian collections, plus other related Web resources.

URL:
http://its-idsc.gov.eg/cult-net/home.htm

Egyptological Bibliography

Books and more books on Egyptology. This database includes categories searchable by subject including general, script and philology, texts and philology, history, art and archaeology, religion, society and culture, and Nubian studies.

URL:
http://www.leidenuniv.nl/nino/aeb92/intro92.html

Giza Plateau Mapping Project

When all around you are sand, rubble, and eroding ruins, the power of the computer can help piece together the puzzle of ancient architecture. This Oriental Institute project is researching the geology and topography of the Giza plateau, the construction and function of the Sphinx, the Great Pyramids and associated tombs and temples, and the Old Kingdom town in the vicinity. Computer graphics and remote sensing technology model the ancient configuration of the Giza plateau in stunning relief.

URL:
http://www-oi.uchicago.edu/OI/PROJ/GIZ/Giza.html

Computer rendered model of the Giza Plateau.

Institute of Egyptian Art and Archaeology

As it turns out, Memphis, Tennessee and Memphis, Egypt aren't so very far apart these days, thanks to the Web. Besides famous kings living in both places, you'll also learn about the Institute of Egyptian Art and Archaeology, and you can visit the exhibit of Egyptian artifacts residing at the University of Memphis. Take a short color tour of Egypt. Uh, thank you. Thank you very much.

URL:
http://www.memphis.edu/egypt/main.html

Kelsey Museum Online

A wooden door, a light green chalice, and second century coins are some of the ancient items you'll discover in the Kelsey Museum's sponsored excavations at Karanis, Egypt. Dust the sand off your mouse and explore exhibits, images, and maps of the ancient Mediterranean world.

URL:
http://www.umich.edu/~kelseydb/

Mansoor Amarna Collection

The Amarna period in Egypt was both a renaissance and a disaster. It almost ruined the country, but it certainly revitalized Egypt's art in a more candid, naturalistic sense. See what it all means at this well-thought-out web page.

URL:
http://www.amarna.com/

Mansoor Amarna Collection

*There once was a pharaoh named Akhnaten
Who tossed the religion he'd gotten
Nefertiti and he tried monodeity
But the rest really thought it was rotten*

And that's exactly what happened. Find out more at the
Mansoor Amarna Collection.

Murmuring Rumors About Mummies

Talk like an Egyptian. News and gossip from the world of Egyptology, plus announcements of conferences, seminars, and exhibitions.

URL:
http://www.newton.cam.ac.uk/egypt/

Museo Gregoriano Egiziano

Typeset in HTML by the Pope himself (who now uses a Sun SPARCstation after realizing that Galileo may have been right about the center of our galaxy), this site lets you see photographs in the Vatican Museum's collection of Egyptian sculpture and art, from the later dynasties to the Roman period.

URL:
http://www.christusrex.org/www1/vaticano/
EG-Egiziano.html

This image of a golden hawk from the late Dynastic period (656-332 B.C.) is among the many stunning pictures you can view and download from the Museo Gregoriano Egiziano.

Nubia Salvage Project

"Nubia: Its glory and its people" and "Vanished Kingdoms of the Nile: The Rediscovery of Ancient Nubia" are two exhibits of the land and peoples of ancient Nubia. Documentation and images depict the region and history of Nubia as seen through tomb paintings, structures, maps, early writing, and artifacts.

URL:
http://www.oi.uchicago.edu/OI/PROJ/NUB/
Nubia.html

Kids, check out Kids and Amateurs.

A B C D E F G H I J K L M N O P Q R S T U V W X Y Z

Oriental Institute Museum Highlights

Excellent photographs accompany this online exhibit of Egyptian artifacts, including a model of a butcher shop, a Book of the Dead, the colossal statue of Tutankhamen, a mummy mask, the coffin of Ipi-Ha-Ishutef, funerary stela, relief from the Tomb of Mentuemhat, and last but not least, a reconstructed predynastic burial. Invite a friend over to share in the excitement.

URL:

    http://www-oi.uchicago.edu/OI/MUS/HIGH/
    OI_Museum_Egypt.html

Papyrus Archive

Duke University's hefty collection of 1,000 papyri is available online, with a searchable database and all sorts of categories to meander through, plus a general section on the fine science of papyrology.

URL:

    http://scriptorium.lib.duke.edu/papyrus/

The Papyrus Archive

Ah, yes, the papyrusless office. It was the technological goal of ancient civilizations. Unfortunately, the study of history necessarily means that humans have to read about it in order to interpret it. But before email, before the National Enquirer, *and even before* Mad Magazine *(not to mention the printing press), people wrote and read on papyrus: a pithy Nile delta plant suitable for being pounded, separated, dried, assembled, scribbled on, and rolled into scrolls. A tremendous amount of grant money has been poured into Duke University's Papyrus Archive so that you can read more about the early beginnings of greenbar. And, if you somehow choose not to explore these ancient bytes of history and wisdom, at least you now know where the word "paper" comes from. Keep that in mind the next time you're in the loo searching for a scroll.*

Reconstructing an Egyptian Mummy in 3-D

If you're interested in getting up close and personal with a mummy, but blush at taking off its wrappings, here's the site for you. Using progressive radiological technology, an Egyptian mummy named Tjentmutengebtiu is reconstructed in 3-D. The details of Tjentmutengebtiu will not disappoint you. No 3-D glasses required.

URL:

    http://www.pavilion.co.uk/HealthServices/
    BrightonHealthCare/mummy.htm

Secrets of the Lost Tomb

The discovery of Tutankhamen's tomb is considered to be one of the major archaeological finds of the century, but for even more information, the tomb of the fifty sons of Rameses II may far surpass Tut. Its sheer size alone is staggering. If Tut's tomb were the size of a matchbook, the new tomb would be the size of a coffee table. Archaeologists are anticipating that the wealth of the tomb will come not from gold, but from new information. The *London Daily Mail* called it, "The mummy of all tombs." Read all about it here, courtesy of *Time* magazine.

URL:

    http://www.pathfinder.com/time/magazine/
    domestic/1995/950529/950529.cover.html

ELECTRONICS

Circuit Cookbook Web Page

Here's a collection of documents relating to the repair of circuitry and electronics. Topics include computers, audio, digital, power, telecommunications, video, and many others.

URL:

    http://www.ee.ualberta.ca/~charro/cookbook/

"Mmmm, honey, what's that delicious aroma? Are you baking transistors again?"

"No, dear, for a change, I thought I'd fry a few electrodes tonight."

If you've sauteed your semiconductors, learn how to repair them at the *Circuit Cookbook* web page.

Diagnosis and Repair of Video Monitors

If you can't decide whether to keep your old computer monitor or sacrifice it to the scrap heap, do the smart thing and find out if it can be repaired! This FAQ list will show you the way, and even if your monitor is just fine, it's still a great place to simply find out how monitors work.

URL:

http://www.paranoia.com/~filipg/HTML/FAQ/
BODY/F_mon_repair.html

Electronics Information

Link up with lots of different sites related to electronics here. This page points the way to information resources on schematics, online magazines, electronics parts, utilities, and other electronics sites.

URL:

http://www.geopages.com/SiliconValley/1164/

Electronics Repair FAQ

Learn to repair anything—from CD players to computer monitors. Here's a great series of documents to help get you started. Also included are links to other electronics repair sites.

URL:

http://www.paranoia.com/~filipg/HTML/FAQ/
BODY/Repair.html

Home Automation

Find hundreds of automation, security, surveillance, audio/video control, and other unique products to make your home or business more comfortable, secure, energy efficient, and fun!

URL:

http://www.techmall.com/smarthome/

HP Calculator Chat Channel

If you love the HP48 and just have to talk to other enthusiasts before you burst, visit the IRC's HP48 Calculator home page. This page describes the HP48 and how to use the IRC to enjoy fellowship with other users of this popular calculator.

URL:

http://twws1.vub.ac.be/studs/tw45639/hp48.htm

Lucasfilm THX Home Page

The audience is listening . . . and surfing Lucasfilm's THX Home Page, too. Find information about film sound reproduction technology, technical details about THX, and how to turn your living room into a thundering home theater. The neighbors will love you (as long as you invite them over to watch the Star Wars trilogy in surround sound).

URL:

http://www.thx.com

Motorola Semiconductor Products

Looking for a microprocessor or a digital signal processor? Search the Motorola online data book with over 32,000 data sheets and see if you can find that chip that causes your VCR to blink 12:00.

URL:

http://motserv.indirect.com/home2/
mothome.html

Museum of HP Calculators

Drop by the Museum of HP Calculators to reminisce about those great HP calculators from the past. Whether you're just missing your old 9100A or are trying to find someone to repair a relic from the early '70s, this is the place. This page has links to detailed information and trivia about each HP calculator model ever produced, a section for buying and selling calculators, and links to other calculator resources on the Net.

URL:

http://www.teleport.com/~dgh/hpmuseum.html

Excerpt from the Net...

(from the Lucasfilm THX Home Page)

"Let me suggest the following progression: Mozart, Beethoven, Mahler, Terminator 2. What we see (or hear, really) is a widening of the frequency and dynamic ranges, increasing spatial effects, and a downward movement of the frequency range in which the acoustic energy is greatest." —Tomlinson Holman, originator of the THX standard, on the importance of subwoofers

A
B
C
D
E
F
G
H
I
J
K
L
M
N
O
P
Q
R
S
T
U
V
W
X
Y
Z

Repair of Audio Equipment and Other Miscellaneous Stuff

Is your old turntable throwing a tizzy every time you put on your old Montovani albums? Don't ditch that dinosaur or give it to some high-priced electronics geek—fix it yourself! This page will show you how, with lots of great tips and tricks to revive your record player, tape recorder, Sony Walkman and Discman, answering machine, touch tone phone, and even your modem.

URL:
http://www.paranoia.com/~filipg/HTML/FAQ/
BODY/F_Audio_n_Misc.html

Sony Camcorders

Compare the specs for Sony's entire line of video cameras and handycams. These spec sheets include everything you could want to know about Sony's cameras including their suggested retail prices, connection options, and all of their features. Before you shop, zoom into the Sony camcorder page for a close-up.

URL:
http://www.sel.sony.com/SEL/consumer/
ss5/camcorder.html/

TI Calculators

Welcome to the unofficial TI calculator page. While this page may be unofficial, it contains a mind-boggling amount of information about TI calculators. You will find links to Texas Instruments' own summaries of various popular TI calculators, including pictures, features, and specifications; FAQs; news; links to FTP archives, Gophers, and other web sites; and information on programming TI calculators by connecting them directly to a PC or Macintosh.

URL:
http://dnclab.Berkeley.EDU/~smack/ti.html

VCR Failure Diagnosis and Repair

If you've already figured out how to set up a VCR, congratulations! Perhaps your next step is to learn how to fix one. Peruse this 65K document covering the A-to-Z's of VCR problems and solutions, and you'll be ready for any mishap. Now if you could just figure out how to set the darned clock. . .

URL:
http://www.paranoia.com/~filipg/HTML/FAQ/
BODY/F_vcr_repair.html

ENERGY

Alternative Fuels Data Center

Operated and funded (ultimately) by the U.S. Department of Energy, the AFDC offers volumes of information on alternative fuels and alternative fuel vehicles. Read about Chrysler, Ford, and GM's current alternative fuel vehicles and their plans for future offerings—including electric vehicles. These pages also offer newsletters, detailed reports on alternative fuel utilization, and a clickable image-map database of U.S. refueling sites.

URL:
http://www.afdc.doe.gov

American Hydrogen Association

The AHA is a nonprofit organization that promotes the use of hydrogen as a clean alternative to fossil fuels, hoping to help shift the U.S. economy to one based on hydrogen by the year 2000. (We'd guess they're going to be late.) On their pages, the AHA offers information about hydrogen as a fuel, fuel-cell technology based on hydrogen, FAQs, newsletters, and activities.

URL:
http://www.getnet.com/charity/aha/

Excerpt from the Net...

(from the Alternative Fuels Data Center)

The Alternative Fuels Data Center (AFDC) is operated by the National Renewable Energy Laboratory (NREL) with funding and direction from the Office of Alternative Fuels within the Office of Transportation Technologies at the U.S. Department of Energy (DOE).

Anyone can browse the Web.

American Solar Energy Society

Seeing sunspots before your eyes? Perhaps you should join the ASES, a national community of people who work and play in the field of Solar Energy. Their page tells you about membership, press releases, publications, fact sheets, and subscriptions.

URL:

http://www.engr.wisc.edu/centers/sel/ases/ases2.html

Ames Laboratory

Study chemicals, engineering, materials, and mathematical and physical sciences to help solve your energy problems. This Department of Energy page outlines research, news, career opportunities, educational programs, and general laboratory information.

URL:

http://www.ameslab.gov

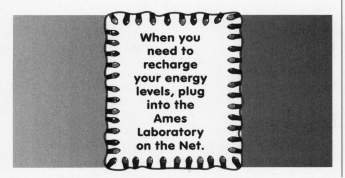

When you need to recharge your energy levels, plug into the Ames Laboratory on the Net.

Argonne National Laboratory

The next time your portable cellular phone starts blinking LO BAT, remember the folks at ANL. They're researching advanced batteries and fuel cells, fission reactor technology, electrochemical treatment for disposal of spent fuel, the decommissioning of aging reactors, and over 400 links to Earth Science-related sources. You'll also find laboratory research results and news on the environment, oceanography, the atmosphere, geology, hydrology, and astronomy.

URL:

http://www.anl.gov

Biofuels Information Network

In the future, those grass clippings from your lawn might go right into your car's gas tank. Well, sort of. The BIN, a branch of the U.S. Department of Energy, researches the domestic production, recovery, and conversion of grasses and fast-growing short-rotation trees to reasonably priced fuels such as ethanol, methanol and biodiesel, which are good to the environment. Time was you put a tiger in your tank. Now you can stuff in a Douglas Fir.

URL:

http://www.esd.ornl.gov/BFDP/BFDPMOSAIC/binmenu.html

Brookhaven National Laboratory

BNL is one of the nation's leading scientific research laboratories with interests in many branches of science related to energy. BNL offers information on laboratory research, events, news, and a visual tour. Prepare to have your neutrons scattered.

URL:

http://suntid.bnl.gov:8080/bnl.html

Excerpt from the Net...

(from the California Energy Commission's pages)

During the next 20 years, California's overall demand for electricity is expected to grow at an average rate of 2 percent per year. However, peak demand for electricity is forecast to grow at an average rate of 2.7 percent per year. If not reduced through energy efficiency improvements, meeting peak demand will require an additional 23,000 megawatts of electricity resources by 2010.

A B C D E F G H I J K L M N O P Q R S T U V W X Y Z

California Energy Commission

Do you live in California? If you do, you're in one of the largest electricity-using and -producing states. Although one third of the West's total electricity-generating capacity is in California, the state has 85 percent of the region's independently produced power capacity. Find out more about these facts and other interesting information, tips, policies, and news on the CEC's web page.

URLs:

gopher://www.energy.ca.gov:70/11/CECIS

http://agency.resource.ca.gov/
    cechomepage.html

CREST's Guide to Alternative Energy

A hypertext guide for and by energy experts, these pages offer direction to alternative energy information on the Web. CREST also includes pointers to Usenet newsgroups, mailing lists, file archives, and databases.

URL:

http://solstice.crest.org/online/aeguide/

EEGADS

Eegads! Your research is so vital, but the funding is a killer. That's why the Eligible Equipment Grant Access Data System (EEGADS) grant program was established by the U.S. Department of Energy to make used equipment for energy-related research available to higher education institutions. There's a whole shopping list of things, from computer equipment to sensors, meters, generators, and other cool stuff.

URL:

http://web.fie.com/web/fed/doe/doemneg.htm

Electric Power Research Institute

The EPRI tries to discover, develop, and deliver new technologies to help the American electric power industry. On their web pages, EPRI outlines their past research and development achievements, presents the full text of *The EPRI Journal*, and provides links to news relevant to the electric power industry.

URL:

http://www.epri.com

Energy Efficiency and Renewable Energy Network

Let's not mince words. You want to decrease your utility bill. So start immediately by running through the house and turning off everything except the computer. Then link to the Department of Energy's page of tips for conserving energy and reducing your electrical expenses. Keep it short—your meter is running.

URL:

http://www.eren.doe.gov

Energy Efficient Housing

Want to know how to save energy on those cold winter nights? This page offers information on the R-2000 model home, the ultimate energy-saving construction for homes. Learn about efficient construction, features, materials, cost, history and background of the home, and related Net resources.

URL:

http://web.cs.ualberta.ca/~art/house/

Energy Information on the Internet

Just for you, this extensive list of annotated links to energy information on the Net is a tremendous resource. With dozens of entries, this is a great page to start your exploration of energy and environmental material.

URL:

http://solstice.crest.org/online/virtual-library/
    VLib-energy.html

Energy Science

If you think cold fusion is a dead end, read **sci.energy** for a while and think again. This Usenet newsgroup is hoppin' with great information from intelligent and knowledgeable people on topics ranging from using vegetable oil for diesel fuel to electric cars and fuel cells.

URL:

news:sci.energy

Looking for a science project? Look in Kids and Amateurs.

Energy Systems Lab

Texas A&M University opens their Energy Systems Lab to you starting with a graphical tour of the facilities. After your tour, sit a spell and look over papers on energy audits, energy conservation, and data acquisition. Next, download software for your computer to help you analyze energy consumption. Before you leave, be sure to get an Energy Systems Update newsletter to take with you. And thank you for not leaving the light on in the kitchen.

URL:
http://www-esl.tamu.edu

Energy Yellow Pages

The Energy Yellow Pages lets your mouse do the clicking through listings of energy companies, trade associations, research organizations, educational programs, and publications.

URL:
http://www.ccnet.com/~nep/yellow.htm

ENERGY YELLOW PAGES
From alternative fuels to wind—and everything in between, including cogeneration, storage, fuel cells, hydroelectric, photovoltaic, and renewable resources. If you need quick information on companies, associations, organizations, programs, and publications in these fields, let your digits do the walking—at the Energy Yellow Pages.

ESnet

ESnet facilitates remote access to major energy research facilities, provides information dissemination among scientific collaborators, and provides widespread access to existing research supercomputer facilities. A hypertext directory, gopher service, white pages, user guides, and information on conferences and meetings can be found online.

URLs:
gopher://gopher.es.net

http://www.es.net

Florida Energy Handbook

What are the best ways to conserve energy throughout the year? This handbook answers this and many other questions about energy efficiency.

URL:
http://hammock.ifas.ufl.edu/text/eh/19898.html

Hydrogen as a Fuel

Ask anyone who has seen images of the Hindenburg disaster (or heard the original radio broadcast replayed over RealAudio) and they'll agree that hydrogen is hot stuff as a source of energy. In this Usenet group you'll find articles on the benefits and risks of using hydrogen as a fuel.

URL:
news:sci.energy.hydrogen

National Renewable Energy Laboratory

Running low on steam? Recharge yourself by tapping into the National Renewable Energy Laboratory for information on alternative fuels, analytic studies, basic sciences, energy systems in buildings, industrial technologies, photovoltaics, and wind technology.

URL:
http://www.nrel.gov

Power Globe Mailing List

This mailing list serves the power-engineering community worldwide by disseminating news and announcements that concern power engineers. This web page serves as an introduction to the list, provides instructions for joining the list, and offers access to the lists' archives.

URL:
http://www.ece.iit.edu/~power/pwrglobe.html

Power Plant Modeling and Visualization

Here's an entire slide show from a presentation at the Delft University of Technology's Laboratorium for Thermal Power Engineering on power plant modeling and visualization. Each slide is accompanied by descriptive text.

URL:
http://www-pe.wbmt.tudelft.nl/ev/mmsugm.html

A B C D E F G H I J K L M N O P Q R S T U V W X Y Z

Power to the People

Cool! Build your own alternative energy machinery such as solar cells, photovoltaic systems, and hydroelectric power! How? This page gives you visual tools, a complete engineering guide, and a shopping list with typical costs for such projects. If it runs on electricity, you'll find a way to build it here—everything from solar ovens to flashlights is covered.

URL:
http://www.nando.net/prof/eco/aee.html

Renewable Energy Education Module

Why is it important to recycle when everyone knows it's in our nature to trash the earth? Well, if you happen to be one of the miscreants asking such an inane question, this page shows you why conservation and recycling help preserve a healthy environment. Learn about biomass as a source of fuel, and solar, geothermal, hydro, and wind energy.

URL:
http://solstice.crest.org/renewables/re-kiosk/
index.shtml

Solar Car Page

Sunrayce, the annual collegiate solar-car competition, heats up every year attracting student engineers from dozens of universities. The car designs are entomologically inspired, some resembling great solar cell-covered cockroaches.

URL:
http://www-lips.ece.utexas.edu/~delayman/
solar.html

Bugged having to use gasoline? Tap into the *Solar Car Page,* and warm up your engine's cockles. You'll encounter a battery of manifold information.

Solar Cells

Within a few years, solar cells that work in the dark may become available. Liquid solar cells, cells that hum when the sun goes down, and cells that work at night are some of the topics on Clark University's Solar Cells page. These silicon cells are known to be an inexhaustible, clean source of energy that does not contribute to global warming.

URL:
http://enuxsa.eas.asu.edu/~rutledge/
solrclls.html

Solar Energy Laboratory

See how the Solar Energy Program shines at the University of Wisconsin-Madison. The lab provides research results and complete source code for study. A ubiquitous list of other solar-related resources can also be found.

URL:
http://www.engr.wisc.edu/centers/sel/sel.html

Solving Power Quality Problems

Electrotek Concepts provides this resource for learning how to evaluate and solve utility power quality problems. These pages describe the methodologies and resources to employ in evaluating problems such as voltage sags, momentary interruptions, transients, and harmonics.

URL:
http://www.electrotek.com/ps_study/utility/
utility.htm

Southface Energy Institute

The Southface Energy Institute promotes developing environmentally sound and enduring energy technologies. On their web pages, Southface provides information on energy-efficient construction, affordable housing, and education and training materials.

URL:
http://www.mindspring.com/~southfac/

Everyone should be on the Web.

U.S. Department of Energy

The DOE reveals their declassified information and research as they try to provide the nation with the technology, policy, and leadership needed to use energy most efficiently. These pages lead you into the deepest corners of this immense department and its many programs, projects, scientists, and institutions.

URL:

   http://www.doe.gov

Waste Isolation Pilot Plant

A new and messy form of solitary confinement at Levenworth? No. This DOE page discusses the safe handling, transportation, and disposal of transuranic waste in salt beds. Included are links to participating agencies and additional details on this program. But what a terrific justice for our worst criminals! A few hours of isolation working in the salt beds and they would suffer from radiation and high levels of sodium at the same time. Who needs a gas chamber?

URL:

   http://www.wipp.carlsbad.nm.us

ENGINEERING

3D Modeling

ICEM CFD is a powerful, state-of-the-art 3D modeling package for generating multi-block structured grids, unstructured tetrahedral and triangular surface grids, quad grids, and Cartesian and refined H-grids.

URL:

   http://icemcfd.com/icemcfd.html

Aluminum Wiring Hazards

Dan Friedman compiles this information and advice for consumers, engineers, and building inspectors on the hazards associated with aluminum electrical wiring. Study a summary of hazard issues and check out links to many public documents.

URL:

   http://csbh.mhv.net/~dfriedman/aluminum.htm

American Nuclear Society

The ANS is an international, nonprofit organization consisting of thousands of engineers, scientists, educators, students, and others interested in nuclear power. On their pages, they offer information about conferences, professional divisions of the organization, and links to student chapters of the ANS.

URL:

   http://neutrino.nuc.berkeley.edu/ans/ANS.html

Analog Circuit Simulation

Automatic Integrated Circuit Modeling Spice (AIM-Spice) is a Microsoft Windows program for simulating analog electrical circuits. On this, the AIM-Spice home page, read an overview about the software and its features and interface, and download the software.

URL:

   http://fulton.seas.virginia.edu/~ty2n/
   aimspice.html

Analog Circuit Simulator Tutorial

Trying to sort out your transistors from your capacitors? Maybe you should take the APLAC tutorial! This tutorial starts at the very beginnings of analog circuit design and shows you how to implement simple circuits with basic analysis modes. Soon, there will also be more advanced tools here covering optimization, noise analysis, and N-Port analysis.

URL:

   http://picea.hut.fi/aplac/tutorial/main.html

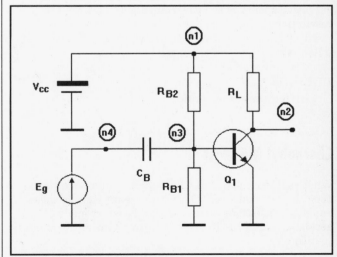

Are your circuits short and your resistance too low? Click on a component of this simple amplifier to find out what it is and what it does. You can find it in the Analog Circuit Simulator Tutorial.

A B C D E F G H I J K L M N O P Q R S T U V W X Y Z

Archive for Power Engineering Education

The University of Wollongong in Australia provides this unique web page for electric power educators. The content here is applicable throughout the world, but may be of particular interest to the Asia-Pacific region. Topics covered here include computational hypermedia issues; teaching and learning resources; multimedia projects in electrical engineering; and regional conferences.

URL:

http://www.uow.edu.au/public/pwrsysed/homepage.html

Ballistic

Ballistic is a layout language for analog circuits. It works with Mentor Graphics' GDT software package and produces technology-independent layouts that are fully parameterizable. This site also provides useful information on circuit layouts and the source code for the language.

URL:

http://www.eecg.toronto.edu/~gdt/

Building Research Establishment

BRE is an organization in the U.K. that carries out research into building and construction—especially as it relates to the prevention and control of fire. This is an executive agency of the government's Department of Environment. You'll find information about BRE—its building research, press releases, literature, events, and consultants.

URL:

http://www.bre.co.uk/

Chernobyl Manifest

What really happened at Chernobyl . . . and what have been the consequences? Get definitive answers to these and other questions about the world's worst nuclear accident here at the Russian Research Center, Kurchatov Institute.

URL:

http://polyn.net.kiae.su/polyn/manifest.html

One nuclear accident can ruin your whole day. Take Chernobyl, for instance. Push a button here, throw a lever there, and BOOM— there goes another autonomous republic . . .

Chuck's Fluids Pages

Chuck Seal is an engineering student with a zeal for fluids. Chuck's so motivated to tell about his work that he's made it available to the world on the Web. Read about Chuck's Transitional Necklace Vortex Regime. If that sounds too material for your mood, why not try the Hairpin Vortex. And of course, there's the ever popular Dynamics of the Vorticity Distribution in Endwall Junctions. Be forewarned: this is cool stuff.

URL:

http://www.lehigh.edu/cvs0/public/www-data/fluids/fluids1.html

CIM Systems Research Center

Arizona State University's CIM Systems Research Center attempts to provide leadership and foster intellectual development for faculty, students, and industry in integrated manufacturing enterprises. Sidle up to a saguaro and check out their research projects, facilities, and their nice collection of links to manufacturing sites on the Web.

URL:

http://enws324.eas.asu.edu

Circuit Cellar

Circuit Cellar is a manufacturing and publishing (odd combination) company involved in home control systems and other home automation tasks. Their main publication is a monthly magazine for design engineers, programmers, and tinkerers who play with embedded processors and electronic circuitry. Check out their informative publication, their line of automation products, and some of the projects they've undertaken.

URL:

http://www.circellar.com

Circuits Archive

This public-access database is a repository for all types of electrical engineering information. Offered here is a brief introduction to the system, instructions on how to submit a design for a circuit or component, circuit diagrams, component data sheets, and so on.

URL:
>   http://weber.u.washington.edu/~pfloyd/ee/

Computational Fluid Dynamics: An Introduction

Intended as an introduction to the science of computational fluid dynamics, this document is targeted at people with diverse scientific backgrounds—from novices to experts.

URL:
>   http://www.nas.nasa.gov/~smurman/
>   cfd_intro.html

Computer Aided Engineering

Computer Aided Engineering (CAE) has as much to do with business in general as it does with computers or computer science. Point your stylus at C.A. Maynard's Computer Aided Engineering page and read about expert systems in manufacturing, Just In Time (JIT) manufacturing, robots in agriculture, and other interesting topics.

URL:
>   http://kernow.curtin.edu.au/cae.html

Did you know that humans couldn't possibly build the computers we use today without the extensive aid of . . . computers?

Next thing ya know, they won't even need us!

If you're not on the Web, you don't exist.

Computer Assisted Mechanics and Engineering Sciences

Computer Assisted Mechanics and Engineering Sciences (CAMES) is a quarterly journal of the Central European Association for Computational Mechanics. It's a great source of information on the field of computational mechanics and related areas of applied science and engineering.

URL:
>   http://www.ippt.gov.pl/zmit/www/CAMES.html

CORE Industrial Design Network

CORE Industrial Design Network is an interesting waypoint for mechanical and industrial engineers doing design work. You'll waste a lot of time here, although you may also come away with some fresh insights.

URL:
>   http://www.core77.com

Design Automation

The Association of Computing Machinery's Special Interest Group on Design Automation (SIGDA) indexes a collection of publications of the ACM. Read about graduate scholarships, conferences, programs, and activities. There's also a SIGDA member list with email addresses, and other items of potential interest.

URL:
>   http://kona.ee.pitt.edu/index.html

dTb Software

dTb Software develops application-specific software tools for electrical engineering. While their systems are commercial products, they also offer five useful sample applications: a standard value resistor selector/calculator, a wire selector/calculator, a divider design program, a units converter, and floating numeric keypad. These are all Windows programs that use DDE to update other applications in real time—for example, a spreadsheet program doing computations based on the outputs of these programs.

URL:
>   http://www.dnai.com/~dtbsware/

Are you on the Web?

A B C D E F G H I J K L M N O P Q R S T U V W X Y Z

Dynamic Stability Lab

Dynamic stability is something we all strive for. Especially after we've had a few too many drinks. Slip into Dr. Mote's lab at U.C. Berkeley where they know more about dynamic stability than any other lab in the western hemisphere. Topics here include vibration control for rotating plates, translating bands, axially moving materials, and high-speed rotating disks. Also of possible interest are optimal designs for guide bearings and ergonomically designed aids to help people with upper extremity musculoskeletal disorders.

URL:

http://mote.berkeley.edu

Electrical Engineering

This enormous page is an excellent starting point for electrical engineering research on the Net.

URL:

http://arioch.gsfc.nasa.gov/wwwvl/ee.html

Engineering Foundation

Since 1914, this organization has provided grants for engineering research and advancement. A large number of projects have been carried out over the years and some of these are described on this page. There are also a conference calendar and a listing of available grants.

URL:

http://www.garlic.com/engfnd/

Engineering Virtual Library

The page lists many links to engineering-related pages on publications, jobs, newsgroups, announcements, standards, manufacturers and vendors, and research institutions, plus upcoming conferences and events, and collaborative projects.

URL:

http://arioch.gsfc.nasa.gov/wwwvl/
engineering.html

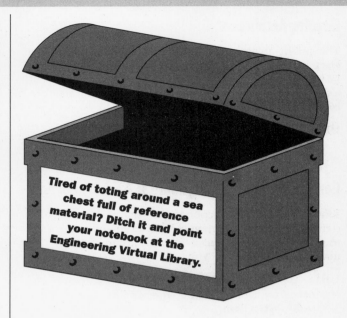

Tired of toting around a sea chest full of reference material? Ditch it and point your notebook at the Engineering Virtual Library.

EZFAB Systems Assembly Service

The EZFAB (presumably pronounced "easy fab") Systems Assembly Service reduces the cost, time, and risks of building systems by using both commercial and university-developed software tools and by developing standard interfaces and design rules. The Service is experimenting with a process for manufacturing and assembling printed circuit boards and multichip modules.

URL:

http://www.isi.edu/ezfab/sacore_info.html

Gas Dynamics Laboratory

The Gas Dynamics Laboratory is part of Princeton University's Department of Mechanical and Aerospace Engineering. Here, you'll find tidbits like a stereoscopic visualization of turbulent boundary layers, high Reynolds number pipe flows, Taylor-Couette flows, combustion and mixing, compressible turbulence, and transition in supersonic flows.

URL:

http://ncd1901.cfd.princeton.edu

Groundwater Modeling Software

The American Society of Civil Engineers (ASCE) offers this page with information, software, and a lot more resources helpful for groundwater and seepage projects.

URL:

http://www.et.byu.edu/~asce-gw/

Housing in Singapore

If you think you do not have any yard space, take a look at the high-rise, high-density environment of Singapore. One-bedroom to executive-room flats are built from a precast system that maximizes space. Call the interior decorator because these flats come with only the essential amenities.

URL:
http://www.arch.unsw.edu.au/subjects/arch/resproj/tan/htm/content.htm

IEEE

The Institute of Electrical and Electronics Engineers (IEEE) is the world's largest technical professional society. This nonprofit organization promotes the development and application of electrotechnology and allied sciences.

URL:
http://www.ieee.org

Institute for Research in Construction

The IRC is a leader in research, technology, and innovation for the Canadian construction industry. The Institute is responsible for coordinating the development of Canada's construction codes. Here you can find information on the IRC, the Canadian building codes, back issues of IRC newsletters, research programs, and information services.

URL:
http://www.nrc.ca/irc/

International Journal of Industrial Engineering

This quarterly journal distributes information about research findings, theories, technical applications, methodologies, and case studies of interest to industrial engineers. On these pages, the journal provides links to current and back issues, subscription information for the hardcopy version, and guidelines for potential new authors to submit their own works.

URL:
http://www.egr.uh.edu/Departments/IE/IJIE/

Laser Focus World

Power up your browser and point it at the Laser Focus World home page for scintillating, focused, and coherent discussions of all that's new and hot with lasers and laser technology.

URL:
http://www.lfw.com/www/home.htm

Laser Programs

As one of the world's preeminent centers for laser science, engineering, and technology, Lawrence Livermore National Laboratory's Laser Programs are applying their expertise to meet national needs in diverse areas. Read about lasers and the ways that LLNL is adapting laser technology to some interesting applications.

URL:
http://www-lasers.llnl.gov

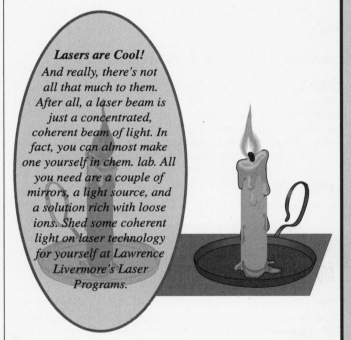

Lasers are Cool! And really, there's not all that much to them. After all, a laser beam is just a concentrated, coherent beam of light. In fact, you can almost make one yourself in chem. lab. All you need are a couple of mirrors, a light source, and a solution rich with loose ions. Shed some coherent light on laser technology for yourself at Lawrence Livermore's Laser Programs.

McCIM

McCIM is an integrated set of tools for developing industrial robotics applications based on MAP 3.0 and allowing demonstration of the CIMOSA concept. To understand this further, McCIM includes papers and helpful diagrams.

URL:
http://iaipc80.iai.kfk.de/mccim/

The Web is too cool.

Nuclear Industry Risk-Based Technologies

This site features discussions that relate to risk-based technologies in the nuclear industry. These include discussions of technologies that work and those that don't, documentation of lessons that have been learned, achievements, questions, answers, comments, and news. Topics include using risk-based approaches, maintenance, quality assurance, online maintenance, and testing.

Mailing List:
    Address: **listserv@peach.ease.lsoft.com**
    Body of Message: **subscribe rbapps** *<your name>*

Nuclear Information Web Server

This page brings together a tremendous collection of links to nuclear resources on the Internet and the Web. From governmental and regulatory agencies to vendors of nuclear power equipment, this page has it all. There's also a bunch of very informative links for learning about nuclear power, and a hypertext list of every nuclear reactor in the world—categorized by continent and subcategorized by country.

URL:
    http://nuke.handheld.com

Nuclear Reactors

The 109 nuclear reactors in the U.S. come in a variety of shapes and sizes. The basic method to generate electricity with a reactor is explained here. Future standardization of designs is discussed, especially in relation to the licensing process.

URL:
    http://www.nrc.gov/reactors.html

Nuclear Regulatory Commission

The U.S. Nuclear Regulatory Commission has a tremendous amount of information it offers to the public on nuclear energy. Their well-illustrated pages describe how nuclear reactors work; maps show the locations of every nuclear reactor in the U.S.; and links lead to detailed information about nuclear reactor sites and address such topics as reactor aging, and licensing and inspection activities and techniques.

URL:
    http://www.nrc.gov

Nuclear Safeguards at Los Alamos

Begun in 1966, the Nuclear Safeguards program at Los Alamos Labs keeps tabs on nuclear safety and personnel training in the U.S. In addition, the program is involved with studies of nuclear systems, treaty verification, and monitoring facilities. Here, you can read up on the Nuclear Safeguards program and its work over the years. You can also get data on nuclear security and follow links to other nuclear resources, such as newsgroups.

URL:
    http://flute.lanl.gov

Nuclear Test News

Nuclear Test News is a mailing list that provides the latest information on nuclear testing, the Comprehensive Test Ban Treaty (CTBT), and disarmament. It is a service of the Comprehensive Test Ban Clearinghouse. The mailing list is also gatewayed to the Usenet group **alt.activism.nuclear-tests.news**.

Mailing List:
    Address: **majordomo@igc.apc.org**
    Body of Message: **subscribe ctb-news**

Patent FAQ

If you're itching to get a patent on your latest gizmo but don't know where to begin, begin here! The Patent FAQ will answer all of your questions, including what a patent really is and what it means, who may obtain patents, how to go about applying for one, and how long the whole process takes.

URL:
    http://www.sccsi.com/DaVinci/patentfaq.html

People Movers

No, this is not about a ride at Disneyland. Mass transportation created segregated working, shopping, and living areas. From trolleys to aviation, intra-urban transportation is shaping our cities.

URL:
    http://www.fileshop.com/apwa/mass.html

Power Engineering

A virtual library on engineering from the Illinois Institute of Technology lists electrical engineering white pages, utilities, corporations, conferences, software, regulatory agencies, news, professional societies, and so on.

URL:
    http://www.ece.iit.edu/~power/power.html

Radioactive Waste Management

How to store and manage radioactive waste is currently of major concern. Coming in liquid, solid, or gaseous form, this stuff can pose high risks to people and the environment if not handled properly. The U.S. Department of Energy provides information on types of waste and hazard levels.

URL:

http://www.em.doe.gov/fs/fs3b.html

Software Technology for Fluid Mechanics

The Group for Software Technology for Fluid Mechanics at the German DLR (Deutsche Forschungsanstalt für Luft-und Raumfahrt) offers a wealth of technologies in aerodynamics and fluid mechanics. Included among their pages are MPEG movies that show tests of a variety of wings and shapes, and still pictures.

URL:

http://www.ts.go.dlr.de/sm-sk_info/STinfo/STgroup.html

Stanford Olympus Synthesis System

This free software package is a vertically integrated design tool for specification and synthesis of digital circuits. Featuring its own hardware design language, high-level synthesis tools, a technology mapping tool, and two simulators, Olympus is a tool that will put the spark back into digital circuit design.

URL:

http://akebono.stanford.edu/users/cad/synthesis/olympus.html

FEELING LEFT OUT OF THE POWER LUNCHES? DROP INTO THE VIRTUAL LIBRARY FOR POWER ENGINEERING, AND SUPERCHARGE YOUR CAREER.

Stimulating Circuit Simulation

How do you construct and measure an electronic device without really building it? Any electrician will tell you it's with circuit simulation, of course. Analyzing this page is not for the beginner.

URL:

http://picea.hut.fi/aplac/tutorial/terminology.html

TEXAS Code Page

TEXAS (Thermal EXplosion Analysis Simulation) is a mechanistic model used for nuclear facts analysis. The TEXAS code is a transient, three-fluid, one-dimensional model capable of simulating fuel-coolant mixing interaction. You'll find samples, code descriptions, definitions of variables, subroutines, and many more details of TEXAS.

URL:

http://trans4.neep.wisc.edu/NSRC/texas/

Thermal Analysis Software

Simulating air flow and analyzing the thermal characteristics of electronic components are problems everyone faces occasionally. Well, at least, maybe electronic engineers do. Icepak is a software package for taking the tedium out of this process and producing consistent, reliable, and documented results.

URL:

http://icemcfd.com/icepak.html

Thermal Connection

This is a resource center for thermal engineers, designers, and analysts. Here, you'll find a cornucopia of links, data, tips, and technologies of interest to thermal engineering and design.

URL:

http://www.kkassoc.com/~takinfo/

VLSI Magic

Magic is fast becoming a popular integrated circuit layout tool, and it is finding its way into several universities and industrial sites. It comes as binary executables for a variety of platforms that include DOS, OS/2, and Linux, but you can get the source code, too. On this, the Magic home page, you can get general information about the program, find out how to use it, download the program and the manuals, and check out the Magic FAQ.

URL:

http://www.research.digital.com/wrl/projects/magic/magic.html

A
B
C
D
E
F
G
H
I
J
K
L
M
N
O
P
Q
R
S
T
U
V
W
X
Y
Z

Yoyodyne

Hosted by Texas A&M University, Yoyodyne is a veritable library of information on object-oriented programming in engineering, adaptive finite element research, and other favorite topics of engineers.

URL:

http://yoyodyne.tamu.edu

ENTOMOLOGY

Academic Research in Entomology

Although they make up over 80 percent of the Earth's life forms, there's still a lot we don't know about bugs. Ohio State's Department of Entomology is helping to fill in the gaps by offering their research through links that range from Acarology and Apiculture to Toxicology and Vector-Virus.

URL:

http://iris.biosci.ohio-state.edu/osuent/
resprograms.html

Arachnology

If you're afraid of spiders, this is one web of the Web that you'll want to avoid. But if you love these creepy eight-legged creatures and scorpions, ticks, and centipedes, you'll find pictures, articles, movies, and more.

URL:

http://sesoserv.ufsia.ac.be/Arachnology/
Arachnology.html

Bee Seeing You

How does a bee see? View the world through the eyes of a bee. A variety of patterns, faces, and insects are shown from the bee's point of view.

URL:

http://cvs.anu.edu.au/andy/beye/beyehome.html

Beneficial Insects

Did you know that stink bugs are good for your garden? Before you step on another bug, check out this handbook and many others offered by the Florida Agricultural Information Retrieval System.

URL:

http://hammock.ifas.ufl.edu/text/bi/19697.html

Bisexual Fruit Flies

The title of this report reads like a headline from the *Weekly World News*: JUMPING GENES YIELD BISEXUAL FRUIT FLIES. Mutations induced by jumping genes can produce marked changes in the courting behavior in male fruit flies, including bisexual traits, according to the report.

URL:

http://www.gene.com/ae/WN/SU/
bisexual_fruit_flies.html

Bug Bodies

The thorax. The antennae. The mandible. All are bug parts. And on this page you'll find each one clearly explained and illustrated. If the mystery of insect anatomy is bugging you, stop squinting at your windshield and take a look here.

URL:

http://www.ex.ac.uk/~gjlramel/anatomy.html

Excerpt from the Net...

(from the Arachnology pages)

Simply letting her know of his intentions is not always good enough, and many males end up cannibalized by the female. In fact, Dolomedes fimbriatus females are extremely aggressive toward courting males, and sexual cannibalism before copulation is common. This may be an adaptive female strategy rather than mistaken identities (Arnqvist 1992). There are several tactics to avoid this outcome by suitors. One is to bring a nuptial gift to the female or to wait at the edge of her web for the arrival of a prey item (Prenter 1994). Metellina segmentata (Clerck, 1757), the autumn spider, does just this. The male waits at the edge of the female's web until a prey item arrives, then he suspends the prey from a nuptial thread on which mating occurs (Prenter 1994). Prenter also discussed the behavior of males of simply waiting to enter the females web until after she had fed upon a prey item, thus gaining a reduced risk of cannibalism.

Bug Books and Insect Literature

This is one heck of a book list. There's the best-selling *Ninety-Nine Gnats, Nits, and Nibblers*. Maybe your interest is more toward music—try *Crickets and Katydids, Concerts and Solos*. Each book has a brief description in this alphabetical listing by author.

URL:

http://www.colostate.edu/Depts/Entomology/
readings.html

Bug Clubs

A listing of mostly British entomological societies and clubs is offered on this page. Some—such as the Glowworm Page—focus on particular insects, while others—including the Royal Entomological Society—take swats at entire swarms.

URL:

http://www.ex.ac.uk/~gjlramel/clublink.html

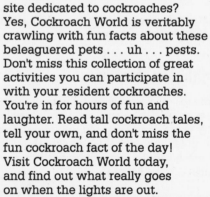

Cockroach World

What could be better than a web site dedicated to cockroaches? Yes, Cockroach World is veritably crawling with fun facts about these beleaguered pets . . . uh . . . pests. Don't miss this collection of great activities you can participate in with your resident cockroaches. You're in for hours of fun and laughter. Read tall cockroach tales, tell your own, and don't miss the fun cockroach fact of the day! Visit Cockroach World today, and find out what really goes on when the lights are out.

Bugs for Breakfast

An incredible (but edible?) collection of recipes, including Chocolate Chirpie Chip Cookies from the students of Iowa State University. The University of Kentucky offers stiff competition for the tastiest treats with the classic Ants on a Log. Yummy!

URL:

http://www.public.iastate.edu/~entomology/
InsectsAsFood.html

Butterfly Gardens

If you want butterflies, the Butterfly Planting Guide will help you attract them and promote their breeding. Choose which species you want in your yard and then find the proper plants to catch their eyes. With a little work you can have beautiful flowers with beautiful butterflies on them.

URL:

http://mgfx.com/butterfly/landscap/
plantgde.htm

Camouflage: Be Seen, Be Heard, Be Dead!

This article demonstrates how insects use camouflage for survival through several simple experiments. The article is geared for teachers of young entomologists. It also touches on methods of defense used by a variety of bugs and includes a number of images.

URL:

http://www.uky.edu/Agriculture/
Entomology/ythfacts/defense.htm

Cockroach World

We know that the Web has blossomed into a universe of information—but if intelligent people create a page devoted to cockroaches, can there be a saturation point in the near future for information? Probably not, so here we go. . . Untastefully presented, "El Mundo de las Cucarachas" will scurry onto your screen with thoroughly disgusting graphics and fun facts. Here's everything you never wanted to know about these varmints—and more.

URL:

http://www.nj.com/yucky/roaches/

Gainesville Air Races

No, there are no airplanes here, but there is a contest to find the fastest flying insect. This page is Chapter 1 of the *University of Florida Book of Insect Records*, entitled "Fastest Flyer." You'll find the results of a lot of research here, but a definitive answer is proving to be flighty. One early favorite—any male in pursuit of a female!

URL:

http://gnv.ifas.ufl.edu/~tjw/chap01.htm

A B C D E F G H I J K L M N O P Q R S T U V W X Y Z

Guide to Beekeeping

You be, me be, we be keepin' bees. Let's compare notes. With varied beekeeping methods, this page lists specific resources for answers to many questions keepers may have. Also included is a list of specialists in entomology and, specifically, bees.

URL:
ftp://rtfm.mit.edu/pub/usenet/
news.answers/beekeeping-faq

Gypsy Moth in North America

With its range continuing to spread, this small pest is one of the most devasting to North American forests. This page reviews the long-range effects of defoliation caused by the moth, management through pesticides, and ongoing research to seek out and destroy gypsy moth populations.

URL:
http://gypsy.fsl.wvnet.edu/~sandy/

Honey Bee Love

We talk about the birds and the bees, but we never stop to think about it literally. That's right, bee sex. This page describes the spectacular mating methods of the honey bee. The drones live and die for love and only get one chance at it. The entire exhausting process takes place in flight. No wonder *Johnny's buzzin' off*.

URL:
http://gnv.ifas.ufl.edu/~tjw/chap08.htm

Insect Olympiads

School is so much cooler today then when we were kids. Remember when you would tie thread to the leg of a big, fat Japanese beetle and fly it like a kite? Well, now they teach this stuff in schools! Really! On this page, teachers demonstrate the many talents of insects by having students put on a Bug Olympics. Take part in the shot put and javelin throw. Set up relay races with ants and watch the air gymnastics of butterflies.

URL:
http://www.uky.edu/Agriculture/
Entomology/ythfacts/olympics.htm

Planning an outing? Check the weather in Meteorology.

Insect Taxonomy

Insect classification is extremely complicated. This page enumerates 29 insect orders in relation to their evolutionary complexity. An estimate of the number of species under each order is given, although the author encourages updates from the Net community.

URL:
http://www.ex.ac.uk/~gjlramel/classy.html

Insect vs. Insect

How many times have you heard that ladybugs are good for your garden? Commercial farmers and home gardeners will find this guide to pest management useful. Next time you see that spider, leave him alone. In your garden, he is a useful predator.

URL:
http://www.nysaes.cornell.edu/ent/biocontrol/

Jumping Spiders

If a spider has four eyes, watch out! It's a jumper. These hunters will actually follow their prey and pounce on it. Jumping Spiders of America North of Mexico offers enough details on this genus to make you itch and squirm.

URL:
http://phylogeny.arizona.edu/salticids/
nasaltshome.html

Mermithid vs. Mosquito

Mermithids (a type of nematode) may prove to be an effective agent for controlling mosquitoes in a biologically friendly way. Jointly created and maintained by the USDA and University of Nebraska, this page explores the use of mermithidae and nemotodes in general as biological control agents.

URL:
http://129.93.226.138/nematode/epn/mermit1.htm

Migrating Monarchs

Unlike most other insects, this butterfly heads south for the winter. Where do monarchs go for the winter? Mexico, of course. Follow the monarch's migration and the history behind the migratory behavior that was only discovered around 1857.

URL:
http://monarch.bio.ukans.edu/migration.html

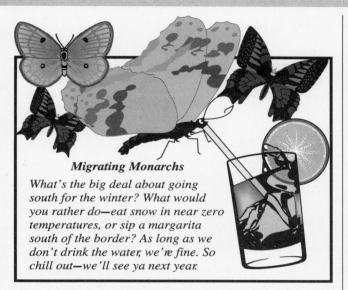

Migrating Monarchs

What's the big deal about going south for the winter? What would you rather do—eat snow in near zero temperatures, or sip a margarita south of the border? As long as we don't drink the water, we're fine. So chill out—we'll see ya next year.

Monarch Butterflies

"King Billy" is a beautiful, common North American butterfly. This page, which is geared toward children, describes the migration of King Billy and other monarchs. It also gives instructions for safely capturing a butterfly, and reminds you to always remember to release them after you have observed them.

URL:

> http://www.parentsplace.com/readroom/ explorer/act995_a.html

Nematodes as Biological Control Agents

One of the most abundant groups of animals, parasitic worms, can be destructive. But they can also be beneficial if you put them to effective use. If you can direct a platoon of nematodes, they can selectively parasitize spiders, leeches, annelids, crustaceans, mollusks, and even insects such as mosquitoes.

URL:

> http://129.93.226.138/nematode/wormepns.htm

Nematodes as Weapons

Two species of nematodes, *Steinernema* and *Heterohabditis*, transmit lethal bacteria to their insect hosts, making them very suitable as biological control agents. They can adapt to a wide range of climates and conditions, and are easily and inexpensively cultured. Study of these and other nematodes has broadened, especially since the ban of some chemical pesticides.

URL:

> http://129.93.226.138/nematode/epn/epnintro.htm

The Pherolist

When those lady moths send out their pheromones, the males come flying. This page that lists sex pheromones of lepidoptera and related attractants is a database of the chemicals produced by each species of moth and how these sex pheromones can be used in pest control.

URL:

> http://www.nysaes.cornell.edu/fst/ faculty/acree/pheronet/pherolist.html

Ticks and Other Blood-Sucking Parasites

Ticks attack people and animals, engorging themselves with their hosts' blood. This is definitely not attractive behavior, and should be discouraged among humans. This short, but informative, article compares the "hard" and "soft" varieties of ticks.

URL:

> http://biomed.nus.sg/PID/animals/tick.html

Wonderful World of Insects

How many bugs are there? Which is the biggest? Which is the smallest? Where do they live? Explore these questions (and find the answers) here. In addition, this page offers links to other bug information on the Internet, including basic bug anatomy, orders and species, and a *Bug Club* just for kids.

URL:

> http://www.ex.ac.uk/~gjlramel/six.html

Wonders and Beliefs

It's time for an interesting collection of myths, superstitions, and medical wonders of bugs. Do bee stings really prevent rheumatism? Want to gain weight quickly? Eat a Churchyard Beetle from the Nile (and pray that someone nearby has the *ant-idote*).

URL:

> http://www.ex.ac.uk/~gjlramel/asides.html

Young Entomologists

But, of course, the biggest mystery is not how to configure your TCP/IP stack, but how to keep crickets from hopping away or how to get more than honey from bees. This is why young entomologists seek the *Bug Club Newsletter*. The articles are easy reading and fun.

URL:

> http://www.ex.ac.uk/~gjlramel/

A B C D E F G H I J K L M N O P Q R S T U V W X Y Z

Hey Kids

When you grow up, wouldn't it be great to have a career where you get to play with bugs all the time? Believe it or not, there really is such a job! It's called entomology. Join the Bug Club and get started today.

Yuckiest Site on the Internet

Your kids may never beg to watch Power Rangers again after seeing this site. The Yuckiest Site on the Internet is a bonanza of creativity, facts, and activities for children of all ages. It has been carefully researched, written, and designed to introduce the fascinating world of insects to the general public in a friendly and unintimidating way.

URL:
http://www.nj.com/yucky/

ETHOLOGY

Animal Behavior News Groups

Follow your natural instinct to learn about how the other four million species on this planet behave. It's monkey-see, monkey-do in the **sci.bio.ethology** newsgroup.

URL:
news:sci.bio.ethology

Like lizards? Check out Herpetology.

Animal Behavior Society

Fish swim, and birds fly, coyotes are wily, and foxes are sly. Why? Check out the Animal Behavior Society to learn why critters do the things they do best. You'll find information about the ABS, a newsletter, a Gopher site, and articles on the significance of animal behavior research.

URL:
http://www.cisab.indiana.edu/
animal_behaviour.html

Applied Animal Ethology Discussion List

Make like Dr. Doolittle and talk to the animals on this mailing list. Exchange information and interesting news items with those working and studying in the area of applied animal ethology. Anyone with an interest in applied animal ethology is encouraged to subscribe and participate.

Mailing List:
Address: **applied-ethology-request@sask.usask.ca**
Body of Message: **subscribe applied-ethology**
<your name>

Center for the Integrative Study of Animal Behavior

Indiana University's CISAB page organizes information and educational programs in animal behavior, including an analysis of neuronal function and connections, modeling behavior in terms of optimization and dynamic systems, field experiments on the behavior of free-living animals, and more. Now you'll know why the cat paws at the cursor on your screen.

URL:
http://www.cisab.indiana.edu

Dolphins: Are They Really Smart?

We all accept the fact that, because they have large brains, dolphins, porpoises and whales are highly intelligent. They can be taught patterns and tricks, and they appear to communicate. However, does brain size alone make an animal smart? Compared to the size of their bodies, their brains are rather small. This page on the intelligence of cetaceans will have you thinking.

URL:
http://tirpitz.ibg.uit.no/WWWW/literature/
especially.html

Ethology Mailing List

If you want to walk with the animals and talk with the animals, subscribe to the Ethology Mailing List. This way, you can talk with the humans first to see how they go about walking and talking with the animals. You can subscribe to this mailing list from the web page, or send mail separately.

Mailing List:
    Address: **listserv@searn.sunet.se**
    Body of Message: **subscribe ethology** *<your name>*

URL:
    **http://cricket.unl.edu/NBBG/Listservers/
    Ethology.html**

Frog-Net Mailing List

For some absolutely ribbeting reading material, try subscribing to Frog-Net at the University of Southern California. You don't have to be amphibious, but it helps.

Mailing List:
    Address: **liaw-request@rana.usc.edu**
    Body of Message: **subscribe liaw** *<your name>*

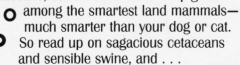
Everybody's always talking about how smart whales, dolphins, and porpoises are. Some even say they spend their free time in deep and meaningful thought and conversation. These folks probably took Flipper altogether too seriously. We don't know much about cetaceans, but we do know this—pigs are among the smartest land mammals— much smarter than your dog or cat. So read up on sagacious cetaceans and sensible swine, and . . . save the pigs!

International Society for Human Ethology

This is an exchange of empirical research on human behavioral science using the full range of methods developed in biology and ethology. Look for human ethology-related electronic publications, archives, discussion forums, multimedia, and digital image-analysis.

URL:
    http://evolution.humb.univie.ac.at/

Ludwig-Boltzmann-Institute for Urban Ethology

Located at the University of Vienna, the Ludwig-Boltzmann-Institute for Urban Ethology focuses on all types of behavior and mass phenomena in big-city environments. Online articles and projects about life in the urban milieu include studies on aggression in anonymous situations, digital image-analysis, impression management, and a study on the use of public space. The Institute claims that by the year 2001, more than 80 percent of the world's population will live in urbanized areas.

URL:
    **http://evolution.humb.univie.ac.at/institutes/
    urbanethology.html**

Nebraska Behavioral Biology Group

Investigate collaborative efforts and research programs in animal behavior and biology among several Nebraska universities. Browse online journals and lists of animal behavior meetings around the world.

URL:
    http://cricket.unl.edu/NBBG.html

Signaling Theory and Animal Communication

A horse is a horse, of course, of course, but this online course at Stanford will appeal mainly to people interested in animal communication, unless you're the famous Mr. Ed. Join the electronic discussions with other Wilburs regarding signal evolution and receiver psychology, territoriality, information theory, sensory systems, and other animal communications topics.

URL:
    **http://leon.econ.lsa.umich.edu/Carl/
    Communication/Comm.html**

The Web will really launch you.

Want to be a doctor? Study up in Medicine.

A B C D E F G H I J K L M N O P Q R S T U V W X Y Z

FORENSICS

Analyzing Entomological Evidence

If the forensic evidence presented in the O.J. Simpson trial didn't bug you enough, this page will certainly have you itching and scratching. Find out what a forensic entomologist looks for at a crime scene, and learn the five vital steps to observing and documenting a crime scene and collecting specimens.

URL:

   http://www.uio.no/~mostarke/forens_ent/
   crime_scene_forens_enty.html

Arthropods and Dead Bodies

Mites are often found in the soil under dead bodies. Winter-gnats will feed on decaying material during the cold months. These and other arthropods are described, along with their contributions to the analysis of dead bodies.

URL:

   http://www.uio.no/~mostarke/forens_ent/
   common_insects.html

Case Histories in Forensic Entomology

Here are over 20 examples of how forensic entomology has been used to solve crimes. The titles of these examples read like the murder mystery section at the bookstore: "The Baby in the Box," "Crime of the Cleaning Woman," and "The Headless Body Case" are just a few. But before you write "murder," you'd better check the evidence—bugs can tell the whole story.

URL:

   http://www.uio.no/~mostarke/forens_ent/
   casehistories.html

The Net is humanity's greatest achievement.

Case Histories in Forensic Entomology

Think what would happen if we didn't have insects all around us waiting for their next meal. The mighty mites come to the rescue in forensics by determining things like time of death in larger animals, such as us humans. The more mites there are chowing down at the crime scene, the longer the dear one has been departed. Lo, how the mighty hath fallen!

Chemistry of Crime

The Ohio University Chemistry Department Forensic Chemistry Page offers extensive information on forensic chemistry. It starts with a basic explanation of what forensic chemistry is. The page details what is offered at Ohio University for those interested in careers in forensic chemistry, and also for preparation for law school and premed majors. This page also provides links to other universities with forensic curriculums.

URL:

   http://quanta.phy.ohiou.edu/intro/forensic.html

Crime Scene Evidence File

Here are the details of a double murder case that was solved in Mississippi. This story has been put on the Internet to demonstrate how cases are solved using forensic evidence. Although many of the pictures are a bit gory and disturbing, it is nevertheless an interesting case study in forensics.

URL:

   http://www.quest.net/crime/crime.html

Crime Scene Photography

What's the best way to shoot a dead person? Explore different techniques of crime scene photography, including infrared, ultraviolet, and florescent. Included is information about courses on photography and forensics.

URL:
http://police2.ucr.edu/photo.htm

Dan's Gallery of the Grotesque

Here's a gallery of forensic photos of homicide, suicide, accidents, and war. This is the real thing. Don't visit here unless you have a strong stomach, because many of these images are quite disturbing.

URL:
http://www.zynet.com/~grotesk/html/
gotg_entrance.html

Footprinting Crime

Applying science to law is the interesting field of forensic sciences. Investigate crimes of murder, causes of fires, medical questions, road accidents, and much more. Work for the prosecution or the defense. This page offers an overview and suggestions on how to pursue a career in forensics.

URL:
http://www.demon.co.uk/forensic/forcareer.html

Forensic Archaeology

Peruse this work in progress by an undergraduate student on Forensic Archaeology. See why this science is more important in the U.S. and Canada than it is in the U.K. Follow the phases of investigation and enhance his information with any that you have to offer.

URL:
http://www.soton.ac.uk/~jb3/lista1.html

Forensic Education

The Council of Forensic Science Educators offers a listing of graduate and masters programs in forensic sciences at national and international universities throughout the world. This Council was created in 1989, and is supported by the American Academy of Forensic Sciences.

URL:
http://www.eskimo.com/~spban/fse.html

Forensic Entomology

Gathering criminal evidence through bugs is explored in this page. See how the life and death of blowflies, spiders, flies, and other arthropods can help determine when and how a victim died.

URL:
http://www.uio.no/~mostarke/forens_ent/
forensic_entomology.html

Forensic List

Not enough death and mayhem for you on the evening news? Settle into that easy chair for some prime time discussions of all issues relating to forensic science.

Mailing List:
Address: **listserv@uabdpo.dpo.uab.edu**
Body of Message: **subscribe forensic** *<your name>*

Forensic Medicine and Science Mailing Lists

This web page provides detailed information on nearly a dozen email mailing lists for discussions on the forensic aspects of several topics, including anthropology, genetic algorithms, pathology, psychology, spectroscopy, expert witnessing, and court evidence.

URL:
http://bionmr1.rug.ac.be/chemistry/list.d-i.html

Forensic Science Society

The FSS is the oldest society for forensics, and the goal of this organization is to advance the study and application of the science and promote communication between people involved.

URL:
http://www.demon.co.uk/forensic/fortop.html

Forensics Web

Link to forensic, law enforcement, FBI, legal, and medical labs from this good starting point for exploring forensics information on the Web.

URL:
http://www.eskimo.com/~spban/forensic.html

Anyone can browse the Web.

A
B
C
D
E
F
G
H
I
J
K
L
M
N
O
P
Q
R
S
T
U
V
W
X
Y
Z

Hair and Fiber Comparison

With 777 hair images to view in TIF format, you can build your own comparison system using a variety of samples and numerous slide images from this site.

URL:
> http://vita.mines.colorado.edu:3857/lpratt/forensics.html

"Hair today, gone tomorrow," might be the motto of the forensic Hair and Fiber Comparison page.

Intercollegiate Forensics

A mailing list designed to promote thoughtful discussion of issues relevant to intercollegiate forensics, specifically the American Forensics Association and National Forensics Association Individual Events communities.

Mailing List:
> Address: **listproc@cornell.edu**
> Body of Message: **subscribe ie-l** *<your name>*

National Fish and Wildlife Forensics Laboratory

This federal law enforcement crime lab is to animals what the FBI is to people. Supporting state, federal, and international wildlife enforcement agencies, the lab focuses on identifying evidence and linking an animal victim and crime scene to a suspect.

URL:
> **http://ash.lab.r1.fws.gov/labweb/for-lab.htm**

United Kingdom Police and Forensic Web

Sherlock Holmes would be envious. The U.K. Police and Forensic Web page is available to British law enforcement officials and the general public. Included is an extensive list of U.K. and U.S. law enforcement agencies.

URL:
> **http://www.innotts.co.uk/~mick2me/mainpage.html**

Zeno's Forensic Page

Zeno offers this complete guide to forensics—from general information to tool marks and shoe prints. This page is a great starting point, covering virtually every forensic subject. A few are: DNA, chemistry, documents/handwriting, hairs and fibers, firearms, linguistics, tool marks, and shoe prints.

URL:
> **http://www.bart.nl/~geradts/forensic.html**

FORESTRY

America's Great Outdoors

Interested in camping in a national forest? Would you like to know the history of a nearby national forest? This page has a variety of features, including a clickable map so you can easily identify forests. Linked pages give more information on the Forest Service's campsite reservations policy, and information on volunteer programs with national forests.

URL:
> **http://www.fs.fed.us/recreation/welcome.htm**

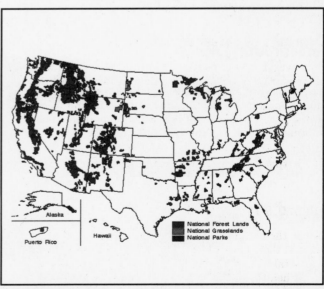

At the U.S.D.A. Forest Service page, you can use your mouse to drill down into this excellent image map to get information about our national forests. Click on a state or region, and you'll get a larger scale image map where you can click on a specific park to read about the forest and recreation activities you can pursue there.

> ## If you're not on the Web, you don't exist.

CSIRO Tropical Forest Research Centre

Learn about the Commonwealth Scientific and Industrial Research Organization (CSIRO) and their dedication to Australia's important tropical forests of North Queensland. Current projects range from examining past disturbance and the process of succession and change to the dynamics of rain forest tree growth.

URL:

http://www.tfrc.csiro.au

Drought Remedies

Drought can cause significant damage in new, and even well-established forests. Although there are few immediate remedies to drought damage, there are strategies to manage forests and minimize drought effects. This page branches out into other topics as well, including seedling selection and care, site preparation, and methods for maintaining a vigorous forest.

URL:

http://www.ces.ncsu.edu/drought/dro-21.html

Dynamics of Forest Biodiversity

This overview of a study started in 1994 looks at methods of measuring biodiversity and new variables used to inventory and analyze land use. A summary of subprojects is included. The main focus of this study is the forests of Finland.

URL:

http://www.metla.fi/projects/vmi/biodiv.htm

Finland's Forest Inventory

At the turn of the century, Finland's forest resources were at their lowest level. The National Forest Inventory was established to produce up-to-date information on forest health, ground vegetation, and growing stock. Discover how new inventory systems help to maintain our forests.

URL:

http://www.metla.fi/projects/vmi/nfi_hist.htm

FireNet Information Network

Australian-hosted FireNet is an international information retrieval and exchange network for rural, landscape, and forest fire management. Links to other fire resources include related institutions, course management, and prevention centers. An interactive software program called IGNITE lets you apply *what-if* scenarios in fire-related models.

URL:

http://online.anu.edu.au/Forestry/fire/
firenet.html

A Forest for a School

The Parkview Center School in Minnesota first created the Green Team and then created a forest. With an on-site, outdoor classroom, the students have been able to experience hands-on education in forestry. With support from adults, students, and agencies, this forest will stay green.

URL:

http://wwwrsl.forestry.umn.edu:10000/parkview/
forest.html

How're ya gonna keep 'em inside four walls after they've built a forest? And how cool to be a kid getting to create your own classroom by planting and watering it? Check out Forest for a School, but don't let your kids know about it, or they may want to move to Minnesota.

A B C D E F G H I J K L M N O P Q R S T U V W X Y Z

Forest Genetics and Tree Breeding

Learn the basics of plant breeding and tree genetics at this attractive and comprehensive page from Finland. You'll find information and links on dendrology, genetics, plant and animal breeding, and plenty of other plant-related topics.

URL:
http://www.metla.fi/~haapanen/breeding.htm

Forest Industry: Quick Facts

Did you know that close to one-third of our land—737 million acres—is forested? Or that last year, over 1.7 billion seedlings were planted in the U.S.? These statistics and more information compiled by the American Forest and Paper Company and Boise Cascade are available here.

URL:
http://www.bc.com/indust.html

Forest Practices Code Guidebooks

If you're a Canadian lumberjack, brave and true, you'll want to log in at this link to environmental regulatory forest-harvesting procedures and practices. Find out what you can and can't cut down, and learn about a variety of silvicultural measures and methods.

URL:
http://mofwww.for.gov.bc.ca/tasb/legsregs/fpc/
fpcguide/guidetoc.htm

Lord knows that loggers have enough problems. People tend to like trees, and a logger's job is to cut them down so that they can be processed into wood pulp to make newspaper to print the news for people to read that too many trees are being cut down! How to reconcile the difference? Log in at the Forest Practices Code page to learn about more environmentally friendly tree harvesting methods than you can shake a stick at!

Having trouble sleeping? See Business Payment Systems.

ForestNet

Before you yell, "Timber!," turn to this page designed for dispensing information within the forestry industry. There are articles on turning wood residues into profit and gaining access to public lands, and national fire and weather reports.

URL:
http://www.forestnet.com

Forestry Extension: NCSU

North Carolina State University provides this comprehensive site for forest resource management information, wildlife management, wood products, and youth- and environmental-education programs. It includes publications that you can download—for example, the Junior Forest Steward Coloring Book. Also included are related pages with information on recreational forest trails, complete with short video vignettes on trail building.

URL:
http://www.ces.ncsu.edu/nreos/forest/
forestext.html

Forestry Links at the Virtual Library

Here's one of the best virtual libraries on forestry materials around. Operated by the Finnish Forest Research Institute, this site has become the standard starting point for anyone interested in locating forestry resources on the Web. The resources here are categorized into working groups and networks; journals; newsletters and proceedings; mailing lists and Usenet groups; bibliographies; research papers and other publications; legislation and international agreements; forest policy; software; databases; entomology; libraries and bibliographies; and conferences and meetings. This site is a must for the virtual forester's research toolkit.

URL:
http://www.metla.fi/info/vlib/Forestry.html

Forestry Media Center at Oregon State University

An online ordering and viewing site for anyone interested in forestry and wood products media materials, this site offers sample video clips and still images. Read from a collection of interesting titles, including Wood Poles: A User's Guide to Inspection and Maintenance, Reproductive Ecology of Broad-leaved Trees and Shrubs, and A Field Guide to Vertebrate Pests of Northwest Conifers.

URL:
http://www.orst.edu/Dept/fmc/newrel.html

Forests in Danger

The Gaia Forest Archives offers some information on rain forests, their uses, abuses, and conservation efforts underway worldwide. This is an extensive resource with links to email lists and archives. Develop a new awareness of what is happening to biodiversity worldwide.

URL:
http://gaia1.ies.wisc.edu/research/pngfores/readme.html

INFOSouth

INFOSouth is a member-supported organization that allows members to conduct literature searches and retrieve documents online. You can also order documents by mail or email. But in addition to member-only services, INFOSouth provides a great many useful resources, such as links to national forests on the Net and forestry-related research sites.

URL:
http://wwwfs.libs.uga.edu

Life in the Redwoods

It is amazing how many animals, insects, and plants depend on redwoods, alive and dead, for food, shelter, and lifestyle. This article will pique your interest for further reading on the redwood forests.

URL:
http://www.northcoast.com/unlimited/tourist_information/what_to_do/life-in-the-redwoods.html

Northeastern Area Forestry

Leaf peepers who enjoy following the annual fall colors will enjoy this site. It includes information on Forest Service activities in the northeastern U.S.—from gypsy moth suppression to fire monitoring and forest health.

URL:
http://www.nena.org/NA_Home/

Plant-It 2000

This is the foundation to contact if you want to organize a tree-planting event. Custom designed for any type of sponsor, for city or forest, the trees are selected for appropriateness to the area. Educational forestry lectures go along with the planting.

URL:
http://atlantis.iits.com/~PLANTIT2000/overview.html

Redwood Fascinating Facts

Did you know part of *Star Wars* was filmed among the redwoods of Northern California? Argentina built its entire railroad system with redwood because the local red ants did not like the flavor of it. These and other interesting factoids are shared here.

URL:
http://www.northcoast.com/unlimited/tourist_information/what_to_do/factoids.html

SmartForest: Forests of the Future

The tool of computer visualization is becoming increasingly important for agency and public projects. SmartForest is an interactive tool with an extensive database. The use of computer modeling to study natural systems is explained in detail. Two case examples are also provided.

URL:
http://imlab9.landarch.uiuc.edu/smartforest/portland.html

The Web will really launch you.

A
B
C
D
E
F
G
H
I
J
K
L
M
N
O
P
Q
R
S
T
U
V
W
X
Y
Z

Southern Regional Extension Forestry

This resource offers information about forest resource management and natural resource education in the southern U.S. A clickable map of the southern U.S. links you to university-based technology transfer and extension information. From here, you can also get to the USDA Forest Service home page. Information contained on these pages includes the mission of the U.S. Forestry Service, as well as details on programs useful to forest landowners, natural resource agencies, youth, and anyone else interested.

URL:
   http://www.uga.edu/~soforext/

Tree and Shrub Seedlings

Each year, Colorado State University grows over 2,000,000 tree and shrub seedlings of over 35 species ranging from Colorado blue spruce to limber pine, green ash, and wild plum. The University offers these seedlings for sale to the public at bargain rates—provided you intend to use them for conservation purposes and meet several other qualifications.

URL:
   http://www.colostate.edu/Depts/CSFS/
   csfsnur.html

Vertebrates and Vascular Plants in National Parks

Search the National Park Service's online Flora/Fauna and species list databases. You can narrow your search to just the lists specific to amphibians, birds, fish, mammals, reptiles, plants, or search the whole thing.

URL:
   http://ice.ucdavis.edu/US_National_Park_Service/

Who Needs Trees?

A few of the many advantages of trees, both in urban and forest areas are presented on this page. Do you realize a tree can reduce your energy bill or increase the groundwater supply? And then there are those riparian trees that the beavers so enjoy.

URL:
   http://atlantis.iits.com/~PLANTIT2000/
   benefits.html

Want to be a doctor? Study up in Medicine.

World Timber Network

Linking the world's forest products community together along with the cyberpublic is the goal of WTN. You'll be "releaved" to find pointers to forest products, organizations, university departments, news about the industry, and connections to other web sites.

URL:
   http://www.transport.com/~leje/wtn.html

Yahoo Forestry Links

The Yahoo directory contains a good-sized list of forestry links categorized by institutes, journals, and paper and pulp. It's a good starting point; however, it's not nearly as comprehensive as some of the other resources listed in this section.

URL:
   http://www.yahoo.com/Science/Agriculture/
   Forestry/

FUTURISM

Brainstorms

Let your mind frolic with ideas that may one day shape our world as the digital revolution evolves into cyber-reality. Your host, Howard Rheingold, presents a multifaceted zeitgeist of discussions regarding global technoculture, humanist futurism, and virtual communities and worlds. There's nothing less at stake than the melding of the Net, telecommunications, electronic markets and agendas, intelligent Snooper-Highways, activism, and democracy. Tomorrow is in your hands, and you may find that this page will help you plan for it.

URL:
   http://www.well.com/user/hlr/

Bubble Culture

The swirling, chaotic, rainbow colors of bubbles fascinate us as they float through the air, landing on one another, joining, popping—almost as if they were alive. Used as a metaphor, these bubbles are an envision of subculture. What is FutureCulture, and is it really in the future or in the here-and-now? Like floating bubbles, is futurism its own predestiny, or its own paradox?

URL:
   http://www.uio.no/~mwatz/futurec/fc-manifesto

Excerpt from the Net...

(from Bubble Culture)

```
You are five years old.  You are lying on a grassy hill,
blowing bubbles up into a clear field of blue sky.  Bubbles.  Right
now, as a five year old child, you look at the bubbles, and words pop
into your head:  "pretty", "oooooo", "float".  To you, the bubbles
are almost like people — at least somewhat analogous to Bugs Bunny
or a Smurf.  Your wide eyes follow the bubbles as they traipse along
the gentle prevailing curves of soft winds, turning, rotating,
revolving endlessly in the air.  A sunray beams its light through one
particular bubble you have been admiring, and within its midst your
eyes become privy to a new world — a heretofore unknown domain of
chaotic rainbows swirling about along the bubble.  The colors, like a
sentient anthill, work at once individually and synergetically to
give the bubble it's unique flavor, an individual identity among the
community of bubbles.
```

Dr. Tomorrow

You may know him as Dr. Tomorrow from his books, articles, and TV shows, but Frank Ogden is a real person living in the future today. On his web pages, find out about Ogden's publications and engagements, peruse the Dr. Tomorrow art gallery, and explore his favorite technology links.

URL:

http://www.drtomorrow.com/drtomorrow

Extropians

Extropians are described as advocating weird technologies and having principles of "spontaneous order." This page describes what extropianism is and is not. Most important, an extropian is **never** dogmatic. Makes you wonder what an entropian is like.

URL:

http://web.gmu.edu/bcox/Bionomics/Extropians/ FAQ.Extropians0.09.html

Nobody sends you email? Join a mailing list.

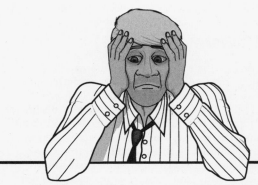

If your dogma just ran over your karma, you'll want to explore the cosmic wrecking yard of Extropian thought.

Futurism

If history, as Konrad Adenauer put it, is "the sum total of all that could have been avoided," then remember the past but concentrate on the future. These words may sound as if they're maxims for the 1990s, but the movement behind them was actually coined and solidified in 1909. Futurism implies energy, speed, vitality, dynamism, and associated characteristics. The movement's most significant historical impacts were in the visual arts and poetry. Rekindle into your vitality by researching the futurism phenomena here at this seminal page.

URL:

http://www.cnam.fr/wm/paint/glo/futurism/

A
B
C
D
E
F
G
H
I
J
K
L
M
N
O
P
Q
R
S
T
U
V
W
X
Y
Z

The Gopher's time is short.

Futurism Menu

Anticipating the surrealists, the futurists declared that discoveries of the subconscious must be brought to the stage. The entertainment would "symphonize" the feelings of the public, exploring and revealing those feelings in every possible way. *Grok* with this compendium of available futurism sites. If it's minimal and postmodern, don't say we did or didn't warn you. Since you're here by now, you're already on your own.

URL:
> http://pharmdec.wustl.edu/juju/surr/futurism/FUT-MENU.html

Futurism Movement

Futurism was the first deliberately organized, self-conscious "art movement" of the twentieth century. It was founded in Italy by the Italian poet, Filippo Tommaso Marinetti. In a series of manifestos designed to shock and provoke the public, the futurists formulated styles of painting, music, sculpture, theater, poetry, architecture, cooking, clothing, and furniture. One Italian critic labeled them "art wiseguys," and their manifestos still provoke, irritate, and amuse us up until this day. Read some of these smarty-pants' manifestos here and decide for yourself.

URL:
> http://pharmdec.wustl.edu/juju/surr/futurism/FUTINTR1.html

> *One man's art is another man's movement. We won't say which is which, but the Futurism Movement page will. Compose your own manifesto, because sometimes no one knows that art stinks until you tell them that it does.*

Futurist Programmers

Programmers of the world, unite! Experience the futurist revolution with this manifesto that attacks "the sluggishly supine admiration for old operating systems, old languages, archaic standards, and . . . the enthusiasm for everything bug-ridden, rotting with code bloat, and eaten away by obsolescence." Brandish your mouse, and join the cause!

URL:
> http://www.sgi.com/grafica/future/futman.html

The Posthuman Body

Will we still resemble human form in two centuries? Can and are our minds and bodies capable of being redesigned? Morph on over to this page that attempts a gaze into future possibilities. There are lots of far-out links to nanotechnology, robotics, and other amazing cyberthoughts.

URL:
> http://www.c2.org/~arkuat/post/

Transhumanism

Delve into the human, or make that posthuman, condition. If you begin with the premise that we must continually strive to reach higher levels, then given imperfect minds and bodies, science and technology can be used to improve human physiology and mental aptitude. Read lots of visionary futurehuman articles excerpted from *Wired* and other print and electronic 'zines.

URL:
> http://www.nada.kth.se/~nv91-asa/Trans/intro_page.html

Traversable Wormholes

"Speeding through the universe, thinking is the best way to travel."—The Moody Blues.

Rather than thinking your way toward distant galaxies, how about using traversable wormholes built with exotic materials that don't even exist at present? Theorize if such interstellar sojourns are possible with methods slower than the speed of light, and read about other possibilities before they ever happen here.

URL:
> http://www.nada.kth.se/~nv91-asa/Trans/wormholes.html

GENEALOGY

Banner Blue Software

Provided by Banner Blue Software—the people who produce *Family Tree Maker*, *Org Plus*, and *Family Archive CDs*—the FTP site offers downloadable demos, screen shots, documents, lists of job opportunities, and press releases.

URL:
  ftp://ftp.best.com/pub/banrblue

Bede Technology

This research-software company provides this page to keep you abreast of their latest research and studies. Bede is currently working on their RoyaList product, which explores 1,200 years of the history and genealogy of the extended royal families of Europe. This page includes examples, features, opinions, and products.

URL:
  http://www.cftnet.com/members/bedetech/

BYU Family History Technology Laboratory

Brigham Young University's Family History pages offer valuable resources for researching your ancestry and family history. Also available here is information about ongoing research projects at BYU and how they're utilizing the latest computer technologies for family research.

URL:
  ftp://issl.cs.byu.edu/famHist/home.html

Everton's Genealogical Helper

Everton Publishers is a publishing company that specializes in publications, products, and services for genealogists—from amateur to expert. On their web pages, Everton introduces the science of genealogy and offers extensive information on genealogy-related products, services, software, news, and links to volumes of other genealogical resources on the Internet. If you subscribe, you can get access to additional information on Everton's catalog and products.

URL:
  http://www.everton.com

Family History Centers

The Family History Library in Salt Lake City provides lists of family history centers around the world, including places such as Australia, New Zealand, the U.S., and Canada. This page also includes information on the services of family history centers and tips to help you when visiting them.

URL:
  http://ftp.cac.psu.edu/~saw/FHC/fhc.html

Genealogy Calendar

The Genealogy Calendar offers updated information on upcoming events in genealogy. Events are categorized chronologically by month and also by location. This page provides information on submitting events for the calendar, and it includes links to other genealogical calendars and resources on the Internet.

URL:
  http://genealogy.org/PAF/www/events/

Genealogy in Scandinavia

Tracing Scandinavian heritage is so much easier with this genealogical resource for Sweden, Denmark, and Norway. Included are a list of surnames (like Johansson), links to other Scandinavia home pages, information on how names changed after emigrating, and parish records.

URL:
  http://www.bahnhof.se/~floyd/scandgen/

Genealogy Newsgroups

Enlist the help of your Internet family while researching your family's genealogical history. All in one place, this page gathers every Usenet newsgroups relating to genealogy, including newsgroups for French, German, Jewish, medieval, British, Scottish, Irish, Nordic, and Hispanic genealogy.

URL:
  http://www.yahoo.com/Social_Science/History/
    Genealogy/Usenet

A B C D E F G H I J K L M N O P Q R S T U V W X Y Z

Said the proper young man looking loftily around him at the wild party, "I'm afraid that I simply cannot bear the presence of fools."

"How odd," replied the young lady standing on her head next to him. "Your mother apparently had no trouble."

If you're having trouble researching your roots, check out Genealogy Newsgroups.

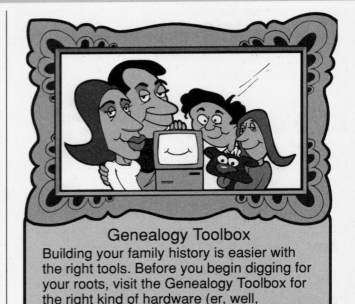

Genealogy Toolbox
Building your family history is easier with the right tools. Before you begin digging for your roots, visit the Genealogy Toolbox for the right kind of hardware (er, well, software) so you can get the job done right.

Genealogy SF

Genealogy SF provides genealogy software, research assistance, and genealogy data for all those interested in researching their family's history. This program can help beginners and expert genealogists with research tips and the biweekly Ask Glenda genealogy tutorials.

URL:

    http://www.sfo.com/~genealogysf/

Genealogy Toolbox

Open this toolbox to access hundreds of genealogy resources and links on the Internet. These links are categorized by guides and indexes; family data; histories; geographical and area-specific resources; commercial and experimental resources; library resources; software; heraldry; and surnames.

URL:

    http://uxl.cso.uiuc.edu/~al-helm/genealogy.html

The Web will set you free.

GenWeb

GenWeb is yet another page collecting resources and pointers to genealogical databases and information. This page offers links to other web pages related to research in genealogy and includes a mailing list and file archive. It also demonstrates a great way to organize your research and information on your family tree. Visual and audio features are included.

URL:

    http://www.vest.sdata.no/skrivervik/employees/
    birger/genealogy.html

Global Heritage Center

The Global Heritage Center offers genealogy software and CD-ROM databases with data from U.S. Census records, marital records, Social Security records, and many other sources. The page includes an order form, a price list, and a master name index.

URL:

    http://www.mindspring.com/~sledet/genealogy/
    ghc.html

The Net is humanity's greatest achievement.

Help for Adoptee Genealogical Research

It's hard enough to trace a family tree. But disjointed branches due to adoptions can be a tangled mess. This brief list of helpful information includes organizations, researchers, books, articles, and magazines relating to adoption and adoptees.

URL:
http://www.everton.com/GENEALOG/
GENEALOG.ADOPTEE

Holler For Janyce

If finding your family's root is driving you crazy, holler for Janyce. She has constructed an incredible page of genealogy information and research tools. Check out her "Root Diggin' Department" for links to other genealogist's home pages and research projects. Regularly updated and always extensive, Janyce offers help on family names from around the world. You can easily look up your own last name to see if you have any distant relatives on the Internet. And remember, it's OK to holler for Janyce, but don't holler at her.

URL:
http://www.wolfe.net/~janyce/

Intro to Genealogy on the Web

If you can't find your name in this list of surnames, you don't exist! These pages have a searchable index of more than 50,000 surnames, submitted by over 3,000 genealogists. One option allows you to type in a location, and a list of people researching that location will be returned. This page also provides a list of other genealogy sites on the Web and updated news.

URL:
http://pages.prodigy.com/UT/fhl/

Joe's Italian Genealogy Page

Joe provides information on Italy and tracing your Italian ancestry, links to the Italian electronic yellow pages, *Italia Online*, and other Italian resources on the Internet.

URL:
http://www.phoenix.net/~joe/

Kellogg Family

Tracing back to the earliest ancestor of the American Kelloggs, Nicolas Kellogg, this family history is extensive. Historical information is given from the Kellie Clan in Scotland on through to the Kelloggs of England. Today there are over 16,462 American Kelloggs.

URL:
http://www.winternet.com/~rlkelog/
KelloggHistory/History.html

... and you thought that Kellogg just made corn flakes ... Check out the Kellogg Family page.

Lineages

Lineages, Inc. is a genealogical research firm, designed to help researchers with brand new and comprehensive genealogical services. By using their wide variety of sources, such as city directories, naturalization records, computer databases, and handwritten records, they can assist you in your genealogical search. They also have a Lineages Research Club you can join, and lists of helpful reference books, CD-ROMs, and floppy disks that you can order through their company.

URL:
http://www.cybermart.com/lineages/

Pomeranian Page

This is not about those cute little dogs. Pomerania was a land that once existed along the Baltic seacoast. Many German Lutherans in the U.S. are descendants from nineteenth century immigrants. Included on this page is a history of the old country and a map. You'll also find a link to Pomeranian literature.

URL:
http://www.execpc.com/~kap/pommern1.html

A B C D E F G H I J K L M N O P Q R S T U V W X Y Z

Some Famous Kelloggs

Yes, this is the family that corn flakes came from. That was the inspiration of Will Keith Kellogg. His flakes changed the way breakfast was eaten in a major way. The first botanist in California was also a Kellogg. Others listed were authors, clergymen, and politicians.

URL:
> http://www.winternet.com/~rlkelog/
> KelloggHistory/Famous.html

Cereal Genealogists
There's nothing flaky about the Internet when it comes to raw, wholesome goodness, especially if you're a cereal genealogist. Dig into a hearty URL of Kellogg's Family History. (Prize inside!)

GENETICS

Cat Genetic Catalog

Here's a veritable meow mix of all the kitty chromosome loci you'll need—assuming you're God—to produce a lovable furball. Assembly instructions are not included.

URL:
> http://www.siumed.edu/ob/fca.html

Cattle Genome Mapping Project

Head 'em up and mooove 'em out at this bovine project. You'll find genetic cow data—including marker tables, type and number of alleles, and a chromosome map with clickable markers.

URL:
> http://sol.marc.usda.gov/genome/cattle/
> cattle.html

Chicken Genome Mapping Project

ChickMap provides access to chickGBASE, a dynamic generator of genomic poultry maps, in response to the user's criteria. You'll find information on the project and chicken images, karyotypes, and additional data used in the mapping project.

URL:
> http://www.ri.bbsrc.ac.uk/chickmap/
> ChickMapHomePage.html

Dog Genome Project

You and your best friend can learn a few new tricks from this genetic project that is producing a map of all of the chromosomes in dogs. See how this program isolates genes that cause disease and control canine morphology and behavior.

URL:
> http://mendel.berkeley.edu/dog.html

FlyBase

Find everything you ever wanted to know, genetically speaking, about fruit flies, including an encyclopedia of *Drosophila*, fly images, and a software archive. Also posted are news items, details of meetings, and community announcements.

URL:
> http://morgan.harvard.edu

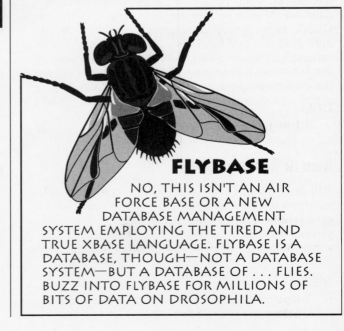

FLYBASE
NO, THIS ISN'T AN AIR FORCE BASE OR A NEW DATABASE MANAGEMENT SYSTEM EMPLOYING THE TIRED AND TRUE XBASE LANGUAGE. FLYBASE IS A DATABASE, THOUGH—NOT A DATABASE SYSTEM—BUT A DATABASE OF . . . FLIES. BUZZ INTO FLYBASE FOR MILLIONS OF BITS OF DATA ON DROSOPHILA.

Fungal Genetics Stock Center

The next time somebody calls you a fun guy or a fun gal, make sure that they're not pulling your leg by checking out this fungi genetic resource. It offers biological and molecular information, different fungus catalogs, microbiological reviews, *Fungal Genetics Newsletter* articles, and last—but not yeast—abstracts from the latest Fungal Genetics Conference.

URL:
http://www.kumc.edu/research/fgsc/main.html

FlyBase Riddle
Q: How do you tell the sex of a fruit fly?
A: Pull down its genes and take a look!
This old joke comes to life at FlyBase, a genetic and molecular database for *Drosophila*. Buzz on over.

GDB Human Genome Data Base

The GDB Human Genome Data Base is an ambitious project for cataloging and mapping human genes. It also includes an online version of the human genetics database, *Mendelian Inheritance in Man*. Use the online GDB Browser to conduct complex searches involving multiple fields and relationships through genetic object categories. Demonstration queries are provided to help you figure out how to use it. Now if they could just isolate the gene responsible for producing bad sitcoms.

URL:
http://gdbwww.gdb.org

Genetic Resources Page

Pull down a vast list of genetic sites on this server that amalgamates genetic database searching and human genome and chromosome centers. Specific sites point to genome-wide linkage, meiotic map construction, and major indices of molecular biology and genetic science links.

URL:
http://www.genlink.wustl.edu/otherlinks/
otherresources.html

Genetics Education Materials

Benefiting teachers and professionals in the field of education, this resource list contains frequently updated materials—including teaching curricula, books, pamphlets, brochures, computer programs, hands-on materials, newsletters, and videotapes—as part of an ongoing genetics education project.

URL:
http://www.kumc.edu/instruction/
medicine/genetics/resource.html

Genetics Glossary

From Adenine to Yeast Artificial Chromosomes, you'll find that this excellent genetics glossary will serve to spice up your conversations the next time that you have your friends over for margaritas. Portions of the glossary text are taken directly or modified from definitions in the U.S. Congress Office of Technology Assessment document.

URL:
http://www.gdb.org/Dan/DOE/prim6.html

GenLink

GenLink is a multimedia database resource for human genetics that is currently under development. This server provides linkage mapping information and software tools that facilitate integration of physical and genetic linkage data to produce unified maps of the human genome. Genetic researchers interested in identifying genes based on map positions will find this resource helpful.

URL:
http://www.genlink.wustl.edu

GenoBase

The National Institutes of Health invites you to search through their experimental GenoBase server. Through a system of tables and various query capabilities, this server provides access to an NIH copy of GenoBase—an object-oriented molecular biology database written in Prolog.

URL:
http://specter.dcrt.nih.gov:8004

Kids, check out Kids and Amateurs.

GenotypesDB

Query a gigantic human genotype database by chromosome, families, and loci. This Washington University at St. Louis server facilitates genetic search information with a great online help tutorial. It incorporates many HTML 3.0 features, so your browser should support these features to view the pages correctly. There is user-friendly, built-in, available help online, with explanations and examples that make using this database a breeze.

URL:

http://hmgmac164.wustl.edu/gtypesdb/index.html

Horse Genetics

Saddle up and head on over to this page on horse genetics and horse-typing services at the Veterinary Genetics Laboratory. It provides general links of interest to genetics and other horse-related items. Use the experimental search engine to search a horse genetics database, and find information on horse genetic disorders and blood types.

URL:

http://vgl.ucdavis.edu/~lvmillon/

A horse is a horse, of course, of course. But what can you do with a great reference on horse genetics? Learn about horse genetics, of course. Find out what makes a horse's coat the color it is, and other exciting tidbits about our equine friends.

Human Chromosome-Specific Web Servers

Delve into your favorite chromosome, from 3 to 22, including the X chromosome, with this clickable list of chromosome servers. Genetic data aplenty await you. Just make sure that you don't mutate into anything in the process!

URL:

http://www.sanger.ac.uk/~marvin/
HGP.resources/HCSWWWsites.html

Human Genome Program

This ambitious project is a 15-year effort to map all human genetic material and determine the complete sequence of DNA subunits in the human genome. The ultimate goal of the project is to discover all of the more than 50,000 human genes and render them accessible for further biological study. You'll find the project's online newsletter, the *Human Genome News*, and the latest project updates.

URL:

http://www.er.doe.gov/production/oher/
hug_top.html

Human Mitochondrial DNA Database

Interested geneticists can view the entire human mitochondrial DNA sequence here at MITOMAP. Explore DNA functional maps, mitochondrial DNA variations, mutations, and aging and degenerative diseases. You'll also find tables and figures of mitochondrial DNA polypeptide assignments and polymorphic MtDNA restriction sites.

URL:

http://infinity.gen.emory.edu/mitomap.html

Interactive Genetic Art

Join Internet users around the world in creating a piece of artwork. Using an approach similar to the genetic process of natural selection, users rank nine graphics images on a scale of 0 to 9. The votes are used to determine the "fitness" of the pictures in the current generation. The more desirable pictures are likely to be used in the creation of the next generation. Unfortunately, all the images tend to end up looking like Arnold Schwarzenegger and Linda Hamilton.

URL:

http://robocop.modmath.cs.cmu.edu:8001/htbin/
mjwgenformI

Maize Genome Database World Wide Web Server

Here's an amazing page with hundreds of genetic maps and thousands of mapped loci, locus variations, genes and gene candidates, probes, and genetic/cytogenetic stocks and stock pedigrees. You'll also find extensive bibliographic references connected to other database objects.

URL:
http://teosinte.agron.missouri.edu/top.html

Mendel

Pull up your genes with Mendel, a database of designations for sequenced genes in individual plant species and plant-wide families. The purpose of these databases is to provide a common system of nomenclature for substantially similar genes across the plant kingdom.

URL:
http://probe.nalusda.gov:8000/plant/
aboutmendel.html

MendelWeb

Gregor Johann Mendel was quite energetic in genetics, and MendelWeb is dedicated to his pioneering spirit. This educational resource for teachers and students focuses on the origins of classical genetics, introductory data analysis, elementary plant science, and the history and literature of the science. You'll find Mendel's 1865 paper "Experiments in Plant Hybridization" in English and German. Students can contribute online annotations to both versions.

URL:
http://www.netspace.org/MendelWeb/

Mouse Genome Database

Are you a man or a mouse? Compare your genes with the ones here at MGD and decide for yourself. This extensive page features the Mouse Locus Catalog, mammalian homologs, genetic maps and mapping data, and combined mouse/human phenotypes. If you're not sure where to begin, there's also a section on mouse nomenclature rules and guidelines.

URL:
http://www.informatics.jax.org/mgd.html

Pig Genome Database (PiGMap)

Learn how to make bacon with all the genetic markers you'll need to produce a pig. While this complete list of amino acid, peptide, and chromosome indicators is substantive, you'll still have to combine the ingredients and supply the oink.

URL:
http://www.ri.bbsrc.ac.uk/pigmap/locus.hlp

Primer on Molecular Genetics

It's too bad that Doctor Frankenstein didn't have a web connection to this molecular genetics primer, or his monster might have turned out better. Get the lowdown on DNA, genes, chromosomes, mapping strategies, sequencing technologies, model organism research, and more at this great genetic information source.

URL:
http://www.gdb.org/Dan/DOE/intro.html

Geneticists Isolate Internet Gene

SAN DIEGO (AP) - Scientists at the Center for Technology Suffers (CTS) have isolated a gene believed to cause the following disorders: insomnia, caffeine and fast food addiction, twitching index finger, and a compulsion to sit at a computer for hours on end. Named *The Internet Gene*, it is mostly inherent in people who use the world's largest information network. It is responsible for family members refusing to talk to one another, except by electronic mail. In extreme cases, people with the gene chant unintelligible babble and acronyms, such as "By Kibo, one day ISDN will save us all, IMHO!" and "Don't NNTP with FTP because the TCP/IP in HTTP will fritz your SMTP and cause your SLIP to POP!" Scientists are baffled by the bizarre behavior, and you should check the genetic FAQs to make sure you're not at risk.

**Looking for a science project?
Look in Kids and Amateurs.**

Rat Genome Database

All you cool cats can find out what's up with rats here at RATMAP. Scurry through lots of literature references sorted by rat gene symbol, rat genetic nomenclature, and rat loci sorted by chromosome number with comparative mapping data. Say "Cheese!"

URL:
http://ratmap.gen.gu.se/

Saccharomyces Genome Database

Commonly known as baker's or budding yeast, *Saccharomyces cerevisiae* makes it possible for you to hold onto a peanut butter sandwich without getting all messy. Scientists at Stanford University have known this for years, but have attempted to protect their tenured positions (as well as their love for PB&J) by persuading us of the yeast's importance in the field of genetic biology. Yeah, sure. Come see the extensive page they've created to perpetuate this myth. Got milk?

URL:
http://genome-www.stanford.edu

Silkworm Genome Mapping Project

This program at the University of Tokyo focuses on insect molecular genetics—specifically, silkworm analysis. You'll find lots of information on silkmoth phylogenetics, including nucleotide and amino acid sequences, and detailed mapped chromosomes.

URL:
http://papilio.ab.a.u-tokyo.ac.jp/shimada.html

Swine Genome Mapping Project

Find swine genetic data, including marker tables, type and number of alleles, and a chromosome map with clickable markers. Mneah, mneah, mneah—that's all, folks!

URL:
http://sol.marc.usda.gov/genome/swine/swine.html

GEOGRAPHY

Alexandria Digital Library

Maps, photographs, atlases, and gazetteers are generally found in non-digital form and are costly to access. Plus, you need to consult a map just to find the geographical reference section in your local library. The Alexandria Digital Library provides easy access to large and diverse collections of digitized maps, satellite images, and other graphical materials.

URL:
http://alexandria.sdc.ucsb.edu

Bodleian Library

The world's seventh largest collection of maps is contained at the Bodleian Library. Online collections include early examples of cartography, such as the fourteenth century Gough Map; charts; estate maps; and many atlases. The Map Case contains an assortment of important cartographic images.

URL:
http://www.rsl.ox.ac.uk/nnj/maproom.htm

Canadian National Atlas Information Service

The Canadian National Atlas Information Service (NAIS) is responsible for the development and maintenance of an authoritative synthesis of the geography of Canada. NAIS makes available both digital and traditional products: base maps; geographical names; and thematic maps that reflect the social, economic, environmental, and cultural fabric of Canada. This text is in both English and French.

URL:
http://www-nais.ccm.emr.ca

Test your knowledge of Canadian geography by completing an interactive quiz at NAIS. Download the questions along with the answer sheet for classroom use from the Canadian National Atlas Information Service. Cool, eh?

Defense Mapping Agency

DMA provides support to military branches, the Office of the Secretary of Defense, and the Chairman of the Joint Chiefs of Staff. Reputedly the largest employer of cartographers in the U.S., DMA maintains the GEOnet Names Server, which provides access to its database of foreign geographic-feature names. Check out the Terrain Modeling Project Office computer and cartographic-modeling simulation page.

URL:

http://www.dma.gov

Digital Elevation Data Catalogue

A 100K file of up-to-date Geography Information Systems (GIS) elevation and bathymetric data. This is a simple text document.

URL:

http://www.geo.ed.ac.uk/home/ded.html

Digital Relief Map of the U.S.

What a relief! Click on one of the table cells that shows latitude and longitude, and you'll be presented with two images. One is a line map, and the other is a highly detailed and colorful relief map of a section of the U.S.

URL:

http://www.zilker.net/~hal/apl-us/

Earth Sciences and Map Library

The University of California at Berkeley's Map Room contains over 310,000 maps. Their online museum has clickable floor plans for a tour of the facilities. Examples include early Sanborn topographic maps, nautical charts, facsimile maps, and aerial photography. Okay, now how to get there from here?

URL:

http://www.lib.berkeley.edu/EART

Environmental Interactions

Here is a thorough analysis of topography, climate, development, and vegetation in the San Francisco Bay and Sacramento regions using the conterminous U.S. Advanced Very High Resolution Radiometer (AVHRR) dataset series. Say that ten times fast. It hurt just typing it.

URL:

ftp://resdgs3.er.usgs.gov/public/mosaic/proj/ofr/
interactions.html

EROS Data Center

The Earth Resources Observation Systems (EROS) Data Center holds the world's largest collection of space and aircraft acquired imagery of the Earth. The center has over two million images acquired from satellites and over eight million aerial photographs, including formerly classified images of Soviet airfields and terrestrial strata.

URL:

http://sun1.cr.usgs.gov/eros-home.html

Federal Geographic Data Committee

The Federal Geographic Data Committee is a clearinghouse on information regarding the National Spatial Data Infrastructure, a project that focuses on the production and sharing of geospatial data.

URL:

http://fgdc.er.usgs.gov/fgdc.html

Geographic Information Systems

GIS technology can be used for scientific investigations, resource management, and development planning. A GIS is a computer system capable of assembling, storing, manipulating, and displaying geographically referenced information, such as data organized by location. Practitioners also regard the total GIS as including operating personnel and the data that go into the system. This site includes relevant software, standards, and publications.

URL:

http://www.usgs.gov/research/gis/title.html

Geography

An excellent launch point for geographical links into the Net.

URL:

http://www.icomos.org/WWW_VL_Geography.html

Geography Quiz

What could be better than winning prizes while answering geography questions? Every few days, there's a new question to test your knowledge of the world. Wouldn't Alex Trebek be proud?

URL:

http://pages.prodigy.com/NJ/geochamp/
geochamp.html

A
B
C
D
E
F
G
H
I
J
K
L
M
N
O
P
Q
R
S
T
U
V
W
X
Y
Z

GeoWeb

Thanks to the Geographic Information Retrieval group, digital geographic information and data are easily available to GIS researchers and the general public. GIR has a mailing list, sponsors online geographical projects, and maintains a geographic data archive.

URL:
    http://wings.buffalo.edu/geoweb/

GIS Dictionary

What's in a name? You're never sure when it comes to scientific terms. That's why the Association for Geographic Information provides this searchable database of around 300 terms used in association with GIS. It is sponsored by the University of Edinburgh.

URL:
    http://www.geo.ed.ac.uk/root/agidict/html/
    welcome.html

GIS FAQs and General Info

What is a GIS? This frequently asked question (FAQ) compelled scientists to create this page so they wouldn't have to answer this same question over and over again for everyone who wanders by. Standards for Geographic Information Systems, online subscriptions, and mailing list information are featured here.

URL:
    http://www.census.gov/geo/gis/faq-index.html

Global Land Information System

The Global Land Information System (GLIS) is an interactive computer system developed by the U.S. Geological Survey for scientists seeking sources of information about the Earth's land surfaces. GLIS contains metadata—that is, descriptive information about data sets. Through GLIS, scientists can evaluate data sets, determine their availability, and place online requests for products.

URL:
    http://sun1.cr.usgs.gov/glis/glis.html

Anyone can browse the Web.

Are you on the Web?

History of Cartography Project

Perhaps the author of *Tom Jones* didn't need a map, but lots of others over the course of history have had reason to use one. Wars are fought to this day over lines drawn on maps, and they reflect not only a visual communicative representation of a region or of the world, but a sense of presence and perception akin to the period, the society, and the cartographer. Lest we forget how important cartographers are, just remember that America was named for its mapmaker. Great images complement the history of slicing and dicing the globe.

URL:
    http://elvis.neep.wisc.edu/~cdean/

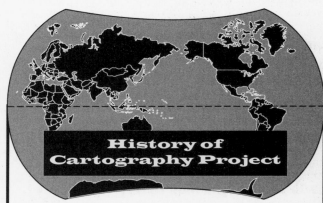

"Map me no maps, sir, my head is a map, a map of the whole world." When Henry Fielding wrote these words in the first half of the eighteenth century, the world was a bigger place. Find out just how much it was expanding by visiting the History of Cartography Project.

How Far Is It?

A friend of ours was moving to Pittsburgh from San Diego. He wanted to know how far it is between the two cities. We immediately plugged into the Web and went to this distance-calculation server. The answer was 2,117 miles, as the crow flies. Packed and ready to leave, he waved as he drove off, wishing he could turn his truck into a crow.

URL:
    http://gs213.sp.cs.cmu.edu/prog/dist

The Web is too cool.

The Journey North

An experimental venture created by students, this page embarks on a journey into the northern arctic region. It incorporates different themes: the environment, wildlife migration, and different cultures discovered along the way.

URL:
http://ics.soe.umich.edu/ed712/IAPIntro.html

Latinworld

Find Internet resources and everything you want to know about Latin America and the Caribbean at Latinworld. This handsome, extensive, and well-organized page takes you on a virtual tour of Latin news, business, economy, politics, culture, and education.

URL:
http://www.latinworld.com

NAISMap GIS

With NAISMap, you're the mapmaker! Whaddya say we make our own custom map of Canada, eh? We'll view and manipulate the National Atlas spatial data layers, and throw in some rivers and forests and a few wetlands. Let's see where the western wood-peewee's range lies. How about the habitat of *Sorex fumeus*, the smokey shrew?

URL:
http://ellesmere.ccm.emr.ca/naismap/
naismap.html

NOAA

As you'll discover, the National Oceanic and Atmospheric Administration is a lot of things to a lot of people. There is a wealth of information on protected species, global warming, weather forecasts and warnings, coastal ecosystems, and global positioning and navigation. Also linked here are the National Environmental Satellite, Data, and Information Service; the National Marine Fisheries Service; the National Ocean Service; and the National Weather Service.

URL:
http://www.noaa.gov

Perry-Castañeda Library Map Collection

If Christopher Columbus had discovered this web site before he left Spain, it would have saved everyone a whole lot of trouble. Now you can discover the New World with detailed maps, all from the comfort of your computer.

URL:
http://www.lib.utexas.edu/Libs/PCL/
Map_collection/Map_collection.html

The Perseus Project

This is a coin, vase, and sculpture database, plus *The New Perseus Atlas* that includes several new forms of geographic information taken from satellite photographs.

URL:
http://medusa.perseus.tufts.edu

The Peters Projection

This is the world as it really is, drawn by Arno Peters in 1974. Arno wanted a map that would take us into the twenty-first century without reproducing existing distortions and biases of other maps. Spatially, the map is a flat projection on which one square inch represents an equal number of square miles. The Peters Map Tutorial Program is a free, MS-DOS-based program that you can download easily from this site.

URL:
http://www.webcom.com/~bright/petermap.html

Ryhiner Map Collection

It helps to understand a little Latin and German at this site, but the effort is well worth it. The Ryhiner Map Collection is an extremely valuable compendium, consisting of more than 15,000 maps, charts, plans, and views of regional and global maps from the reigning countries of the sixteenth to the eighteenth centuries. This page is for the serious student of cartography.

URL:
http://ubeclu.unibe.ch/stub/ryhiner/
ryhiner.html

Timezone Converter

The forms on this page allow you to choose a time zone, the time and date to convert from, and the time zone to convert to. There are instructions on using the converter and configuring it, as well as technical details on how it works.

URL:
http://hibp.ecse.rpi.edu/cgi-bin/tzconvert

A
B
C
D
E
F
G
H
I
J
K
L
M
N
O
P
Q
R
S
T
U
V
W
X
Y
Z

Excerpt from the Net...

(from the Peters Map pages)

Five thousand years of human history have brought us to the threshold of a new age—an age typified by science and technology, by a growing awareness of the interdependence of all nations and all peoples.

Such a moment in history demands that we "look" critically at our understanding of the world. This understanding is based, to a significant degree, on the work of map makers of an earlier age.

Tourist Expedition to Antarctica

Take a tour of Antarctica in this first-hand account of the Grand Antarctic Circumnavigation cruise of the *MV Marco Polo*. Visit sites within the Antarctic Peninsula and the Ross Sea, traveling from Punta Arenas, Chile to Christchurch, New Zealand.

URL:

   http://http2.sils.umich.edu/Antarctica/
   Story.html

Travels with Samantha

"This book is about the summer I spent seeing North America, meeting North Americans, and trying to figure out how people live," writes author Greenspun in this electronic version of a popular travelogue of North America. Come face to face with examples of the stunning ethnic, scenic, and cultural richness of the continent. Excellent photography and engaging storytelling won this page a Best of the Web award in 1994.

URL:

   http://www-swiss.ai.mit.edu/samantha/
   travels-with-samantha.html

U.S. Gazetteer

The U.S. Gazetteer is a searchable database of U.S. place-names. After entering your city's name, click on the Tiger Map link to build a map of your city. Use the mapping controls to zoom right into your neighborhood down to the street level.

URL:

   http://tiger.census.gov/cgi-bin/gazetteer

Do you really know where you live? Not just the location, but population, geographic coordinates, census information, and ZIP codes? You should. (After all, the terrorists with the ICBMs know *exactly* where you live.) Find your hometown in the U.S. Gazetteer. Then move into cyberspace, where it's safe.

Virginia County Interactive Mapper

The Virginia County Interactive Mapper provides Web-available customized digital maps of every Virginia city and county. Different features may be displayed, zoomed, and customized, allowing greater functionality.

URL:

   http://ptolemy.gis.virginia.edu:1080/tiger.html

Xerox PARC Map Viewer

Jules Verne would have named this resource *Around the World in 80 Clicks*, though it takes fewer than that. You don't need a balloon—just a mouse—to explore this digital world map. The Xerox PARC Map Viewer contains a plethora of features to display your world by type of projection, in color, zoom in/out, and show or hide rivers and borders. Where do you want to click today?

URL:

   http://www.xerox.com/map

Where in the World is San Diego? Find out on the Xerox PARC Map Viewer.

GEOLOGY

Active Tectonics

If you're all shook up about geology, then you'll want to shift, slide, or otherwise collide into the Active Tectonics page. Frequently updated information regarding the Active Tectonics initiative encourages research in tectonic environments. Included are research projects, images, and news about earthquakes, explosive volcanism, and more.

URL:
> http://www.muohio.edu/tectonics/
> ActiveTectonics.html

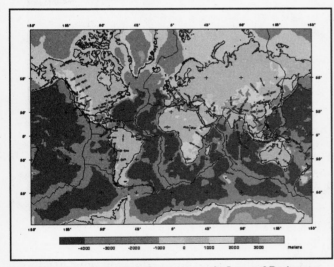

Having trouble finding geological resources on the Internet? Don't stress out. With pointers to extensive information, you'll be seduced into the hottest subduction zones, just like those shown here on the *World Stress Map*.

Alaska Volcano Observatory

The next time your flight to Nome is delayed, it could be due to volcanic activity over the Northern Pacific around the Aleutian arc. This and other interesting information on Alaskan volcanoes and the hazards they generate can be found here at AVO.

URL:
> http://www.avo.alaska.edu

Ant-earth

Here's a mailing list for dissemination of information and discussion regarding the earth sciences division of the U.S. Antarctic Program.

Mailing List:
> Address: **listproc@mcfeeley.cc.utexas.edu**
> Body of Message: **subscribe ant-earth** *<your name>*

Cascades Volcano Observatory

When a volcano erupts, it seems, literally, that all hell breaks loose. At CVO you'll find information on volcanic hazards, the mission of CVO, recent publications, and plenty of color images of Cascade volcanoes.

URL:
> http://vulcan.wr.usgs.gov/home.html

The CREWES Project

The CREWES Project is an applied geophysical research group concentrating on the analysis and interpretation of multicomponent seismic data. Their goal is to obtain improved 3-C and 3-D geological images of the subsurface.

URL:
> http://www-crewes.geo.ucalgary.ca

Diagenesis

The next time someone asks you what you've been up to, tell them that you've been investigating *thermochemical sulphate reduction* and *anthropogenic sulphur incorporation*, two areas of great interest to those in the geological field of diagenesis. In case you don't know what diagenesis is all about . . . hey, wait a minute, that's what the Web is for! Look it up yourself!

URL:
> http://earth.geo.ucalgary.ca

Earth Resources Observation Systems

The EROS Data Center is a data management, systems development, and research field center of the U.S. Geological Survey's National Mapping Division. Download digital elevation models, digital line graphs, land use and land cover information, declassified CIA images, global land information system photos, and more.

URL:
> http://edcwww.cr.usgs.gov/eros-home.html

A B C D E F G H I J K L M N O P Q R S T U V W X Y Z

Earthquake Discussion

The western U.S., Japan, Mexico City, Indonesia—nobody knows where a quake will hit next. Talk about earthquakes on Usenet. Topics range from software and prediction tools to recent seismic events.

URL:

news:sci.geo.earthquakes

Excerpt from the Net...

```
(from Diagenesis)

Real geologists...

don't eat quiche. They don't even know what it is. Real geologists like raw
meat, beer, and tonsil-killer chili.

don't need rock hammers. They break samples off with their bare hands.

don't sit in offices. Being indoors makes them crazy. If they had wanted to
sit in offices, they would have been geophysicists.

don't need geophysics. Geophysicists measure things nobody can feel or
see, make up a whole lot of numbers about them, then drill in all the wrong places.

don't go to meetings, except to point at a map and say "drill here!" And leave.

don't work 9 to 5. If any real geologists are around at 9:00 A.M. it's because
they're going to meetings to tell the managers where to drill.

don't like managers. Managers are a necessary evil, for dealing with
bozos from human resources, bean counters from accounting, and other
mental defectives.

don't make exploration budgets. Nervous managers make exploration
budgets. Only insecure mama's boys try to stay within exploration
budgets. Real geologists ignore exploration budgets.

don't use compasses. That smacks of geophysics. Real geologists always know exactly
where they are and the nearest place where beer is available.

don't make maps. Maps are for novices, the forgetful, managers, and
pansies who like to play with colored pencils. A real geologist will only
draw a map to show the ill-informed managers where to drill.

don't write reports. Bureaucrats write reports, and look at what they're like.

don't have joint venture partners. Partners are for wimpy bedwetters
who are unable to think big.
```

Earthquake Hazard Maps

You left your heart in San Francisco, and now you owe it to yourself and your loved ones to find it among the rubble from the last quake that hit the beautiful City by the Bay. You can view earthquake-hazard maps by city, scenario, and fault lines. Jittery about living in the Bay Area? Don't forget, L.A. is just a few miles south!

URL:
http://www.abag.ca.gov/bayarea/eqmaps/
eqmaps.html

Earthquake Information

This page offers updated information on highly seismic areas such as Japan and California. Here at the UC Davis Department of Geology you'll find detailed charts and maps.

URL:
http://www-geology.ucdavis.edu/eqmandr.html

Envision

Envision is a free software package for the storage, management, manipulation, and display of large multidimensional arrays of numerical geophysical data. Envision runs on most Unix workstations using X/Motif, C language, and interprocess communication.

URL:
http://www.atmos.uiuc.edu/envision/
envision.html

Federal Geographic Data Products

The Manual of Federal Geographic Data Products describes federal geographic data products that are national in scope and commonly distributed to the public. Geographic data products include maps, digital data, aerial photography and multispectral imagery, earth science, and other geographically referenced data sets.

URL:
http://info.er.usgs.gov/fgdc-catalog/title.html

Flagstaff USGS Center

Browse the solar system or look at color-shaded relief maps of the western states on the U.S. Geological Survey's Flagstaff Field Center home page.

URL:
http://flgsvr.wr.usgs.gov/Flagstaff.html

Geo Exchange

This list of applied and commercial geoscience resources includes news, newsgroups, mailing lists, government sites, and geographic information systems. A good starting point for geoscience research and information.

URL:
http://giant.mindlink.net/geo_exchange/

Geoed-l

Get your rocks off and join the mailing list of the Geology and Earth Science Education Discussion Forum.

Mailing List:
Address: **listserv@uwf.cc.uwf.edu**
Body of Message: **subscribe geoed-l** *<your name>*

Geological Society of America

The GSA provides updated earth-science information for the benefit of educational, governmental, business, and industrial purposes.

URL:
http://www.aescon.com/geosociety/

Geological Survey of Canada

Like most geological-survey field centers, this GSC page includes the links to many geoscience resources, libraries, projects, news, and more.

URL:
http://www.emr.ca/gsc/

Geology Materials

Pictures of Earth, as seen from various space shuttle missions, highlight this page at the University of Oregon. It also offers maps of recent earthquakes throughout the world; volcano information and images; and animations of ocean, hurricane, and sea surface temperatures.

URL:
http://zebu.uoregon.edu/geol.html

Geophysics Organizations

Around the world in eighty (plus) links. Here is an extensive list of geophysics organizations around the world.

URL:
http://www.crewes.ucalgary.ca/VL/html/
gp-orgs-by-location.html

A B C D E F G H I J K L M N O P Q R S T U V W X Y Z

Michigan Tech Volcanoes Page

Michigan Tech's Volcanoes Page provides information on current global volcanic activity, remote sensing, links to government agencies and research institutions, and even some volcano humor. Their near-real-time data on eruptions is a model of how such information is readily available on the Net.

URL:

http://www.geo.mtu.edu/volcanoes/

Mount St. Helens

Before, during, and after. An easy-to-read page which gives information on the geological history of Mount St. Helens before its eruption on May 18, 1980. Read on to find out about the various aspects of the eruption and the immediate, as well as far-reaching, results.

URL:

http://volcano.und.nodak.edu/vwdocs/msh/
msh.html

Mount St. Helens lets loose on May 18, 1980 with an explosion and devastation that only Mother Nature could conjure. View dozens of breathtaking images of Mount St. Helens and other Lava Domes at the Mount St. Helens web.

NASA Earth Observing System

EOS is a series of polar-orbiting, remote-sensing satellites planned for launch starting in 1998, and spanning a period of at least 15 years. These satellites will collect long-term data sets about the Earth and its climate. Samples of images of volcanic activity are available online now.

URL:

http://www.geo.mtu.edu/eos/

Ralph Kugler's Geology and Geography Page

Ralph Kugler is seriously into Earth Science. On his home page, Ralph provides links to dozens of great geology resources, geology software, interactive geographic information systems, and excellent images of clay minerals from scanning electron micrographs. If you're interested in geology, this should be one of the first pages you check out.

URL:

http://www.onramp.tuscaloosa.al.us/~rlkugler/

Ridge Inter Disciplinary Global Experiment

Learn about the topography of volcanic ridges in the RIDGE project at Woods Hole. See how a geographically dispersed, multidisciplinary project is managed in cyberspace.

URL:

http://copper.whoi.edu

School of Ocean and Earth Science and Technology

SOEST's presentation is of interest to those in the fields of geology, meteorology, oceanography, and ocean engineering. You'll find the *Volcano Watch News Letters* covering the ongoing Kilauea eruption in Hawaii, seismic hazards of the Hawaiian chain, other eruptions around the world, and other volcanology issues.

URL:

http://www.soest.hawaii.edu

Smithsonian Global Volcanism Program

The Smithsonian's Natural History Web includes the Global Volcanism Program, devoted to documenting volcanic eruptions on Earth during the last 10,000 years—the interval of geological time known as the Holocene. The GVP plays a central role in the rapid dissemination of information about ongoing volcanic activity on Earth and in the archiving of data, maps, and photographs for Holocene volcanic eruptions.

URLs:

gopher://nmnhgoph.si.edu/11/.gvp

http://nmnhgoph.si.edu/gopher-menus/
SmithsonianGlobalVolcanismProgram.html

Discover 10,000 years of geologic time in only a few seconds at the Smithsonian Global Volcanism Program.

U.S. Geological Survey

As the nation's largest earth science research and information agency, the USGS maintains a long tradition of providing "Earth Science in the Public Service." The USGS, a bureau of the U.S. Department of the Interior, was established to provide a permanent federal agency to conduct the systematic and scientific "classification of the public lands and examination of the geological structure, mineral resources, and products of the national domain."

URL:

http://www.usgs.gov

Society for Sedimentary Geology

The Society for Sedimentary Geology is dedicated to spreading scientific facts and data in sedimentology, stratigraphy, paleontology, environmental sciences, hydrology, and other areas of earth science. This page includes extensive information on membership, publications, resources, and additional outside links.

URL:

http://dc.smu.edu/sepm_sp/home.html

UCD Department of Geology

The Department of Geology at the University of California at Davis provides departmental and administrative information and services, as well as research data. These pages include information on UCD research, activities, related societies, links to other institutions, and course and faculty information.

URL:

http://www-geology.ucdavis.edu

USGS Earthquake News

This is *the* center for earthquake information, preparedness, bulletins, and updates for earthquake studies and earthquake-related disaster control in Southern California. Among other important information, you can find out what seismologists think about earthquake risk, and you can listen to a seismogram. If a whole lotta shakin's goin' on, duck and cover, then link here after you get the backup generator going!

URL:

http://quake.wr.usgs.gov

Hey, we said *sedimentary*, not *sedentary*. So don't just sit there—check out the Society for Sedimentary Geology!

Space Physics Data Access and Display System

Defense Meteorological Satellite Program (DMSP) Data Archive is located at the National Geophysical Data Center (NGDC). This web site includes images, satellite data, and links to other related topics such as geomagnetism, hurricane images, and data.

URL:

http://web.ngdc.noaa.gov/dmsp/dmsp.html

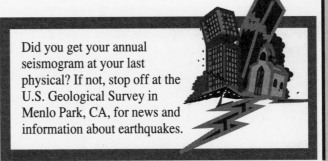

Did you get your annual seismogram at your last physical? If not, stop off at the U.S. Geological Survey in Menlo Park, CA, for news and information about earthquakes.

Volcanic Eruption Modeling

This page is part of an ongoing project to model volcanic eruptions using a computer. All stages are covered, from simple petrology to complete eruptions. Photos and reference materials are also included. The project is funded by the Japanese Science and Technology Agency, the Geological Survey of Japan, and the National Science Foundation.

URL:

http://www.aist.go.jp/GSJ/~jdehn/v-home.htm

Volcanic Hazards Mitigation

Volcanic activity can be one of the most spectacular events in nature, but the damage inflicted on life and property can be cataclysmic. Here is where hard scientific technology and economics are making a difference. This extensive presentation illustrates how enhanced satellite photography is used to map changes in activity at five major volcanoes. Computer models predict economic loss by applying cost of living indexes against the area's volcanic risk.

URL:

http://www.geo.mtu.edu/volcanoes/hazards.html

Volcano Systems Center

Among news, information, and related links on the University of Washington's VSC page, you'll find interesting research projects. For example, here is detailed topography of the Juan de Fuca Ridge, a spreading center in the northeastern Pacific where the RIDGE Observatory Experiment (ROBE) and research on the CoAxial Segment have recently been conducted.

URL:

http://www.vsc.washington.edu

Volcano World

Join Rocky the Volcano on a tour of some of nature's most dramatic phenomena. Experience volcanic activity around the globe and throughout the solar system. Tour such famous fumaroles as Mount St. Helens and the Hawaiian volcanos. These pages also include news updates on eruptions, and online vulcanologists who can answer any burning questions you may have. You'll lava it!

URL:

http://volcano.und.nodak.edu

GRAPHICS AND CAD

3D Artist Magazine

3D Artist is a hard cover magazine that also publishes an electronic magazine, *The Tessellation Times*. You can read "Tess" current and back issues, and look for additional information concerning 3D events, classes, and galleries. There are sections for job listings in the 3D field and posting ads for artists, and an FTP site with many resources that relate to desktop 3D graphics.

URL:

http://www.3dartist.com

3D Graphic Engines

Here's a great overview and compendium of over 100 sources for texture mapping, landscape rendering, gouraud and flat shading, wireframe, non-realtime, and Doom/Wolfenstein-type graphic engines. In many cases, the original source code, extensions, features, and FTP source are provided.

URL:

http://www.cs.tu-berlin.de/~ki/engines.html

Excerpt from the Net...

```
(from the Volcanic Hazards Mitigation pages)

Formula to determine volcanic risk...
Value = Standard of Living × Population Density + Infrastructure + Landuse
   (dollars/hectare)
Vulnerability = Measures a proportion (0-100%) of the value likely to be lost in a
   given event
Hazard = The probability of a given area being affected by a potentially destructive
   process within a given period of time

VOLCANIC RISK = Value × Vulnerability × Hazard
```

Did you ever wonder how great games like Doom and Dark Forces actually work? Find out on the **3D Graphic Engines** page.

ACM SIGGRAPH Online

The Association for Computing Machinery Special Interest Group on Computer Graphics is a professional organization that brings together computer graphics professionals from all disciplines. SIGGRAPH's page offers sections on sponsored conferences and workshops, educational resources, art and design resources, publications, and an archival area. Also find information on SIGGRAPH Professional Chapters and regional contacts.

URL:

http://www.siggraph.org

Get Graphic.
Visit the Association for Computing Machinery Special Interest Group on Computer Graphics.

AHPCRC Graphics Software

The Army High Performance Computing Research Center (AHPCRC) provides an array of computer graphics software for Silicon Graphics workstations at this site. This software includes volume rendering, movie-making tools, Motif control panels, and special-purpose X software.

URL:

http://www.arc.umn.edu/html/gvl-software/gvl-software.html

ARL Scientific Visualization

The Army Research Laboratory High Performance Computing Division explores software and hardware technology related to visualization and produces a variety of animation projects. Among other resources, you'll find discussions of selected visualization projects, technical papers, and online simulations of Detonation Interaction and Finned Projectile Aerothermal Analysis.

URL:

http://frontier.arl.mil/SciVis/scivis.html

CICA Graphics 3D Object File Formats

The Center for Innovative Computer Applications provides a page that contains information on various 3D object file formats and how to view them from Web browsers. There are overviews and links to a variety of 3D Object and Virtual Reality Modeling Language (VRML) file formats. Included is a sample 3D object in Object Oriented Graphics Library (OOGL) format.

URL:

http://www.cica.indiana.edu/graphics/3D.objects.html

Computer Animations at LSI

Ideas are the images of imagination, and at the University of São Paulo Laboratory of Integrated Systems, ideas are not in short supply. Learn more about their programs in artificial intelligence and automation, digital systems, and microelectronics materials and processing. There are local software programs available for workstations that include RTV, a volume visualization package with tools for segmentation, classification, and animating medical images. For fun, take a peek at several student art projects.

URL:

http://www.lsi.usp.br/~dsd/animate/animations.html

A B C D E F G H I J K L M N O P Q R S T U V W X Y Z

Computer Graphics at Berkeley

Cruise on over to Berkeley, where there's always something interesting afoot. Take a virtual walk through the campus graphics building or sit in on a weekly lunchtime lecture on graphics. Then get a glimpse of what Cal Berkeley is doing with optics and topography; radiosity; and geometric and spline modeling.

URL:
> http://http.cs.berkeley.edu/projects/graphics/

Computer Graphics at Stanford University

Palo Alto is no slouch when it comes to computers and graphics. At this Stanford page, you'll find cool demos, technical papers, research projects, and course information. Downloadable software packages developed by members of the Stanford research group include the VolPack Volume Rendering Library and ZipPack Polygon Mesh Zippering Package, both for SGI workstations.

URL:
> http://www-graphics.stanford.edu

Computer Graphics at University of Manchester

Artists and programmers at the University of Manchester not only produce nice computer art, but also provide high-performance multimedia and image-processing facilities and services. Browse their online gallery, or learn about their graphics and visualization education activities.

URL:
> http://info.mcc.ac.uk/CGU/CGU-intro.html

Computer Graphics Group at Caltech

Read up on Caltech's wide-ranging endeavors in the fields of generative modeling using interval analysis; magnetic resonance imaging; model extraction; and morphogenesis and developmental biology. Downloadable software includes the DBF device-independent graphics package. There are abstracts available for selected books and theses published by Caltech.

URL:
> http://www.gg.caltech.edu

Planning an outing? Check the weather in Meteorology.

The Gopher's time is short.

Computer Graphics Research Group at Sheffield

Computer graphics students at Sheffield focus on 3D computer animation. On their demo page, you can see the results of some of their projects, including modeling and animation control of deformable objects and human beings (not that the two go together, of course!). They're also interested in flame simulation, so fire up your browser and take a look.

URL:
> http://www.dcs.shef.ac.uk/research/groups/
> graphics/

Data Visualization Group

This site in Italy contains samples of visualization and animation systems applied in the areas of geographic databases, meteorological data, computational fluid dynamics, oceanographic modeling, digital animation, and ray tracing in FLC format.

URL:
> http://esba-www.jrc.it/dvgdocs/dvghome.html

Digital Illusions

Digital Illusions publishes the online *Digital Studio Magazine*. This magazine contains new articles each week on the latest in computer graphics and animation written by industry professionals. There are interviews with graphics experts on topics ranging from special effects to 3D modeling and animation. Separate sections feature tutorials, job offers, technical production, software FTP sites, and Usenet conferences.

URL:
> http://www.mcs.net/~bcleach/illusions/

DIVE Home Page

This Swedish site contains sample graphics and movies of DIVE, a fully distributed heterogeneous virtual reality system. Users navigate in 3D space and may see, meet, and collaborate with other users and applications in the environment. This Distributed Interactive Virtual Environment software is available free of charge for SGI, IBM RS6000, and SUN4 SPARCstation platforms. Included are links to other DIVE-related pages.

URL:
> http://www.sics.se/dce/dive/dive.html

Electronic Visualization Laboratory

EVL at the University of Illinois at Chicago researches interactive techniques in computer graphics through a blend of engineering, science, and art. Interdisciplinary studies here include virtual reality, multimedia, paradigms for information display, and televisualization (distributed graphics over networks). There is an electronic art gallery and a list of related computer graphics resources.

URL:

    http://www.ncsa.uiuc.edu/evl/html/ homePage.html

Graphics FAQ Collections

Here you'll find FAQs for the Usenet newsgroups: **comp.graphics.algorithms, comp.graphics.animation, comp.graphics, comp.graphics.apps.gnuplot, comp.graphics.opengl, comp.graphics.api.pexlib,** and other subjects including binary space-partitioning trees, color and gamma, graphics file formats, and raytrace graphics.

URL:

    http://www.cis.ohio-state.edu/hypertext/faq/ usenet/graphics/top.html

"The significant problems we face cannot be solved by the same level of thinking that created them."—Albert Einstein Solve or create your own significant problems based on frequently asked questions on Computer Graphics.

Graphics File Format Page

Here's an invaluable source of information for many of the different formats encountered in computer graphics. You'll find detailed documentation on virtually every major format flavor from BMP to JPEG and TIFF, as well as 3D and CAD. Several sections even offer conversion programs and source code.

URL:

    http://www.dcs.ed.ac.uk/home/mxr/gfx/

The Graphics Forum

As part of the Kai's Power Tips & Tricks for Adobe Photoshop web, the Graphics Forum is a place you can discuss all types of computer graphics with experts and newcomers. The forum is divided into categories—a general discussion area for any graphics topics, a web graphics forum for discussion of web-specific topics, and other subjects, including Adobe Photoshop and prepress graphics.

URL:

    http://the-tech.mit.edu/KPT/Forum/

Graphics Newsgroups

Here's a list of general and specific discussion groups relating to many aspects of computer graphics and animation for a variety of computer platforms.

URLs:

    news:alt.3d
    news:alt.graphics.pixutils
    news:comp.graphics.algorithms
    news:comp.graphics.animation
    news:comp.graphics.animations
    news:comp.graphics.data-explorer
    news:comp.graphics.explorerlorer
    news:comp.graphics.openglerlorer
    news:comp.graphics.packages.3dstudio
    news:comp.graphics.packages.aliasdio
    news:comp.graphics.packages.lightwave
    news:comp.graphics.raytracingightwave
    news:comp.graphics.rendering.renderman
    news:comp.graphics.researchg.renderman
    news:comp.graphicspixutils
    news:comp.sys.amiga.graphics.renderman
    news:comp.sys.mac.graphicscs.renderman
    news:comp.sys.sgi.graphicscs.renderman
    news:rec.arts.animationicscs.renderman
    news:sci.virtual-worlds.apss.renderman
    news:sci.virtual-worldsicscs.renderman

Going shopping on the Web? Remember to pick up some ecash.

A B C D E F **G** H I J K L M N O P Q R S T U V W X Y Z

Graphics, Visualization, and Usability Center

The GVU Center's mission at Georgia Tech is to make computers as accessible and usable as automobiles, stereos, and telephones. Here, you'll find projects on topics such as animation, virtual environments, medical informatics, scientific visualization, and user-interface software.

URL:
   http://www.cc.gatech.edu/gvu/research.html

The Gravigs Project at ITTI

Thanks to the Information Technology Training Initiative, workstation users may now take online courses on the use of computer graphics and scientific visualization. The four courses currently available are *Standards for Computer Graphics*; *Color in Computer Graphics*; *Visualisation 1: Graphical Communication*; and *Geometry for Computer Graphics*. Additional courses are under development and all include teaching packs and exercises—and they are yours—free of charge.

URL:
   http://info.mcc.ac.uk/CGU/ITTI/gravigs.html

GWeb

Graphics Web is an electronic trade journal for professionals in the computer animation industry and for those pursuing careers in computer animation. Highlights include interviews with various award-winning graphic designers and animators, an up-to-date job listing section, and a large list of computer graphics schools and departments from around the world.

URL:
   http://www2.cinenet.net/GWEB/

Human Modeling and Simulation

Dubbed "Jack," a University of Pennsylvania 3-D interactive demo features a detailed human model with realistic behavioral controls. The center investigates computer graphics modeling, animation, and rendering techniques.

Note: A Silicon Graphics workstation is required to use the demonstration software.

URL:
   http://www.cis.upenn.edu/~hms/

How to use a Gopher: type the URL into your Web browser.

Image Gallery

Even if you can't afford a new (or even used) Silicon Graphics Crimson Elan, you can still see the kinds of amazing artwork they can produce. There are excellent renditions of general and scientific images. All entries were created on SGI visual computing workstations.

URL:
   http://www.sgi.com/Fun/free/gallery.html

KPT Online Graphics Gallery

This gallery is a part of Kai's Power Tips and Tricks Online, and it shows some excellent examples of artists pushing the graphics software available today to its extreme. Two of the incredible images here have the titles *Alien Desert* and *Ice Planet*.

URL:
   http://the-tech.mit.edu/KPT/Gallery/

Medical Imaging in 3D and Computer Graphics Lab

For research in advanced computer architectures for virtual environments and advanced distributed simulation, recon this Air Force Institute of Technology's Graduate School of Engineering page. Their virtual environments and distributed interactive simulations include Virtual Cockpit (a manned flight simulator of an F-15E), Synthetic BattleBridge, Satellite and Solar System Modeler, and the Virtual Emergency Room.

URL:
   http://www.afit.af.mil/Schools/EN/ENG/LABS/
   GRAPHICS/GRAPHICS.html

Fascinated with fishes? School yourself in Ichthyology.

MIRALab

This Swiss site is involved in computer graphics and animation, particularly in the areas of the human body and facial animations using synthetic actors. MIRAlab's projects also explore cloth animation, hand deformation, medical informatics, and virtual reality.

URL:
http://cuisg13.unige.ch:8100/HomePage.html

NCSA Digital Information System

The National Center for Supercomputing Applications displays a stunning selection of images and videos in computer graphics, astronomy, biology, chaos, chemistry, geosciences, math, meteorology, and technology.

URL:
http://www.ncsa.uiuc.edu/SCMS/DigLib/text/
overview.html

Nifty Raytraced MPEGs

This site not only has nifty raytraced MPEGs of 3D Hilbert space-filling curves, but also descriptions of how they were created and pointers to FTP sites where you can get the programs that the author used. Then you can create your own raytraced MPEGs of 3D Hilbert space-filling curves!

URL:
http://www.santafe.edu/~nelson/mpeg/

Perceptual Science Laboratory

The PSL at the University of California at Santa Cruz studies perception and cognition, especially facial animation and speech perception from visual cues. Separate sections with text and graphics cover various historical paradigms in the studies of facial animation, facial analysis, and lipreading. There are speech and model-fitting examples, and links to other related computer graphics resources.

URL:
http://mambo.ucsc.edu

Headed for the city? Get the subway map from the Web.

Publish it on the Web.

Photoshop Techniques

Photoshop Techniques is a publication of Swanson Tech Support—a company that has graciously given permission for students at MIT to publish helpful articles from the magazine as part of the Kai's collection. Articles include "How to use Quick Mask to create special photo edges," "How to create a neon object," and "How to create an embossed object." Many other tremendously useful tips can be found here.

URL:
http://the-tech.mit.edu/KPT/Techniques/

Program of Computer Graphics

Cornell University is the home of CU-SeeMe Internet videoconferencing software. So it should come as no surprise that their computer graphics program is top-notch. Established back in 1974—when computers were mainframes, expensive, and scarce—Cornell's program is one of the most advanced in the U.S. There is an excellent tutorial on computer graphics for beginners that covers such techniques as color, shading, object rendering, ray tracing, and radiosity—a feature discovered by the university that led to the creation of the Cornell Box. There is also information on current research and graduate studies.

URL:
http://www.graphics.cornell.edu

Ray Tracing

Probably the mother of all ray tracing pages. . . You'll find general information and resource material, with links to virtually every major software site and organization associated with ray tracing. Each FTP source contains additional links to notes, user's guides, and mirror sites.

URL:
http://arachnid.cs.cf.ac.uk/Ray.Tracing/

A B C D E F G H I J K L M N O P Q R S T U V W X Y Z

Scientific Applications and Visualization Branch at NASA

If workstations could work anywhere they chose, they'd probably want to work at this NASA facility. SAVB uses high-performance graphics workstations to produce superb videos and animations of space, lunar, and Landsat imagery. Take a virtual MPEG ride, visit the stills gallery, or consider downloading several pixel-restoration and 3D mapping programs that are available—for workstations only, of course.

URL:
http://sdcd.gsfc.nasa.gov/SAVB/SVS/SVS.html

SciViz

Sardinia isn't the first place one thinks of for high-tech Internet graphics, but it may be from now on—thanks to this innovative group at SciViz. At this site there are voluminous documentation and videos of many SciViz inventions including i3D, a high-speed, three-dimensional scene viewer. You can download a large number of MPEG movies produced in the fields of medical imaging, fluid dynamics, environmental modeling, meteorological animations, aerodynamics, and semiconductor devices.

URL:
http://www.crs4.it/~zip/group_homepage.html

SimTel Index:Graphics

At this site are more graphics programs for PCs than you can shake a paintbrush at. You'll find everything from 3D wireframe display programs and toolkits to morphing and conversion programs—all zipped up and ready to go.

URL:
http://www.sinica.edu.tw/
simtel/simtel_index_graphics.html

Swiss Federal Institute Graphics Lab

The Computer Graphics Lab (LIG) at the Swiss Federal Institute of Technology (EPFL) in Lausanne is mainly involved in creating computer animation and virtual reality. There are downloadable MPEG demos of morphing (from the Mona Lisa to Cindy Crawford) and simulated actors playing tennis. This site lists computer-generated, film-related Ph.D. dissertations, books, papers, events, and conferences.

URL:
http://diwww.epfl.ch/w3lig/indexeng.html

Syracuse University's Art Media Studies—Computer Graphics

Here's a good starting point for Internet art experiments. Featured at this site are collaborative art projects, present and previous exhibitions, and undergrad projects. The university also makes computer graphics courses available online.

URL:
http://ziris.syr.edu

Thant's Animations Index

This index contains over 200 animation sources from many different universities, companies, and individuals. Most are related to the hard sciences (math, chemistry, and physics) in one way or another. There are links to such sample titles as "3D Volume Morphing" and "Wind Vectors Over Maui," and animations of biochemical processes such as enzymology, heme metabolism, and lipid metabolism.

URL:
http://mambo.ucsc.edu/psl/thant/org.html

World Wide Web Virtual Library: Virtual Reality

Fraunhofer Institute in Stuttgart, Germany, maintains this VR page of FAQs, overviews, software application referrals, and other 3-D subjects.

URL:
http://www.iao.fhg.de/Library/vr/
OVERVIEW-en.html

Yoshiaki Araki's Home Page

Yoshiaki is heavily influenced by the artwork of M.C. Escher. On his pages, Yoshiaki offers a number of original renderings and animations done in Escher's style, and musings on Escher's philosophy and artwork. There are also hyperbolic animation MPEGs, several HotJava references, and links to FTP sites for related graphics software.

URL:
http://www.sfc.keio.ac.jp/~t93827ya/

Be a star-gazer. Scope out Astronomy.

HARDWARE

Amiga Video Card

Still trying to breathe life into your Amiga? Maybe you need a new video card! The Picasso II is a multiple-format graphics board that integrates into AmigaDOS 2.04 and higher, allowing you to use 24-bit graphics modes up to a resolution of 1600x1200. Of course, the Picasso card is a commercial product, but this isn't an official page—it's an informal, but very informative, review of the Picasso II card.

URL:

> http://www.phone.net/home/mwm/picasso/

The AWE32 Page

The unofficial home page for the AWE32 is an amazing resource for owners of Creative Labs' top-of-the-line sound board. Created and maintained by Jesper Nordenberg and Johan Nilsson, this page sports the most information, utilities, and tools for the AWE32 (and AWE32 Value and SB32) sound boards outside of Creative Labs. You'll find a chat page dedicated to talk of the AWE32; pointers to drivers, utilities, and programs; hints; bug fixes; the AWE32 FAQ; and, of course, direct links to Creative Labs and their tech support department.

URL:

> http://www.edu.isy.liu.se/~d93jesno/awe32.html

CD Information Center

The CD Information Center is a resource provided to the Web by The CD-Info Company. This company promises to keep you up to date on CD and CD-ROM technology, and it delivers. On this web, you can find information about premastering; recordable CD technologies; a glossary of CD and CD-R terminology; and sections on the CD industry, CD applications, and the history and future of CD and CD-ROM technology.

URL:

> http://www.cd-info.com/cd-info/
> CDInfoCenter.html

Need help getting connected? See Rick's book The World Wide Web Complete Reference.

CD-ROM Drive Specs

Thinking of upgrading that old 1x-speed CD-ROM drive with a fast new one? Don't do it without checking the specs and comparing features first. This page has links organized by interface type and speed. If you're interested in 6x-speed SCSI drives, just click the link and you're presented with a table comparing manufacturers' models features and specifications.

URL:

> http://www.cs.yorku.ca/People/frank/
> cd_specs.html

Don't put up with waiting minutes for that next level of Doom to load or a stuttering and stammering Carmen Sandiego. Upgrade that ancient 2x CD-ROM drive to something actually developed in this decade. But don't just dig into your pocket for the first flashy box you find at the computer store . . . arm yourself with the specs. Check out the CD-ROM Drive Specs page.

Computer Monitors: Separating Fact from Fiction

If you've been thinking about getting a new hi-res monitor for your PC, you definitely want to read this page before you reach into your pocket. This informative article explains how it's nearly impossible for monitors to truly resolve high-resolution signals at the specified dot pitch within the advertised active display area. The page also has a link to a very useful table of monitor specifications that provides the active display areas; dot pitch; bandwidth; horizontal and vertical sync rates; and maximum resolution.

URL:

> http://hawks.ha.md.us/hardware/monitors.html

Want to see some cool resources? Check out Internet Resources.

Disk Controller Manufacturers

This page lists the names of literally hundreds of manufacturers of floppy and hard drive controllers and related products. Each list entry links to a page presenting the physical address and phone numbers for the company. Perhaps, in the future, these pages will also have links to those companies on the Web.

URL:
> http://theref.c3d.rl.af.mil/manufacturers/
> m_boca_research.html

Fixing Gamma

If you are not happy with the way your computer displays digital images, find out how to adjust the gamma of your monitor. Because there isn't a standard gamma setting for displays, play it safe and read this FAQ to decide how to optimize the images on your screen.

URL:
> http://photo.net/philg/photo/fixing-gamma.html

Floppy and Hard Drive Cable Diagrams

So you got cocky and thought you could add that second hard drive yourself. Now you've got an open box with so many wires and cables snaking around that it looks like Indiana Jones' snake pit. Don't despair, though; help isn't far away (if you still have access to a Web browser). Pull up the TechTalk page on TheRef for all the nitty-gritty details about connecting controllers, floppy drives, and hard disks. (Maybe you should print the page out before you start that project.)

URL:
> http://theref.c3d.rl.af.mil/tech_talk/
> tech_talk01.html

Hard Disk Drive Manufacturers

This page sports a comprehensive list of hard disk drive manufacturers. Under each manufacturer's link are links to each model of drive the manufacturer makes. Finally, following a link for a drive model takes you to a page with the full specifications of the drive. Information presented includes the physical and logical characteristics of the drive; its rated performance; and other specs, such as M.T.B.F., power consumption, and dimensions.

URL:
> http://theref.c3d.rl.af.mil/hard_drives/
> h_7345s_cheyenne.html

Intel Technology Briefing Index

When you're looking for up-to-date information about computer technology, begin your search at the horse's mouth. Intel offers a collection of excellent outlines on current technology, including topics such as the Pentium and Pentium Pro (P6) processors, mobile Pentium processors, the Intel CPU architecture, PCI local bus, making Plug and Play devices, Indeo video technology, and the CPU manufacturing process. These articles are excellent introductions to these technologies. They're illustrated and written in non-technical terms that anyone can understand.

URL:
> http://www.intel.com/product/tech-briefs/

Matrox Millennium (Unofficial) Page

David Jeske likes his Matrox Millennium video card so much that he's put up this web page to extol the Millennium's virtues and otherwise disseminate information about this excellent 3D video card (at least until Matrox gets its act together and gets onto the Web). David describes the Millennium models, including technical specs; he shows what comes in the box; and he tells of his own experiences with the Millennium card.

URL:
> http://www.cen.uiuc.edu/~jeske/
> MatroxMillenium/

Mwave Technology

Using on-board Digital Signal Processing (DSP) techniques, Mwave technology can combine a variety of telephony and audio functions onto a single card. This introduction describes the Mwave technology and some of its benefits and applications in today's hardware market.

URL:
> http://watson.mbb.sfu.ca

> **Going shopping on the Web? Remember to pick up some ecash.**

PC Hardware Discussion Groups

Check out these pages if you're looking for answers beyond tech support for your video card or wondering if you have what it takes to build your own PC from scratch. The experts in these and other PC hardware fields lurk on these newsgroups regularly.

URLs:

    news:comp.sys.ibm.pc.hardware
    news:comp.sys.ibm.pc.hardware.cd-rom
    news:comp.sys.ibm.pc.hardware.chips
    news:comp.sys.ibm.pc.hardware.comm
    news:comp.sys.ibm.pc.hardware.misc
    news:comp.sys.ibm.pc.hardware.networking
    news:comp.sys.ibm.pc.hardware.storage
    news:comp.sys.ibm.pc.hardware.systems
    news:comp.sys.ibm.pc.hardware.video

PC Hardware Introduction

Buying a new computer can be a confusing and mysterious process. Terms like SX and DX, and acronyms like IDE, SCSI, and SVGA make it all the worse. Before you dig into your pocket, arm yourself with some information. Learn about the basics of PC hardware and the important differences in design and architecture between different computer systems. This illustrated tutorial introduces topics like CPUs and what their different speeds mean to you, memory and virtual memory, disk caches, the buses, video adapters, and disks drives.

URL:

    http://pclt.cis.yale.edu/pclt/pchw/platypus.htm

PCI Local Bus Primer

Read about why the PCI is the premier local bus technology in use today and why it's likely to stay that way for some time to come. This interesting history and non-technical overview of the PCI input/output bus is provided by one of the major players in the development of the PCI bus—Intel.

URL:

    http://www.intel.com/product/tech-briefs/
    pcibus.html

IBM vs. The World
For all its technological superiority and snobbery, where's the MCA bus today? If it were really all that good, wouldn't we all be using MCA cards instead of the blindingly fast plug-and-play PCI cards we use today? Read about the PCI bus and why it's so great on Intel's PCI Local Bus Primer.

RZ1000 IDE Controller Problems with PCI

Some early PCI motherboards that use the integrated RZ1000 IDE controller chip have problems with corrupting data on IDE hard drives, and even possibly damaging disk partitions. On this page, Intel describes the problem, notes affected operating systems, and explains workaround procedures (which mostly involve obtaining updated drivers or patches). Also available on this page is a DOS program that detects the RZ1000 IDE controller, and phone numbers to call for more help identifying and correcting this problem.

URL:

    http://www.intel.com/procs/support/rz1000/

SCSI Frequently Asked Questions

The SCSI FAQ is the standard FAQ for the Usenet newsgroup **comp.periphs.scsi**. The FAQ list is posted to the newsgroup periodically, but you can read it anytime on the Web. This FAQ list is an excellent primer for learning about SCSI technology. It starts at the very beginning—"What Is SCSI?"—and gradually gets as technical as you could hope. The FAQ is divided into two volumes, both of which are a click away from this page.

URL:

    http://www.cis.ohio-state.edu/hypertext/faq/
    usenet/scsi-faq/top.html

A
B
C
D
E
F
G
H
I
J
K
L
M
N
O
P
Q
R
S
T
U
V
W
X
Y
Z

There's nothing scuzzy about SCSI literature. Small Computer System Interfaces are very intricate in their technology, and you don't have to feel scuzzy reading about them either. (Just don't tell your mother!)

Sound Blaster AWE32: The Naked Truth

If you want the truth about Creative Labs' AWE32 sound board, you've got to see this page. Mathias Hjelt has dissected and examined the AWE32 in every possible way. He's taken it right down to the bits and bytes and has put it back together again. Mathias offers an introduction to the AWE hardware; truths and tricks about getting the best sound from (and into) the board; personal advice; and pointers to software archives and other resources of interest to AWE owners.

URL:

http://spider.compart.fi/~mhjelt/awe/

SPEC Benchmarks

SPEC (Standard Performance Evaluation Corporation) is a nonprofit organization formed by a number of computer and technology companies to develop standardized methods for evaluating the relative performances of computers, CPUs, and components. The SPEC benchmark is quickly gaining a foothold as the standard benchmark. From this page, you can read the SPEC FAQ list, read details about the benchmark, peruse SPEC summaries, and read back issues of the SPEC newsletter.

URL:

http://performance.netlib.org/performance/
html/spec.html

TheRef

TheRef (also known as Bob) maintains this incredible web of pages presenting specifications and other information about a wide variety of peripherals. TheRef has cataloged hundreds of disk drive controllers, hard disk drives, floppy drives, and optical drives. He maintains this server on his own time and at his own expense. Nevertheless, this is a first-class repository of technical information on these devices.

URL:

http://theref.c3d.rl.af.mil/

Tommy and Biff's 3D Accelerator Page

So what is all this stuff about 3D video cards? What do they really do? And do they really do it better or faster than standard cards? For the answers to these and any other questions you have about 3D video accelerators, come to Tommy and Biff. Tommy and Biff's pages provide links to the PC 3D Graphics Accelerators FAQ, information on 3D accelerator chip sets, the (unofficial) Matrox Millenium page, and a collection of links to more information about 3D games and resources.

URL:

http://www.cowboy.net/~biff/

Turtle Beach Users Group

Turtle Beach is a manufacturer of sound boards and piggyback "daughter boards" for other manufacturer's sound boards. One of Turtle Beach's popular products is the "Rio" board that adds wavetable synthesis capabilities to Creative Labs 16-bit sound boards. At this page, the home for Turtle Beach users, you can download useful utility files; read the group's FAQ list; check out specific products, such as the Maui sound board; and hyperlink directly to Turtle Beach's web.

URL:

http://www.cs.colorado.edu/~mccreary/tbeach/

Ziff-Davis Benchmark Operation

The Ziff-Davis Benchmark Operation develops and supports the benchmark programs used by all Ziff-Davis publications. (Ziff-Davis publishes computer magazines, including *PC Magazine*, *PC Week*, *MacUser*, *PC Computing*, *MacWeek*, *Windows Magazine*, and so on.) Available here are software packages for testing computers, and components like video cards and disk drives for a variety of computers. The packages are WinBench 95, Winstone 95, PC Bench, MacBench, NetBench, and ServerBench.

URL:

http://www.zdnet.com/~zdbop/

HERPETOLOGY

American Federation of Herpetoculturists

"Reptiles have silly grins, dark eyes, and shiny teeth. Reptiles all have pretty scales, they run real fast and like to bite," sings Oingo Boingo's Danny Elfman. See if this embellishment is true by visiting the AFH page for selected articles from the federation's magazine, *The Vivarium*. There's a herpetological classified ad section, and you'll find a list of upcoming events and online photos of some favorite amphibians and—reptiles.

URL:
http://www.reg.uci.edu/AFH/

American Society of Ichthyologists and Herpetologists

The ASIH is devoted to all things slippery and slithery. Here you'll find information about the organization, news and announcements, and conference updates. There are links to other ichthyological and herpetological societies, both online and off.

URL:
http://www.utexas.edu/ftp/depts/asih/

Bibliography of Crocodilian Biology

Corral your yearning for crocodile learning with this compendium of croc references ranging from paleontology and ecology through molecular biology. The bibliography currently contains over 2,500 references on crocodiles, alligators, caimans, and extinct crocs of various sorts (down to Crocodyliformes).

URL:
http://www.welch.jhu.edu/homepages/mmeers/html/croc.bib.cover.html

California Academy of Sciences Department of Herpetology

Herp on over to the CAS Gopher for information on the department and personnel, grants, and funding. You'll also find an online species catalog and collections database. Best of all, if you're a qualified researcher, see how the department can loan you herpetological specimens from its collections.

URL:
gopher://cas.calacademy.org/11/depts/herp/

The next time your dog mistakenly believes that your carefully taxidermied gecko specimens are chew toys, you may wish to check out the California Academy of Sciences Department of Herpetology loan program for qualified researchers. Of course, if you tell them your sad story, you run the risk of quickly becoming *unqualified*.

Crocodile Specialist Group

Crocodile Specialist: an occupation that'll look great on your business card. CSG is a worldwide network of biologists, wildlife managers, government officials, researchers, and others dedicated to the conservation of alligators, crocodiles, caimans, and gharials.

URL:
http://www.flmnh.ufl.edu/docs/departments/crocs.htm

Crocodilian FAQ

Crocodilians play a valuable role in conserving wetland ecosystems by increasing nutrient recycling, maintaining aquatic refugia for other wetland species during droughts, and keeping waterways open. They are the largest predator in their habitat, and this FAQ will show you—among other things—why alligators, caimans, gavials, and crocodiles are amazing creatures.

URL:
http://gto.ncsa.uiuc.edu/pingleto/crocFAQ.html/

Looking for Earth Science resources? Check Earth Science, Ecology, Geography, and Geology.

Crocodilians on the Net

If you like crocs, gators, and geckos, you'll find nirvana here. The Crocodilians on the Net page is crawling with links for resources on alligators and crocodiles. Consult the Croc FAQ to learn how to distinguish between alligators, caimans, gavials, and crocodiles. Read about their temperaments, eating habits, and how they live. There's even a link to southern Florida's oldest alligator farm.

URL:

> http://geowww.geo.tcu.edu/faculty/
> crocodile.html

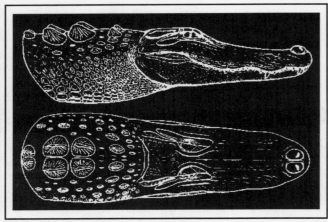

Crocodile or alligator? Find out at *Crocodilians on the Net.*

Genbank Sequence Data Organized by Taxon

Here's a huge, searchable, technical database containing taxonomical information on hundreds of thousands of entries for reptiles and amphibians. You can conduct searches on such subjects as attributes of protein organisms and nucleotide organisms according to journal, author, subject, gene symbol, and many other ways.

URL:

> http://xtal200.harvard.edu:8000/herp/
> genbank.html

Herpetocultural Home Page

The landlord said your lizard had to go. So he ate him. You know, with all the rent money you'll save you can afford that Komodo Dragon you've always wanted. Here's a practical page on the art and science of raising and studying reptiles and amphibians. There's a herpetofaunal life list, a large guide to keeping iguanas in captivity, a large photo gallery, and links to many other related herpetological sources.

URL:

> http://gto.ncsa.uiuc.edu/pingleto/herp.html

> # Like lizards? Check out Herpetology.

Herpetologists' League

Established in 1936, HL is devoted to the study of amphibian and reptile biology. The League plans to have an online newsletter available in the near future; but in the meantime, you'll find information on membership, meeting notices, publications, and League programs.

URL:

> http://www.ecs.Earlham.edu/htmldocs/
> departments/biology/HL/HLmainpage.html

Herpetology Journals

Sssssslither your way to several online herpetological journals with subjects ranging from snakes to salamanders. You won't be ssssorry.

URL:

> http://xtal200.harvard.edu:8000/herp/general/
> journals/journals.html

Herpetology-Texas Natural History Collections

The state of Texas is home to many fine herps, and you can check out approximately 53,000 cataloged specimens in this collection at the University of Texas. At some point in the future, you'll also find skeletal and preserved specimens, and audio recordings of the sounds some of your favorite herps make.

URL:

> http://www.utexas.edu/depts/tnhc/.www/herps/

Herpmed Communications

The folks at Herpmed provide a kind of general clearinghouse for information on reptilian and amphibian research and education. Among many other useful and interesting entries and links, you'll find worldwide snakebite emergency first-aid information, a zoological online ordering center, a venom supplier, and online herpetology booksellers.

URL:

> http://www.xmission.com/~gastown/herpmed/

Excerpt from the Net...

(from the Taming and Training Your Reptile FAQ at Trendy's House of Herpetology)

Questions and Answers

Q: Can reptiles become tame ?

   Short answer: YES

   Long answer: Most will, some won't. Reptiles have not been categorized and bred for gentle temperament, even though some heavily captive bread species, like corn snakes, seem mellower than wild caught specimens.

   Some species/individuals are more likely to get used to than others.
Some species/individuals will get non-shy and eat from your hand but will never let you touch them. Some species/individuals will tolerate handling in return for food, others will hate it.

   Some species/individuals will run, others will bite....and some will be really tame and develop a taste for having their neck scratched.

When buying a reptile, if you want it to become tame, buy and animal that seems calm and friendly. Unless you are prepared to invest a lot of time and effort without promise of success, don't go for the more 'challenging' animal.

Some species start out friendly as babies and will develop a rotten temper when they get older. Reticulated Pythons have that reputation. But several people state that some reticulated pythons are very well tempered, and that maybe they become less mellow not so much as a character trait, but because they are handled less when they get large.

Some species, no matter how tame, can become dangerous when they get bigger. They can hurt you, when they get angry or scared. And they can potentially kill you, if they mistake you for food.

No matter how tame your animal gets, expect to get bitten or scratched, occasionally, if the animal is upset or scared. Use tongs for biters, and gloves.

Remember: ALL REPTILES ARE WILD ANIMALS. Treat them as such, and you can avoid trouble.

Q: Do reptiles like to be petted ?

Some do, some don't. I know several boas who love to be petted, even on their head, after they develop a taste for it. And that is the key: many reptiles don't take naturally to being petted. But once they get used to it, they often start to like it.

Many snakes will like to cuddle in warm, dark places, like T-shirts. If your idea of petting is an animal curled up on your stomach inside your shirt while you are reading a book, then a snake is a good pet for you.

Turtles and tortoises are not cuddly, but many like their belly scratched, i.e. the bottom of their plastron, and they will definitely express their pleasure. "I had a chance to pet some young Galapagos tortoises in June. They go off to Never-Never Land if you scratch the jaws and neck."

Q: How do I know, my animal likes to be petted ?

Your animal likes to be petted if it:

      does not run away given the option
      does not try to avoid your touch
      goes to sleep in your arms
      stays around for more if you stop
      visibly relaxes as you go
      doesn't hiss or bite

Interactive Guide to Massachusetts Snakes

Identify all 14 types of snakes indigenous to Massachusetts on this handy page. A series of questions leads you through a process of elimination to determine which snake you've seen or are seeing. If the snake looks back, have you been snake-eyed?

URL:

http://klaatu.oit.umass.edu/umext/snake/

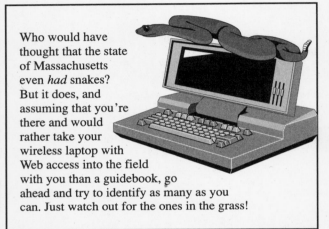

Who would have thought that the state of Massachusetts even *had* snakes? But it does, and assuming that you're there and would rather take your wireless laptop with Web access into the field with you than a guidebook, go ahead and try to identify as many as you can. Just watch out for the ones in the grass!

Seedman's Herpetocultural Hideaway

This page is closely linked to the Iowa Herpetological Society and contains information on herpetology and herpetoculture (reptiles and amphibians and how to keep them). In addition, you'll find the VARANid Information eXchange (VARANIX), an organization devoted to the study and research of monitor lizards.

URL:

http://www.public.iastate.edu/~seedman/

Society for the Study of Amphibians and Reptiles

SSAR was founded in 1958 to advance research, conservation, and education about amphibians and reptiles. This society is recognized worldwide for its diverse program of publications, meetings, and other activities. You can peruse the society's history and preview contents of *Herpetological Review*.

URL:

http://falcon.cc.ukans.edu/~gpisani/SSAR.html

Trendy's House of Herpetology

If you're not afraid of reptiles and amphibians that make your skin crawl or cause your blood to run cold, enter Trendy's House for interesting reading and excellent photographs of snakes, lizards, scorpions, and more.

URL:

http://fovea.retina.net/~gecko/herps/

World Wide Web Virtual Library—Herpetology

Link up with like-minded lovers of amphibians and reptiles at this page that features a cornucopia of herpetological organizations, discussion groups, books, journals, virtual field guides, online and off-line museums, meetings, and announcements.

URL:

http://xtal200.harvard.edu:8000/herp/

HISTORY: AMERICAN

The 1960s and Beyond

Turn on, tune in, and drop back to the '60s with this timeline of significant events in the counter-culture movement and the Vietnam conflict. Relive the days of baby boomers, bohemians, beatniks, hippies, the Chicago 7, and the whole far-out era.

URL:

gopher://gopher.well.sf.ca.us/11/Community/60sTimeline

American Civil War: Resources on the Internet

Emancipate yourself with this huge collection of pointers to virtually every resource on the Web that pertains to the Civil War. We're not whistling Dixie.

URL:

http://www.dsu.edu/~jankej/civilwar.html

American Memory

Celebrate the land of the free and the home of the brave! This site is comprised of collections of archival and source material pertaining to American culture and history. These exhibits are all part of the Library of Congress's national digital library, and include wonderful text, images, and photographic presentations of the American experience.

URL:

http://rs6.loc.gov/amhome.html

American Studies

Visit the American experience online with this excellent all-purpose directory of American Studies resources. Besides American history, material culture, economy, and politics, this site also has an exceptional selection of race and ethnicity resources, covering African-American, Asian, Latino, and Native American Studies.

URL:

    http://pantheon.cis.yale.edu/~davidp/
    amstud.html

What was so civil about the Civil War? Don't wait 'til Johnny comes marching home to find out! Shoot on over to American Civil War: Resources on the Internet.

The Cuban Missile Crisis

Don't blink, but this resource documents the famous crisis in the early 1960s between the U.S. and the U.S.S.R. from a British perspective. Read a contemporary analysis of the crisis from primary and secondary sources; and find references to a select bibliography and a list of additional resources, books, and articles.

URL:

    http://www.paranoia.com/~az/cuba/

Cuban Missile Crisis
America and Russia got into a fight,
Over missiles to Cuba arriving at night.
Kennedy threatened and Kruschev blinked,
Russia backed down and America winked.

The French and Indian War

Explore this somewhat-forgotten war that led to a more important one—America's freedom from British rule. With an impressive compilation of minutiae on the major turning points in the Seven Years War, this page also highlights major battles, soldier lists, troop movements, casualty statistics, and related historic sites to visit.

URL:

    http://web.syr.edu/~laroux/

History of the NW Coast

Read first-person accounts of explorers to the northwest coast of America in this collection taken from journals and ethnographic papers. This history of European/Indian contact spans the years from 1778 to around 1900, and includes Captain Cook's expedition and excerpts from Franz Boas' account of the northwest major Indian groups.

URL:

    http://www.hallman.org/~bruce/indian/
    history.html

Images from the Philippine-United States War

Here are approximately 40 images of the Spanish-American War. You'll find photographs and portrait likenesses of major and minor figures, incidents, and battles. The site contains links to other Philippine and Spanish-American War sites.

URL:

    http://www.msstate.edu/Archives/
    History/USA/filipino/filipino.html

JFK Assassination Fascination

The Kennedy assassination has haunted America for over 30 years, and many people seem to believe that the only person who didn't kill the president was Lee Harvey Oswald. You can download sound files from that terrible day in 1963, and read this conspiracy theory that tags the CIA, the FBI, the Mafia, paramilitary groups, anti-Castro Cuban exiles, the French Secret Army Organization, Israeli intelligence, and Corsican assassins. How come no one ever suggested that it might have been Lyndon Johnson?

URL:

    http://smartnet.net/jfk/snake.html

A B C D E F G H I J K L M N O P Q R S T U V W X Y Z

Martin Luther King, Jr., Papers Project

Do you have a dream? Visit this online civil rights worker's page to learn more about his life, his times, and his vision. You'll find a comprehensive bibliography of King-related materials and images, and access to full texts of King's speeches and related audio and video resources.

URL:

http://www-leland.stanford.edu/group/King/
mlkpp.htm

National Trust for Historic Preservation

Saving America's diverse historic environments and preserving the livability of communities nationwide is the goal of the National Trust. Find out about travel to various American historic destinations, historical research, a calendar of events, and other historical resources.

URL:

http://www2.nthp.org/trust/

The Papers of George Washington

Old George may have made some horrendous military decisions (and had wooden dentures), but he still crossed the Delaware, won the Revolutionary War, and is more famous than his British counterpart, General Cornwallis. Here you can read about America's first president and the "father of his country" in selected papers and letters written by him and to him.

URL:

http://poe.acc.Virginia.EDU/~gwpapers/
GWhome.html

Prohibition Materials

Curl up with a nice ginger ale and relive the arid days of axe-toting, saloon-smashing Carry Nation and that dry time known as Prohibition. Drink in images and text from this 1920s movement, including the Ohio Dry Campaign of 1918 and the Alcohol and Temperance History Group. Cheers!

URL:

http://www.cohums.ohio-state.edu/history/
projects/prohibition/

We cannot tell a lie. Once, when George Washington was sitting with his back to a fire, the heat became so intense that he had to move away from it. Someone called out that a general should be able to withstand fire. "But it doesn't look good if he receives it from behind," was General George's reply. Get fired up and peruse the Papers of George Washington.

Vietnam-era U.S. Government Documents

Peruse the Senate Select POW-MIA Affairs Report, a State Department White Paper on Vietnam, and the Tonkin Gulf Message and Resolution. There are documents from the Paris Peace Talks, as well as other papers on the Vietnam conflict.

URL:

gopher://wiretap.spies.com/11/Gov/US-History/
Vietnam

Women's Suffrage

It's hard to believe that less than a century ago, women in America did not have the right to vote. In these enlightened times, you can learn more about the pioneers of the women's movement and the history of the suffragettes. Elizabeth Cady Stanton and Susan B. Anthony are highlighted on this informative page.

URL:

http://www.history.rochester.edu/class/suffrage/
home.htm

HISTORY: WORLD

1492 Exhibit

Explore the pros and cons of Columbus' famous voyages to the New World. This exhibit examines Native American and European ways of life before and after contact, and it discusses the roles of other European explorers, conquerors, and settlers from 1492 to 1600.

URL:

http://sunsite.unc.edu/expo/1492.exhibit/
Intro.html

The Aboriginal Page

If the beginning of history is defined as the debut of written language, then this definition pales in comparison to the Australian aborigines. Perhaps 60,000 years before the dawn of conventional "history," Australian aboriginal people were creating paintings unmatched until those of Lascaux Cave in France—some 35,000 years into the future! Read and learn more about their history, society, culture, and phenomenal artwork here at this server.

URL:

http://www.vicnet.net.au/vicnet/COUNTRY/
ABORIG.HTM

Archives of the Soviet Communist Party and Soviet State

Throughout most of its history, the Soviet state was a one-party monolith, run by the Communist Party. This catalog of documents and "finding-assistants" is part of a joint project of the Russian State Archival Service and the Hoover Institution to provide approximately 25 million sheets of records and archival documentation on microfilm.

URL:

http://hoover.stanford.edu/www/archives/
front.html

Comrade, relive those salad days of Lenin, Trotsky, and Stalin at the Archives of the Soviet Communist Party and Soviet State. Better hold the salad, though, and pass the cabbage instead!

Armenian Research Center

Experience the dramatic and often-tragic history of Armenia and the struggles of the Armenian people by visiting this site at the University of Michigan. It contains information about the sociocultural makeup of Armenia and Karabaugh; provides details regarding the Armenian genocide; and explains the present-day struggles of the Nagorno-Karabaugh region.

URL:

http://www.umd.umich.edu/dept/armenian/

British Columbia History

Research British Columbian history and historical associations, organizations, and history publishers on this gateway page. You'll find online periodicals, magazines, exhibits, and selected British Columbiana.

URL:

http://www.freenet.victoria.bc.ca/bchistory.html

Canadian History Exhibition

Oh, Canada! Study the formations and foundations of this world power. A rich historical timeline from the era of the Vikings up until the 1890s is presented in the Phase 1 project of this impressive Web exhibit, sponsored by the Canadian Museum of Civilization.

URL:

http://www.cmcc.muse.digital.ca/cmc/
cmceng/canp1eng.html

The Chechen Homeland

Read numerous articles and book excerpts on the history of Chechnya, an autonomous republic of the U.S.S.R., and decide for yourself if the Russians had a right to invade the country. There is a great deal of information pertaining to historical travel accounts and the history of the Chechen people.

URL:

http://www.smns.montclair.edu/~chechen/

**Need some great music?
Lend your ear to the Music:
MIDI section.**

A B C D E F G H I J K L M N O P Q R S T U V W X Y Z

Constantinople and Russia in the Early Eleventh Century

The Dark Ages weren't so dark after all—and thanks to this excellent treatise, you'll be further enlightened about the Byzantine Empire and Russia.

URL:
   http://www4.ncsu.edu/eos/users/s/sfcallic/SCA/
   PERSONA/Reality.html

Cybrary of the Holocaust

Visit this online memorial to the Jewish Holocaust during World War II. In addition to historical perspectives and permanent sections on Hitler's "final solution," there are changing monthly exhibits and excerpts from recently published books. *Shalom.*

URL:
   http://www.best.com/~mddunn/cybrary/

Eighteenth Century Resources

This site is the closest vehicle to a time machine that you'll encounter on the Net for eighteenth century studies. You'll find resources for eighteenth century history, literature, philosophy, art, architecture, music, and more.

URL:
   http://www.english.upenn.edu/~jlynch/18th.html

Eighteenth Century Studies

Link up here with hundreds of eighteenth century works, including novels, plays, memoirs, and poems of the period. There is information on subscribing to the eighteenth century mailing and discussion lists.

URL:
   http://english-www.hss.cmu.edu/18th.html

Empires Beyond the Great Wall: The Heritage of Genghis Khan

Saddle up and ride with the Mongol Horde toward conquest and history with this online exhibit from the Royal British Columbia Museum. There are many photographs of artifacts, and excellent supporting text of the life and culture of Genghis Khan.

URL:
   http://vvv.com/khan/

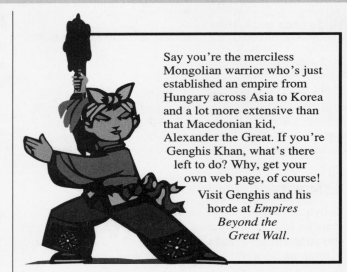

Say you're the merciless Mongolian warrior who's just established an empire from Hungary across Asia to Korea and a lot more extensive than that Macedonian kid, Alexander the Great. If you're Genghis Khan, what's there left to do? Why, get your own web page, of course!

Visit Genghis and his horde at *Empires Beyond the Great Wall.*

English Literature: Renaissance

Much as the mythological Phoenix bird arose from its own ashes, the Renaissance allowed humankind to be reborn from the Dark Ages. Read from this collection of English and Italian Renaissance works, and marvel at your own rebirth.

URL:
   http://humanitas.ucsb.edu/shuttle/eng-ren.html

The Fall of Singapore

The greatest military defeat in British history probably occurred at Singapore on February 15, 1942. Eighty-thousand British Empire and Commonwealth troops surrendered in humiliation to the Japanese army in a campaign that captured the "impregnable" fortress in only 70 days. Explore the timelines by following day-to-day events of the Malayan Campaign; read an authoritative account of the war from a reference textbook; view an image archive of the Battle of Malaya; and hear a sound archive of WWII radio broadcasts and excerpts of wartime speeches.

URL:
   http://www.ncb.gov.sg/nhb/dec8/war.html

H-LATAM

Become a lover of Latin American history. There is an extensive amount of book reviews, topically pertinent journals, Latin American information, and bibliographies at this site. You'll also find a list of scholarly organizations that you can join.

URL:
   gopher://gopher.uic.edu:70/11/research/history/
   hnetxx/40227007

Historical Text Archive

"History is the sum total of everything that could have been avoided."—cheery hindsight commentary from Konrad Adenauer. Start yourself up by learning more about the vast compilation of human foibles, facts, and fantasies. Click on a continent to learn how that part of the world came to be, in spite of itself.

URL:
http://www.msstate.edu/Archives/History/

History Index

Search the World Wide Web over, and you may never come up with a better index than this pointer to virtually everything of historical interest in world history. Learn it, do it, say it, write it, and keep up the faith.

URL:
http://kuhttp.cc.ukans.edu/history/index.html

History of Computing

Point your Pentium over to this site for a sobering moment to learn just how computers came to be. The machine that changed you and the world has a history of its own. Among a multiplex of bits and bytes in computer archaeology presented here, see the Babbage machine's original nemesis—the first actual computer bug on exhibit at the Smithsonian.

URL:
http://ei.cs.vt.edu/~history/

History of Marrakesh

Won't you take me on the Marrakesh Express? The city of Marrakesh conjures up magnificence and mystery in the mind's eye, and Morocco adapted its own name from it. Come visit the history of this imperial city and meet some of its kings, sages, poets, and beauties.

URL:
http://www.dsg.ki.se/maroc/marrakesh/

History of Science, Technology, and Medicine

Here's a thoroughly entertaining site. Its interdisciplinary nature makes for enjoyable browsing, and it can be categorically searched by subject arrangement. The variety ranges from the history of cartography to perspectives of the Enola Gay and a study of Greek scientists. Live it up!

URL:
http://www.asap.unimelb.edu.au/hstm/
hstm_ove.htm

History Reviews Online

Have you heard about that new book on Scottish monasteries in the Late Middle Ages, but want to read what others think of it first? Look no further than this site that reviews selected books and other works on world history.

URL:
http://www.uc.edu/www/history/reviews.html

Jesuits and the Sciences

The Jesuits may have been the instigators of the Inquisition, but they were also very involved in the sciences. See what they did on this well-planned page with excellent text and graphics that covers the history of the order from 1540 to the present.

URL:
http://www.luc.edu/libraries/science/jesuits/
jessci.html

Joan of Arc

Joan of Arc is synonymous with the history of France and its medieval struggle to defeat the English. At her death, those who were in attendance reported that a white dove was seen flying out of the flames in which she perished. Read about the life of this mystic female warrior and the lives and deaths of two of her contemporaries—William Rufus and Saint Thomas Becket.

URL:
http://www.cs.cmu.edu/afs/andrew.cmu.edu/
org/Medieval/www/src/contributed//
yapm97a@prodigy.com/jnofarc.html

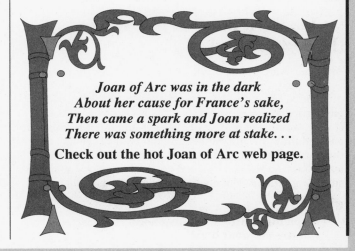

*Joan of Arc was in the dark
About her cause for France's sake,
Then came a spark and Joan realized
There was something more at stake. . .*
Check out the hot Joan of Arc web page.

A
B
C
D
E
F
G
H
I
J
K
L
M
N
O
P
Q
R
S
T
U
V
W
X
Y
Z

Korea 1935-1955: An Eyewitness Account

Read a large and mesmerizing account of one Korean's personal history and life throughout the Korean War and during its aftermath. This resource contains photos and excerpts from recently declassified U.S., Chinese, and Russian documents. It is a remarkable document that anyone interested in the Korean War should access. The page also contains a great deal of information about North Korea spying operations and the CIA.

URL:
http://www.kimsoft.com/korea/eyewit.htm

Korean War Project

Learn about the Korean War through this dedication project that encompasses Korean War history, timelines, resources, photographs, maps, and veterans groups. Relive the war from Panmunjom to the Punchbowl to Pork Chop Hill. This resource even includes elaborate tourist information on travel to North Korea.

URL:
http://www.onramp.net/~hbarker/

Labyrinth

You don't have to be a minotaur to roam through Labyrinth—linking medieval history servers around the world. Labyrinth includes a searchable index of professional directories, online bibliographies, and information on specific subjects grouped by geographic region. Read full-text versions of medieval works in Latin, French, Italian, and Old and Middle English. Arthurian Studies, Manuscripts, Medieval Music, Vikings, and Runes and Norse Culture are part of the experience.

URL:
http://www.georgetown.edu/labyrinth/
labyrinth-home.html

Les Tres Riches Heures du Duc de Berry (The Book of Hours)

Ah, what more apt example of *noblesse oblige* than the Duc de Berry, who simply commissioned his very own illustrated book while his humble serfs toiled endlessly to pay for it. This very rich book of hours is just so—*the* classic example of a collection of text and immaculately painted illustrations wedded to each hour of the day for the entire year. View some of its famous paintings and learn how the fruits of the serfs made this book one of the surviving pinnacles of medieval masterpieces.

URL:
http://humanities.uchicago.edu/images/
heures/heures.html

Mil-Hist

War is hell, so experience it from the comfort and safety of your computer screen with this site that covers world military history from ancient times to the present day. You'll find a great deal of military images and primary source material consisting of speeches, letters, diaries, and official reports. The material is grouped by periods, including pre-nineteenth century; the Civil War; history through 1919; the inter-war years and World War II; Vietnam; and 1973 to the present.

URL:
http://kuhttp.cc.ukans.edu/history/milhst/
m_index.html

Operation Desert Storm Debriefing Book

Relive the recent Gulf War with this Desert Shield and Desert Storm page. Meet all the major players, including former President Bush, Colin Powell, Arnold Schwarzkopf, and the infamous Iraqi despot, Saddam Hussein. Elaborate sections cover operation summaries, the armies, the aircraft, and the amazing weapons that ultimately won the war and freed Kuwait. There are also links to other Desert Storm pages.

URL:
http://www.nd.edu/~aleyden/contents.html

Plague and Public Health in Renaissance Europe

There was virtually no aspect of European society that was not affected by the plagues that ravaged the continent and beyond during the Middle Ages. The Black Death, as it was known, consumed perhaps one-third to one-half of Europe's population in only two years. Read more about this calamity on this page that presents a hypertext archive of narratives, medical information, governmental records, religious writings, and images that portray this dreadful scourge.

URL:
http://jefferson.village.virginia.edu/osheim/
intro.html

Russian and East European Studies

REESweb provides a convenient compendium of links to many pages of historical and political information on Russia and its former satellite republics. Many of these emerging nations have established home pages, and this site offers one of the easiest ways to take the pulse of activities in parts of the world that are rarely covered by standard news services.

URL:
http://www.pitt.edu/~cjp/rees.html

Excerpt from the Net...

(from Plague and Public Health in Renaissance Europe)

Bubonic Plague In Renaissance Europe

The coming of the Black Death, when in just two years perhaps one third to one half of Europe's population was destroyed, marks a watershed in Medieval and Renaissance European History. Bubonic plague (Yersinia pestis) had been absent from Western Europe for nearly a millennium when it appeared in 1348. The reaction was immediate and devastating. Up to two thirds of the population of many of the major European cities succumbed to the plague in the first two years. Government, trade and commerce virtually came to a halt. Even more devastating to Europeans, there was hardly a generation which did not experience a local, regional or pan-European epidemic for the next two hundred years. There was virtually no aspect of European society that was not affected by the coming of plague and by its duration. At the most basic level, recurrent plague tended to skim off significant portions of the children born between infestations of plague, dampening economic and demographic growth in most parts of Europe until the late seventeenth century. The responses of Europeans are often treated as irrational or superstitious. Yet medical tracts, moral treatises and papal proclamations make clear that for most Europeans there were, within the medieval world view, rational explanations for what was happening. Plague stimulated chroniclers, poets and authors, and physicians to write about what might have caused the plague and how the plague affected the population at large the framing story of Boccaccio's Decameron is merely the most famous of the writings. Nonetheless, in the wake of the first infestations there were attacks on women lepers and Jews who were thought either to have deliberately spread the plague or, because of their innate dishonor, to have polluted society and brought on God's vengeance. The violence against outsiders demonstrated, in a tragically negative manner, the nature and the limits of citizenship in Europe. This was a society which defined itself as Christian and recurrent plague changed religious practice, if not belief. Christians had long venerated saints as models of the godly life and as mediators before God, in this case an angry and vengeful one. A whole new series of "plague saints" (like St. Roch) came into existence along with new religious brotherhoods and shrines dedicated to protecting the population from plague. The recurrence of plague also affected the general understanding of public health. Beginning in Italy in the 1350s there were new initiatives aimed at raising the level of public sanitation and governmental regulation of public life. And, finally, by the sixteenth century a debate over the causes of plague spread in the medical community as old corruption theories inherited from Greece and Rome were replaced by ideas of contagion. The story of plague in Renaissance society is not merely a medical, religious or economic subject. To properly understand the impact of plague it is necessary to consider almost all aspects of society, from art and music to science.

The Siege and Commune of Paris, 1870-1871

This page presents an enormous amount of reference and research material on the Paris commune and the disastrous events that befell it. There are links to over 1,200 digitized photographs and images recorded during this period, along with 1,500 caricatures, newspaper articles, and referrals to hundreds of books and pamphlets.

URL:
http://www.library.nwu.edu/spec/siege/

Reach for the stars. Gaze into Astronomy and Physics: Astrophysics.

A
B
C
D
E
F
G
H
I
J
K
L
M
N
O
P
Q
R
S
T
U
V
W
X
Y
Z

Significant Events from 1890 to 1940

Travel back in time by clicking your favorite year and seeing who was born, who died, who created what, and what happened in general.

URL:

http://weber.u.washington.edu/~eckman/
timeline.html

Slovak Document Store

Wouldn't you know it? Just when you had finally learned how to spell Czechoslovakia, the country split up into two separate entities. This is a home page that focuses on educating visitors about the cultural distinction between Slovakia and the more prominent Czech Republic. It provides text and images of historic milestones for the region and a history of Bratislava; maps; and country, culture, and tourist information.

URL:

http://www.eunet.sk/slovakia/slovakia.html

Stories of the Irish Potato Famine

The Great Potato Famine in Ireland changed the course of world events, both in Europe and in America during the middle part of the last century. Find out why, with this large collection of accounts taken from a period journal of the times, the *Illustrated London News*. Included are approximately 40 contemporary engravings.

URL:

http://userwww.service.emory.edu/~ussjt/
Famine/

The Titanic

Take a tour of the most famous boat that ever went to the bottom of Davy Jones's Locker. View historical and current deep-sea photographs of the ship, contemporary accounts of the disaster, and passenger biographies. There are also links to other Titanic-related pages.

URL:

http://iccu6.ipswich.gil.com.au/~dalgarry/

It's been the butt of jokes for decades, but going down on your honeymoon is no laughing matter. (All right . . . we'll try to keep from sinking to new depths.) What really happened with the Titanic? Were the lookouts still intoxicated from the maiden celebrations? If questions such as this float your boat, nose into a slip at the Titanic Home Page.

World War II: The World Remembers

Link up with many World War II resources at this site created by the students and faculty of Patch High School. There is a large array of World War II material collected from government and military archives and *Stars and Stripes* newspaper, plus famous speeches from the National Archives, and maps and battle plans from the Center for Military History.

URL:

http://192.253.114.31/D-Day/
Table_of_contents.html

HOLOGRAPHY

Counterfeit Proof Hologram and Machine Readable Technology

See how scientists are using holograms to thwart thieves and counterfeiters through various applied technologies in the field of holography. Topics covered include currency, passports, drivers licenses, and brand-name products.

URL:

http://hmt.com/holography/hdi/counterfeit.html

Generalized Geometry for ISAR Imaging

Peruse this technical paper on microwave holography and tomography to learn how both are being applied to inverse synthetic aperture radar (ISAR) imaging. You may view images from the resulting geometry, which allows full rotational images in 2D and 3D.

URL:

http://www.cquest.com/pubs/amta.html

Hologram FAQ

If you're wondering how scientists and photographers construct 3-dimensional pictures on a 2-dimensional surface, consult this FAQ. Learn what holograms are, how they work, how they are made, and why they look so spectacular.

URL:

http://hmt.com/holography/hdi/hdiFAQ.html

What are holograms—really? Are they those shiny stickers of birds on credit cards and Microsoft software packages that flap their wings when you move them, or are they 3-D images that you can walk around and look at from any angle, like on the holodeck of the Starship Enterprise? Find the answer in the Hologram FAQ.

Holography News

View highlights and news excerpts from this international business newsletter of the holography industry. Keep informed about new products, applications, and developments for holography; plus corporate news, industry analysis, and company profiles.

URL:

http://hmt.com/holography/hnews/hnhome.htm

Holography Page

Check out what's new in holography on the Web with this compendium of links to articles and publications, artists, galleries, and commercial and educational sites. You'll also find news and marketplace information.

URL:

http://www.hmt.com/holography/

Interactive Holographic Pattern Generator

Create your own holographic image with Dimensional Arts' Interactive Holographic Pattern Generator. You can design your own unique patterns interactively by choosing a design type and entering your parameters. To get yourself started, choose "Fan" type, and enter **1.5** for the FanWeight, **.5** for the FanTwist, and **2** for FanCycles. Then click "Do It!" and sit back and watch it generate a beautiful pattern.

URL:

http://www.holo.com/cgi-bin/holoform.cgi

Internet Museum of Holography

Holography is a process of recording the image of an object using the special properties of light from a laser to create a true three-dimensional picture. Take an online interactive tour at this site to learn about the science of lasers, holograms, and optics. You'll find a wealth of images and information; and if you have a question about holography, you can submit your question to the museum and receive a reply within 24 hours.

URL:

http://www.enter.net/~holostudio/visitor.html

Practical Holography

This document is an excellent introduction to holography. Written by Christopher Outwater and Van Hamersvel, it's truly every detail you could want or need to know to begin creating your own holograms. The text is highly illustrated, and great examples abound.

URL:

http://www.holo.com/holo/book/book1.html

Search the Light—Holography

Find biographies from some of the world's pioneering holography artists and photographers, and view some stunning images and details about their work. You can peruse online articles about the scientific aspects and techniques of holography, and link to research sections, news, and exhibitions.

URL:

http://www.holo.com/search/search.html

A B C D E F G **H** I J K L M N O P Q R S T U V W X Y Z

ICHTHYOLOGY

Aquatic Conservation Network

Just say no to fish sticks. This nonprofit corporation is dedicated to conserving aquatic life, specifically freshwater fish. This page presents research projects, a bulletin, membership and registration access, and a discussion group.

URL:

http://www.ncf.carleton.ca/freeport/
social.services/eco/orgs/aquat-con/menu

Free the Fish!
Find out how at the Aquatic Conservation Network.

Captain Strong Salmon Project

Since they are so delicious, salmon are an important link in the northwest U.S. food chain. Not only humans, but all kinds of animals rely on salmon for nourishment; and when the fish are threatened, so is everything else in the ecosystem. The salmon project at Captain Strong Elementary School has a large online salmon exhibit with pictures, text, sound, and movies of many salmon species.

URL:

http://www.bgsd.wednet.edu/doc/salmon/
salmon1.html

Cichlid Home Page

The CHoP is for those popular freshwater fish known as cichlids—beloved by aquarium owners everywhere. Each entry contains information on appearance, origin, care, and the proper environment for each kind of cichlid. You'll find a bibliography of cichlid books and references, information on treating cichlid diseases, and lots of cich pics.

URL:

http://trans4.neep.wisc.edu/~gracy/fish/
opener.html

Columbia River Salmon Passage Model

The Columbia River Salmon Passage model (CRiSP) is an interactive, multiple-window program that assists managers of water districts, hydroelectric power plants, fisheries, and recreation areas in seeing the impact of their decisions on fish populations in the Columbia River. You can access databases of fish species counts by the Army Corps of Engineers and the Fish Passage Center, and there is a large ichthyological literature index. You'll need X Window or an emulator to successfully run the model demos.

URL:

http://www.cqs.washington.edu/crisp/

Desert Fishes Council

Yes, Virginia, there *are* fish that live in the desert, and even in—of all places—Death Valley! Find out more at the Desert Fishes Council page, which includes DFC meeting proceedings; high-resolution fish images; and assorted information about desert fishes and their ecology, biology, conservation, distribution, and management.

URL:

http://www.utexas.edu/depts/tnhc/.www/fish/
dfc/dfc_top.html

Dr. Gunnar Jonsson's Huge Searchable Dictionary of Marine Life

If you've been looking for that one special database that translates a school of fish names into English, Danish, French, Latin, and Icelandic, then your ship has come in. We'll bet you didn't know that the Abyssal Armed Grenadier is also called the Deepwater Whiptail and is known as *Brynhali* in Icelandic.

URL:

http://www.hafro.is/cgi-bin/ordabok/7

The Fish Division of the Texas Natural History Collection

Perhaps the next time you think of Texas, your thoughts will turn to fish after you visit this online searchable fish collection database overflowing with information about the University of Texas's collection, fish-distribution maps, and selected links to other Net resources in the areas of ichthyology, fish and aquatic biology, hydrology, and ecology.

URL:

http://www.utexas.edu/depts/tnhc/.www/fish/
main.html

Fish FAQ

Learn about whether life is found at all depths in the ocean, how many fish species there are, which fish is the oldest, what the world's largest and smallest fishes are, and what the differences are between anadromous and catadromous fish, plus a whale of a lot more.

URL:

http://www.wh.whoi.edu/homepage/faq.html

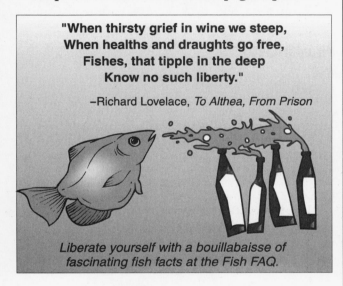

"When thirsty grief in wine we steep,
When healths and draughts go free,
Fishes, that tipple in the deep
Know no such liberty."

–Richard Lovelace, *To Althea, From Prison*

Liberate yourself with a bouillabaisse of fascinating fish facts at the Fish FAQ.

Fish Information Service Index

Tired of staring at TV? Get an aquarium. They're much more interesting, and you never have to change the channel. The FINS page maintains a large archive of information about freshwater and marine tropical and temperate aquariums. You'll find aquaria FAQs, and catalogs of marine and freshwater fish and invertebrates that include information, and common and scientific names. There are pictures, aquarium software, and information on disease diagnosis and treatment—plus plans for various aquarium do-it-yourself projects, movies, and aquaria images.

URL:

http://www.actwin.com/fish/

Fish Virtual Library

Holy mackerel! Hook lots of fish resources for ichthyologists and hobbyists here with information and links to pages on aquariums, aquatic environmental science, biology, and sport fishing. Don't let this whopper of a resource get away!

URL:

http://www.actwin.com/WWWVL-Fish.html

Fish-Ecology List

Read the latest about just how slippery or swimmingly our fine-finned friends are doing. A handy hypertext link from your browser makes subscribing to this mailing list as easy as shooting fish in a barrel. Now, who'd want to do that?

Mailing List:
    Address: **listserv@searn.sunet.se**
    Body of Message: **subscribe fish-ecology** *<your name>*

URL:

http://cricket.unl.edu/NBBG/Listservers/
    Fish-Ecology.html

From Tuna to Wahoo

Holy mackerel, what a site for references on tuna, bonito, wahoo, and mackerel! Each genus and species is well-documented with fun fish facts and is beautifully illustrated to boot. Visit to learn that the bluefin tuna is the world's most expensive fish, and that mackerel body shapes have influenced the design of ships, submarines, and even airplanes!

URL:

http://www.mackerel.com/fish/T2Wmenu.html

Great White Shark

Yes, the *Carcharodon carcharias* has pretty teeth, dear, and if you want to see how it keeps them that way, sail on over to find out more about great white sharks; what their Greek name means; and a lot of myths, legends, and fallacies, plus some cold hard facts. You'll find lots of great white physiological information.

URL:

http://www.netzone.com/~drewgrgich/shark.html

The Web will set you free.

A
B
C
D
E
F
G
H
I
J
K
L
M
N
O
P
Q
R
S
T
U
V
W
X
Y
Z

The Great White

Something about the name of this shark just doesn't seem right. Oh, they're big—but they're neither great, nor white. So why do we so revere these denizens of the deep? Probably because if one catches you in the surf, you'll agree to anything.

LeanFish

The LeanFish page describes a process for producing high-quality fish-meal protein from marine and aquatic offal and by-catch. There is analysis of recent developments in the fish-meal industry and proposals given to meet current demand.

URL:

> http://www.ingvar.is/Engineering/LeanFish/LFtechno1

National Marine Fisheries Service

The Northeast Fisheries Science Center Headquarters at Woods Hole, Massachusetts, has a 125-year history of conserving U.S. coastal and ocean resources. Learn more about its research ships and laboratories, and find out about fisheries' information-gathering techniques and fishstock assessments.

URL:

> http://www.wh.whoi.edu/noaa.html

NEODAT Fish Biodiversity Gopher

Here's a whopper of a fish resource complete with information on specimens from fish collections; species and bibliographic databases; a who's who in neotropical ichthyology; information on fish mapping and collection management software; and pointers to related Gophers, news, and information services.

URL:

> gopher://fowler.acnatsci.org

Pacific Fishery Biologists

Established in 1936, the Pacific Fishery Biologists organization promotes fisheries science through research, cooperation, and the free exchange of ideas. Here, you'll find a newsletter, notices of meetings, and information on the organization. There are also a few original "fish pics" worth checking out.

URL:

> http://www.teleport.com/~tfish/pfbhome.htm

Piranha-Cam

If this page only featured downloadable shots of live-action piranha in somebody's aquarium (which it does), it wouldn't have qualified as an entry. If it also had lots of information on piranha and fish biology (which it does), this would make it a real humdinger of a resource! This page is pure piranha paradise, with movies and an actual project for kids on how to build their own piranha tank out of Jello!

URL:

> http://www.floater.com/strength/

Where else but on the Web can you view a nearly real-time image of an aquarium full of piranha? There are three camera views to choose from—one on each side, and one at the bottom peering upward. Catch them at feeding time for some great live action. We can't wait to catch it when the owner's cat sticks it's paw into the tank to catch a fishy. (This should be better than a Chuck Norris movie!)

Salmon FAQ

Did you know that salmon can weigh over 100 pounds and can live to be 8 years old? Find out more fascinating salmon facts, including how they migrate and where they travel once they reach the ocean, with this FAQ for salmon afishionados.

URL:

> http://www.wh.whoi.edu/faq/fishfaq2d.html

Salmon Fisheries Information

Salmon have been very much an integral part of the culture and heritage of the Pacific Northwest from time immemorial to Native Americans and fishermen of all kinds. Learn more about different salmon species, and see the progress that people are making to ensure that salmon will survive *throughout* time immemorial.

URL:

http://kingfish.ssp.nmfs.gov/olo/unit12.html

Shark Images from Douglas J. Long, UCMP

Fourteen images and some excellent documentation about sharks, especially great whites, accompany this sharp page. These toothy masters of the deep have existed as a group for over 350 million years. There are some great great white pics; and, since you're here, sink your teeth into these sharks and some of their most endearing qualities, especially their feeding habits.

URL:

http://ucmp1.berkeley.edu/Doug/shark.html

South African White Shark Research Institute

Since many more people eat sharks than sharks eat people, *they're* the endangered ones—not us. South Africa declared the White Shark an endangered species in 1991, and you can learn a great deal here about shark conservation efforts, shark physiology, and research activity.

URL:

http://www.cru.uea.ac.uk/ukdiving/orgs/wsri/index.htm

Status of Fisheries Resources Off Northeastern United States

The Conservation and Utilization Division of the Northeast Fisheries Science Center (NEFSC) publishes an annual assessment of fish and shellfish resources in the northeastern coast areas of the U.S. The report summarizes the general status of these regional resources, from Cape Hatteras to Nova Scotia. Excerpts of this full report are available here online.

URL:

http://www.wh.whoi.edu/library/sos94/sos.html

West Coast Salmon News

If salmon are important to you, you'll want to keep abreast of what people are doing to keep them around. Read monthly news and updates from the Congressional Research Service on recent laws and enactments regarding marine and freshwater salmon-fishing regulations.

URL:

http://kingfish.ssp.nmfs.gov/salmon/sal-news.html

Zebrafish Server

Zebrafish aficionados of the world, unite at this site that features more information on z-fish than you can shake a fishing pole at. You'll find data on laboratory use, genetics, molecular DNA, reference listings, news, and the *Zebrafish Science Monitor*—a bulletin that reports current methodologies and research news.

URL:

http://zfish.uoregon.edu

IMAGING

Center for Complex Systems and Visualization

The mathematics and computer science projects of CeVis at the University of Bremen include cellular and linear automata, fractal patterns, and time analysis, as well as more practical endeavors like medical visualization and diagnostic systems. Download a full-feature version of ImgLab, an interactive image-processing program for SGI platforms. This page is also available in German.

URL:

http://www.cevis.uni-bremen.de

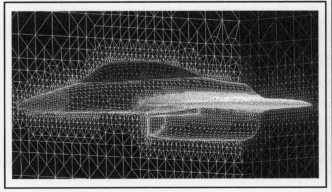

Learn how to visualize your data on the Net.

DesignSpace

Scamper on over to Stanford for a visit to DesignSpace at their Center for Design Research (CDR). View and read about a few of their conceptual models of simple surface and solid geometric shapes used in mechanical design. Then get into deeper discussions of virtual space, design theory and constraints, and comparisons of CDR's programs to virtual reality and teleoperation.

URL:
> http://gummo.stanford.edu/html/
> DesignSpace/home.html

Laboratory of Eidomatics at the University of Milan

It may be Greek to you, but *eidos* means image or idea. Combine *eidos* with *matics* as a suffix (as in *infomatics*), and you have an interdisciplinary area that transcends image processing, computer graphics, vision, and the cognitive aspects of computer imaging. Faced with a name like that, you're encouraged to explore MPEG examples of photometric modeling and solid simulation, and optimization/parallelization of ray tracing and radiosity algorithms. Among other wonders, you can visualize a bit of history with a computer simulation of Edwin Land's Polaroid Retinex Model.

URL:
> http://cube.sm.dsi.unimi.it/Users/imaging/
> HOME.html

Scientific Visual Analysis Lab

SVAL at Virginia Tech offers movies on the lab's research in areas of 3D X-rays, MRI tomography, Doppler velocity field measurement, and supercomputer simulations. You can learn more about the lab's projects and technical publications, and follow links to resource development. Check out listings of upcoming seminars, view student projects online, and take a tour of other campus labs developing visualization approaches in their research.

URL:
> http://www.sv.vt.edu

Scientific Visualization at NYU's Academic Computing Facility

New York University's Academic Computing Facility has much to offer in the way of visualization software and computer animations. There are workstation packages that include TecPlot and VolVis, and online documentation and tutorials for each program. Download some spiffy graphics and animations of mathematical models of the heart and solar magnetic fields. While you're at it, sample a few of the articles from NYU's *Connect* newsletter.

URL:
> http://www.nyu.edu/pages/scivis/

SHASTRA Collaborative Modeling and Visualization Environ

SHASTRA is the Sanskrit word for *Science* or a branch of knowledge. Purdue University's SHASTRA scientific toolkits are X11-based applications that are supported on Sun, SGI, and HP platforms. They include medical image reconstruction, molecular modeling, geometric design, physical analysis, and animation toolkits available for downloading. Version 1 of each toolkit is in the public domain.

URL:
> http://www.cs.purdue.edu/research/shastra/
> shastra.html

Visualization and Graphics Research

The Institute for Computer Applications in Science and Engineering (ICASE) works with NASA to produce state-of-the-art 3D imaging research in applied mathematics, numerical analysis, fluid dynamics, and computer science. ICASE maintains a visualization and graphics bibliography, and offers MPEG examples of its high-performance workstation rendering.

URL:
> http://www.icase.edu/docs/hilites/
> index.cs.viz.html

> ## The Net is humanity's greatest achievement.

> ## Kids, check out Kids and Amateurs.

INFORMATION SYSTEMS

All-in-One Search Page

All-in-One conveniently provides forms for submitting search requests to a number of search engines, all available from within a single form. Unlike SavvySearch, which queries multiple engines simultaneously, searches are performed on a per-engine basis.

URL:
http://www.albany.net/~wcross/all1srch.html

Directorate of Time

Find out exactly what the time is in 11 different time zones. The U.S. Naval Observatory employs HP9000/747i systems that host Datum VME synchronized generators using IRIG-b timecode from USNO Master Clock 2. The system clocks are synchronized to within a few tens of microseconds of USNO Master Clock 2.

URLs:
telnet:tick.usno.navy.mil:13
telnet:tock.usno.navy.mil:13

http://tycho.usno.navy.mil/time.html

GNU Information Files

If you've ever attempted to navigate through *info*, GNU's Unix-based information system, you'll appreciate this hypertext version. Included here is a comprehensive directory to info files on everything from emacs to zsh. No funky meta-control key sequences to memorize. Just point and click.

URL:
http://www.ai.mit.edu/!info/dir/!!first

Inktomi Search Engine

The Inktomi Search Engine was the first fast web indexer with a large database. Inktomi uses parallel computing technology to build a scaleable Web server using SparcStation 10s. Inktomi is part of the Network of Workstations (NOW) project at the University of California at Berkeley.

URL:
http://inktomi.berkeley.edu

> ## Looking for a science project?
> ## Look in Kids and Amateurs.

Excerpt from the Net...

```
(from the US Naval Observatory Master Clock)

Time from the Master Clock


The time at the tone ...

US Naval Observatory Master Clock, at the tone,

November 3, 1995, 08:12:44 Universal Time

November 3, 1995, 04:12:44 Atlantic Standard Time
November 3, 1995, 03:12:44 Eastern Standard Time
November 3, 1995, 02:12:44 Central Standard Time
November 3, 1995, 01:12:44 Mountain Standard Time
November 3, 1995, 00:12:44 Pacific Standard Time
November 2, 1995, 23:12:44 Alaska Standard Time
November 2, 1995, 22:12:44 Hawaii-Aleutian Standard Time
November 2, 1995, 21:12:44 Samoa Standard Time

November 3, 1995, 18:42:44 Australian Central Daylight Time
November 3, 1995, 04:42:44 Newfoundland Standard Time

beep! [Note: there is no audio, we were just kidding...]
```

A B C D E F G H I J K L M N O P Q R S T U V W X Y Z

Lycos

It's fitting that Lycos gets its name from the first five letters of the species name for the Wolf Spider, because Lycos crawls all over the Web. Sure, the Net gives you access to piles of information, but it doesn't tell you where to find what you're looking for. One of the largest Internet search systems, Lycos catalogs over 95 percent of the Web. (Don't worry, you didn't just throw away good money on this book. You can't take Lycos to the bathroom, and it won't make you laugh as much.)

URL:

http://www.lycos.com

NASA Thesaurus

NASA's hypertext thesaurus presents descriptions of thousands of terms, including an extensive technical collection.

URL:

http://netsrv.casi.sti.nasa.gov/
nasa-thesaurus.html

SavvySearch

SavvySearch is a parallel Internet query engine. After entering your search request, SavvySearch contacts multiple Internet search engines, simultaneously gathering information from each source and returning the linked results. Since many engines are queried, responses can take a while.

URL:

http://www.cs.colostate.edu/~dreiling/
smartform.html

INTERNET CONNECTIVITY

Internet Hyper Glossary

Just getting started on the Internet and confused by the endless terms and acronyms? The Hyper Glossary comes to the rescue with dozens of clearly written descriptions of baffling terms like ARPANET, CHAP, Subnet Mask, MIME, and more.

URL:

http://www.windows95.com/connect/
glossary.html

ARPA, DARPA, OLE
OS, IS, UDP
ISDN, DNS
AUI and FTP

Sound like alien poetry?
Find out what it all means at
the Internet Hyper Glossary.

Native Windows 95 PPP Connectivity to the Internet

Of course, if you can read this page, you might not need it because you're already connected. But it's a great reference nevertheless. These pages take you by the hand and explain—step by step—how to connect a Windows 95 system to an Internet service provider using Win 95's built-in PPP connectivity software.

URL:

http://www.erv.com/w95_ppp.htm

NetWatch

NetWatch, an online magazine, identifies and discusses Net-based enabling technologies and associated applications. While focusing on audio, video, and server technology, the publisher includes marketing gems and editorial pieces. NetWatch is the first Web magazine to provide a RealAudio soundtrack.

URL:

http://www.pulver.com/netwatch/

Need help getting connected? See Rick's book The World Wide Web Complete Reference.

PC Internet Access Tutorial

So you have a PC, and no idea how to get connected to the Internet. The best solution is to run down to the bookstore and pick up a copy of *The World Wide Web Complete Reference*. But if you can't do that, here's a thorough overview of how to connect a PC to the Internet. (Of course, if you can't get to the Web somehow to get here, you've got a real dilemma.) These pages lay down a good outline, giving a general background on network navigation, types of data, windows sockets (winsock), and TCP/IP. The planning section details the hardware required and covers COM ports, modems, and Ethernet cards. There are also guides for installing hardware cards and software, including packet drivers, WINSOCK, OS/2 Warp, and Microsoft's TCP/IP. Finally, there's a section describing Internet services and programs, including FTP, Archie, Gopher, the Web, and Usenet news readers.

URL:

    http://pclt.cis.yale.edu/pclt/winworld/ contents.htm

Web66 Network Construction Set

This tutorial describes Ethernet network components and provides you with all the details you need for setting up LANs and WANs with Internet connectivity. Having completed and understood this tutorial, you should be able to analyze and understand almost any modern computer network. The tutorial is in HTML format and is illustrated with color diagrams.

URL:

    http://web66.coled.umn.edu/Construction/ Construction.html

INTERNET RESOURCES

ASCII Tables

If you're a programmer, you know how aggravating it can be remembering where you put your ASCII table. Pick one of these ASCII tables, put it in your bookmark list, and forget about finding the wily chart on your desk. Several of these charts also list the corresponding HTML codes to reproduce any ASCII character on a web page.

URLs:

    ftp://dkuug.dk/i18n/WG15-collection/ charmaps/ANSI_X3.110-1983
    ftp://dkuug.dk/i18n/WG15-collection/charmaps/ ANSI_X3.4-1968

    http://www.bbsinc.com/iso8859.html
    http://www.bbsinc.com/symbol.html
    http://www.infocom.net/~bbs/symbol.html

Liszt

Liszt is an excellent directory of email mailing lists. You can enter any word or phrase into the web form to search this directory of well over 23,000 Listserv, Listproc, Majordomo, and independently managed mailing lists from 555 sites. There are instructions on using the service, configuration options, and a form to submit new mailing list details.

URL:

    http://www.liszt.com

Liszt
Don't believe those other books that tell you that you have to download and wade through some ugly text files on your computer to find out about mailing lists you might be interested in. Just pull up the Liszt page, type in a keyword, and get the real scoop—the easy way.

Liszt of Newsgroups

Use this web form to enter any word or phrase and search a list of over 13,000 newsgroups. The form (or actually the associated CGI script) will return a list of newsgroups meeting the criteria of your search.

URL:

    http://www.liszt.com/cgi-bin/news.cgi

Want to see some cool resources? Check out Internet Resources.

A
B
C
D
E
F
G
H
I
J
K
L
M
N
O
P
Q
R
S
T
U
V
W
X
Y
Z

JOBS

American Astronomical Society Job Register

There's money and careers to be made by stargazing. Search through this database that lists hundreds of positions offered in the fields of physics, astronomy, and astrophysics.

URL:
http://www.aas.org/JobRegister/aasjobs.html

Career WEB

CareerWeb lists over 150 available positions in various career fields, from finance and sales, to electronics and marketing. Job seekers can search according to discipline, location, specific employer, or keyword. Motorola Inc., Northrop Grumman Corporation, The Virginian-Pilot, and Sprint Cellular are just a few of the employers listed. You can take the online Career Fitness Test and the Readiness Inventory Test. Check out announcements of upcoming career fairs throughout the U.S. This site also provides an online library, articles from the *National Business Employment Weekly*, and links to employers and other career resources.

URL:
http://www.cweb.com

CareerMosaic

Search this CareerMosaic J.O.B.S. database for thousands of the latest opportunities from hundreds of employers in the world's leading companies for high tech, health care, finance, retailing, and media jobs. There are Usenet links and an online job fair. You can view in-depth profiles of prospective corporate employers, and learn tips for networking and how to prepare a proper résumé and cover letter—all this and more at the electronic Career Resource Center.

URL:
http://www.careermosaic.com/cm/

Like birds? Flutter into Ornithology.

Looking for Earth Science resources? Check Earth Science, Ecology, Geography, and Geology.

Environmental Health Perspectives

The EHP Bulletin Board is updated monthly and features job opportunities, fellowships, grants, and awards from mainly universities and non-profit organizations. These online job postings are from *Environmental Health Perspectives*, the monthly journal of the National Institute of Environmental Health Sciences.

URL:
http://ehpnet1.niehs.nih.gov/docs/bboard/bboard.html

Job Openings in Academe

This page at the Chronicle of Higher Education may help you on your way toward a teaching, research, or faculty career in any of the fields of humanities, social sciences, science, technology, and other professional endeavors. You may search using any word or words of your choosing by domestic or international locality.

URL:
http://chronicle.merit.edu/.ads/.links.html

Jobs in Mathematics

How do you make your skills as a mathematician add up to employment? By visiting this page on the academic math job market, that's how! Take a look at job openings advertised on the Internet for undergrads, grad students, and Ph.Ds. Brush up on information about the types of skills that employers are looking for in mathematicians, and view a recommended reading list that may help you reach your employment goals.

URL:
http://www.cs.dartmouth.edu/~gdavis/policy/jobmarket.html

MedSearch America

MedSearch America is a large Internet employment advertising and communications network that focuses specifically on the Healthcare industry. This resource provides detailed employer profiles, job listings, resume postings, and industry and career opportunities. Through MedSearch America you can post your own job ad, which will remain online for two months. The site also features a physician-referral service.

URL:

http://www.medsearch.com

The Monster Board

The Monster Board can post your online resume to Resume City, a vast database that is searched by employers throughout the world. It currently features over 3,100 job listings offered by over 500 companies. Employers are welcome to post job offerings to attract prospective employees. If you are a job-seeker, you can submit your resume to their database, look for posted jobs in Career Search and Job Shop 'Til You Drop, take a virtual tour of hundreds of companies in Employer Profiles, and check out the CIGNA Career Fair and Target Career Fairs Online.

URL:

http://www.monster.com

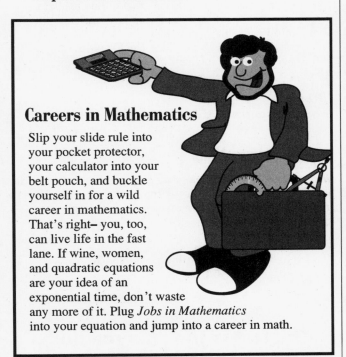

Careers in Mathematics

Slip your slide rule into your pocket protector, your calculator into your belt pouch, and buckle yourself in for a wild career in mathematics. That's right— you, too, can live life in the fast lane. If wine, women, and quadratic equations are your idea of an exponential time, don't waste any more of it. Plug *Jobs in Mathematics* into your equation and jump into a career in math.

Reach for the stars. Gaze into Astronomy and Physics:Astrophysics.

Seagate Technology Employment Opportunities Summary

The Seagate Technology Employment Opportunities site offers a comprehensive list of available positions in many parts of the world in the computer programming and engineering fields. Job seekers can view all current Seagate jobs and can select job descriptions by categories such as entry-level positions, product/company divisions, or location.

URL:

http://www.careermosaic.com/cm/seagate/
seagate7.html

Telecommuting Advisory Council

Save time, gas, energy, and your sanity by working for your company at home. The Telecommuting Advisory Council has put together this page that promotes the economic, social, and environmental benefits of gainful employment via modem. You'll find a wealth of articles on telecommuting, information on tax credit benefits, a book bibliography, and other great links toward simplifying your working life.

URL:

http://www.telecommute.org

Work@Home

Many people dream of SOHO (small office home office) work. If you have a professional or technical job that requires computer skills, consider leaving the rat race behind by telecommuting to work. This page may assist you toward that goal with some excellent links to working-at-home resources and ideas. After all, face the truth—the race *is* over. The rats won!

URL:

http://www.iadfw.net/msmith/

Excerpt from the Net...

(from the Work@Home page)

You can DRAW PEOPLE TO YOUR PAGE IN DROVES. It won't cost you anything, except a little bit of your time.

Don't be fooled into giving money to self-proclaimed experts who will sell you links to their sites, or promise heavy traffic in exchange for hundreds of dollars. They don't know half as much as I'll show you, and I'll show you how to do it all yourself. You can and you will get visitors to your site, if you simply use the right technique.

KIDS AND AMATEURS

Activities for Kids

Adventure through scads of science activities with your young'uns. Experiments abound in subjects such as bioscience, communications, computers, Earth science, energy, mathematics, oceanography, physical sciences, and space. Build an ocean in a bottle, or check out "Static Electricity—A Hair Raising Phenomenon."

URL:

http://sln.fi.edu/tfi/activity/act-summ.html

Ants in a Bottle

Who said ant farms are for nerds? They're fun and easy to make. This informative page shows you step-by-step how to build an ant farm from common household items. All you need is a large peanut butter or pickle jar, a small bottle, some clay, sand, bread crumbs, and . . . oh—ants, of course. What this page doesn't tell you is how to go about getting the ants.

URL:

http://www.parentsplace.com/readroom/
explorer/act_1.html

Drawing Holograms by Hand

This science hobbyist presents a method for creating holograms without lasers or any other fancy equipment. All you need is a compass, some Plexiglas scraps, and his instructions. The story of how he discovered this phenomenon is every bit as interesting as how to make holograms.

URL:

http://www.eskimo.com/~billb/amateur/
holo1.html

Experimental Science Mailing List

"Experimental Science" may sound redundant, but you'll enjoy this brand of *weird science* that's always interesting. To appreciate this mailing list, you may need to have some proficiency with mathematics, physics, chemistry, and basic shop tools. Discussions have included such nerdy things as building mirrored telescopes, Tesla coils, and vacuum pumps.

Mail:

Address: **tesla@einstein.ssz.com**

Franklin Institute Science Museum

Explore the Franklin for several great online exhibits, including "The Heart: A Virtual Exploration," and "Benjamin Franklin: Glimpses of The Man." Then take an electronic field trip by reading the Franklin's monthly online magazine for "inQuiring" minds, the *inQuiry Almanack.*

URL:

http://sln.fi.edu/tfi/welcome.html

Green Eggs and Ham

Why are some egg yokes a gross gray-green color? Why do some of them have that disgusting rotten egg smell? This informative page offers the answers and a simple experiment to prove it.

URL:

http://www.parentsplace.com/readroom/
explorer/act_5.html

The Net is the new medium.

Helping Your Child Learn Science

Learn more about science with your kids by pushing straws through potatoes and making concoctions out of flour, egg whites, and gelatin cubes. Here are a series of fun experiments guaranteed to really get the kitchen messy and everyone in the mood to read up on Einstein's theory of relativity.

URL:
http://www.ed.gov/pubs/parents/Science/

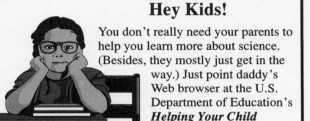

Hey Kids!

You don't really need your parents to help you learn more about science. (Besides, they mostly just get in the way.) Just point daddy's Web browser at the U.S. Department of Education's *Helping Your Child Learn Science* page and read about all the stuff the government wants your parents to teach you.

Miami Museum of Science

This science museum offers dozens of interesting links to follow. The Exploratorium, the Boston Museum of Science, and the Florida Marine Aquarium Society are just a few of the highlights. Fly into a hurricane, learn about instruments used to gather information on hurricanes, and study other types of killer storms. You can also explore acids, bases, and pH factors through a graphical database interface.

URL:
http://www.miamisci.org

Corrections? Changes? Additions? See our web page at http://www.iypsrt.com.

Microscopes and Microscopy

If you squint too hard, your eyeballs might pop out. So here's a rich assortment of subjects related to microscopy, with lots of images, information, and indexes. A monthly section called "In Focus" is devoted to special topics and microscope minutia. You can jump to news, events, reference libraries, workshops, mailing lists, links, and related FTP sites.

URL:
http://metro.turnpike.net/jefferie/

Murray, the Thespian Microbe

For a bit of microbial mirth and wisdom, try Murray, the bacterium with an identity crisis. You and your kids can visit this page that describes Murray's bold journey of self-discovery. Learn interesting facts about both good and bad bacteria.

URL:
http://www.demon.co.uk/scotcal/murray/murray.html

Museum of Science, Boston

Explore an online exhibit of fractal patterns in nature at "The Dance of Chance," and learn how these patterns arise in nature from the conflict between random and constraining forces. You can also take a tour of the museum, and read about the programs, courses, and travel opportunities the museum has to offer.

URL:
http://www.mos.org

NCSA Education Program

The National Center for Supercomputing Applications is known for its Web browsers, servers, and network of massive supercomputers. But if you're an educator, you'll want to visit NCSA's SuperQuest for Teachers, a modeling workshop and year-long support program for high school science and mathematics teachers. There are a number of related education-oriented applications in scientific visualization that teachers will find helpful.

URL:
http://www.ncsa.uiuc.edu/Edu/EduHome.html

The Web is the Net's GUI.

A B C D E F G H I J K L M N O P Q R S T U V W X Y Z

Newton's Apple

Here's the real reason why teachers love to use the Net! Newton's Apple assists grade school teachers by providing lesson plans for science studies. There are 26 lessons in all that cover the major topics explored on the popular television program, *Newton's Apple*. Each lesson includes vocabulary examples, scientific insights into each fun topic (hang gliding, karate, brain mapping), and questions to stimulate student discussion—as if they'd need any!

URL:

> http://ericir.syr.edu/Newton/welcome.html

Nikola Tesla Page

This fascinating man was part scientist and part sorcerer. He outdid Thomas Edison by lighting the 1893 Chicago World's Fair, and by inventing fluorescent lighting, new forms of generators and tranformers, and his famous coil that is widely used today. Nikola Tesla held more than 700 patents. Check out this page of related links to other enlightening Tesla resources.

URL:

> http://www.eskimo.com/~billb/tesla/tesla.html

Nikola Tesla

Another nineteenth century quack? Or a brilliant electrician and inventor who rivaled the likes of Thomas Edison? Plug into the consummate collection of Nikola Tesla information on the Web at the Nikola Tesla page and decide for yourself.

Nobody sends you email? Send us a note—we'll answer you.

NSF MetaCenter Science Highlights Repository

Imagine having access to over 10,000 scientific research project reports with images, animations, and sound that range in scope from astronomy to zoology! Just plug into this National Science Foundation page and be enthralled by all of the possibilities. Happily, there is a searchable database either by keyword or through a list of topics.

URL:

> http://www.ncsa.uiuc.edu/SCMS/Metascience/ Home/welcome.html

Ontario Science Centre

Take a virtual tour of OSC through its various levels, such as The Information Highway and The Human Body exhibits. You can also explore displays on space, transportation, technology, and communications, and visit a science arcade.

URL:

> http://www.osc.on.ca

Optics for Kids

What is light and how fast does it travel? Light can be controlled by blocking, bending, or reflecting it. This page explains some of the basic properties of light and how they are used with optical devices such as cameras, binoculars, and scanners. This covers a lot of areas—including lasers, lenses, and careers working with optics.

URL:

> http://www.opticalres.com/kidoptx.html

Questacon

At this science center in Australia, you can sample exhibits in the Questacon galleries and try hands-on science activities using simple, everyday materials. Check out science news and events, and explore a gallery of "science snippets." Learn about improving student science programs, and investigate the array of educational science products that are available. Check out Questicon's hands-on activities for schools and families, and find info about upcoming programs for the general public.

URL:

> http://actein.edu.au/Questacon/

Get Virtually Real.

Quick Thoughts

Inspire your child to look at the world through a scientist's eyes. Here, you'll find a collection of thoughts and questions about the world around us. There are suggestions for projects and field trips. Why does wood float and rocks sink? What are the differences between grasshoppers and crickets? These and other interesting topics are covered here.

URL:
http://www.parentsplace.com/readroom/
explorer/act_10.html

Science at Home

Treat yourself to a collection of hands-on science experiments that the whole family can enjoy, using easy-to-find and inexpensive materials from around the house. These well-tested activities are geared for students in grades 4-8, but they're great fun for everyone else, too.

URL:
http://education.lanl.gov/RESOURCES/
Science_at_Home/SAH.welcome.html

Science Friday Kids Connection

This site features archives of *Science Friday Kids Connection* with RealAudio program recordings of shows, behind-the-scenes information on how the show is put together, science discussion forums for kids and teachers, and links to related resources on the Net.

URL:
http://www.npr.org/sfkids/

**Draw the bridge!
Man the bastion!
Read Security and Firewalls.**

Science Hobbyist

There's nothing like a good jolt to spark your interest in science. So put a bookmark on this very cool page with pointers to amateur science projects, experiments, and demos you can experience right on the Internet. Be sure to check out the Weird Science page for wacky links on energy, antigravity, crazy inventions, and "dangerous capacitor bank" experiments!

URL:
http://www.eskimo.com/~billb/

Science Hobbyist

No, you probably won't find out how to make a beautiful robot that looks like Cindy Crawford and follows your every command, but there's a lot of other fun stuff. Check out the weird science and crazy inventions on the Science Hobbyist page.

Get your doctorate in Ufology.

A B C D E F G H I J K L M N O P Q R S T U V W X Y Z

Science Museum of London

Several exhibits featured at this venerable London museum begin with descriptions and images of scientific apparatuses. View the earliest complete Renaissance medicine chest; Babbage's Calculating Engine (the original "computer"); James Joule's Paddle-Wheel Apparatus, with which he confirmed the First Law of Thermodynamics; and the history of flight from Leonardo da Vinci to the present, with over 70 images, photographs, and drawings.

URL:
http://www.nmsi.ac.uk/Welcome.html

Smithsonian Computer History

Enjoy an excellent slide show and online descriptions of the Smithsonian exhibition, "Information Age: People, Information & Technology." There are visual and interactive displays on how electrical information technology has changed our society and our lives over the past 150 years.

URL:
http://www.si.edu/perspect/comphist/computer.htm

Smithsonian Gem and Mineral Collection

See and learn more about gems, gemology, and mineralogy on this page that features the Smithsonian's exquisite collection of fabulous stones. Examples range from small yellow wulfenite crystals and Zuni turquoise to the Hope diamond.

URL:
http://galaxy.einet.net/images/gems/gems-icons.html

Society for Amateur Scientists

Take part in the great scientific issues of our time by becoming an amateur scientist. The Society of Amateur Scientists is dedicated to helping people explore the outer reaches of their own genius by developing their scientific skills and removing the roadblocks that today make it nearly impossible to do research without a Ph.D. degree.

URL:
http://www.thesphere.com/SAS/

Society for Scientific Exploration

SSE and its companion publication, the *Journal of Scientific Exploration*, report on topics that are either ignored or studied inadequately by mainstream science. Read articles from earlier editions of the *JSE*, including titles such as "The Nature of Time" and "Experiments in Remote Human-Machine Interaction."

URL:
http://valley.interact.nl/av/kiosk/SSE/home.html

Society of Amateur Radio Astronomers (SARA)

Sixty-five percent of our current knowledge of the universe originates from radio astronomy, including discoveries of the Big Bang, black holes, pulsars, and quasars. SARA members include optical astronomers, ham radio operators, engineers, teachers, and nontechnical persons. Check out the full text of SARA's newsletters and observational programs right here.

URL:
http://irsociety.com/0c:/sara.html

Uncle Bob's Kids' Page

Step right up, boys and girls! Children from 2 to 92 are welcome at Uncle Bob's place. Uncle Bob has collected the neatest scientific sites around and put them all within easy reach. Among many other great topics, learn about volcanos, dinosaurs, outer space, and math.

URL:
http://gagme.wwa.com/~boba/kids1.html

UT Science Bytes

See what scientists at the University of Tennessee are up to in Science Bytes. Each byte uses text and pictures to describe the particular work or interest of a UT scientist. Topics include "On the Wings of a Dragonfly," "Rhinos and Tigers and Bears—Oh My," and "Search for Antarctic Spring."

URL:
http://loki.ur.utk.edu/ut2kids/science.html

Take flight in Aerospace and Space Technology.

Virtual Science and Mathematics Fair

From the "Effect of Soap on Bubble Bath" to "Manatee Tracking," students' entries for the 1995 Virtual Science and Mathematics Fair are presented here. Enter your Rube Goldberg contraption that takes you into the Virtual Science and Mathematics Fair, which takes place *virtually* on the World Wide Web. Each presenter plans a scientific "poster" with several components: an abstract; a presentation of data with images, tables and the like; a discussion; and a hypothesis. Entries are evaluated online by peer review.

URL:
http://www.educ.wsu.edu/fair_95/

Virtual Science and Mathematics Fair

Who needs plaster of paris volcanoes? How about elemental density computations illustrated with ray-traced graphics presented on a web page? Okay, plaster of paris volcanoes are still cool. But why not a web page with a Java applet that simulates a volcano blowing its top and spewing ash into the stratosphere?! You've got to see some of the science fair entries that today's Net-kids are coming up with to believe them.

Is music art or science?

Virtual Science Center

Singapore Science Center shows off its online science exhibits, plus the latest scientific news updates and projects. Download QuickTime movies of interesting experiments involving gyroscopes, chicken egg embryos, parallel lines, and thunder balls.

URL:
http://www.ncb.gov.sg/vsc/vsc.html

Weird Research and Anomalous Physics

If scalar xmitters, gravity wave detectors, resonance and caduceus coils, or magnetic resonance amplifier devices trip your trigger, check out this page loaded up with weird technology and funny physics. This page also has links to even weirder pages.

URL:
http://www.eskimo.com/~billb/weird.html

You Can

Here's a lively, colorful collection of questions and answers about space and planets, and other diverse topics. There are great photos from the Hubble telescope. But you can also find out why feet smell, learn how to make up a batch of fake snot, and discover how eyeglasses correct problems with eyesight. Quite a wide variety of topics are covered here on Beakman & Jax's You Can page. Pure fun for children and adults alike.

URL:
http://www.nbn.com/youcan/

Seeking stimulating circuit simulations? See Engineering.

Hug a tree. Hang out in Forestry.

A B C D E F G H I J K L M N O P Q R S T U V W X Y Z

LINGUISTICS

Aquarius Directory of Translators and Interpreters

Having trouble reading the signs on the Autobahn? Find yourself looking for pictures on the cans in the supermarket? Do you need an interpreter—fast? Aquarius is a directory of willing interpreters organized by area and language. This organization also offers a newsletter and an international graffiti page for interesting scribblings, notes, and requests.

URL:

http://www.xs4all.nl/~jumanl/

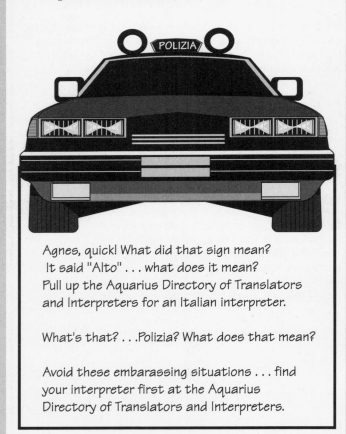

Agnes, quick! What did that sign mean?
It said "Alto" . . . what does it mean?
Pull up the Aquarius Directory of Translators and Interpreters for an Italian interpreter.

What's that? . . . Polizia? What does that mean?

Avoid these embarassing situations . . . find your interpreter first at the Aquarius Directory of Translators and Interpreters.

Take a break. Visit a zoo in Zoology.

Find out what's shakin' in Seismology.

AsTeR—Audio System For Technical Readings

What do you get when you run Faa de Bruno's formula through an ordinary text-to-speech synthesis program? Without AsTeR, you get a computer's impression of a tongue twister. AsTeR was developed to render technical documents in spoken audio. This site contains many samples of AsTeR's output demonstrating its ability to handle increasingly complex formulas.

URL:

http://www.cs.cornell.edu/Info/People/raman/aster/demo.html

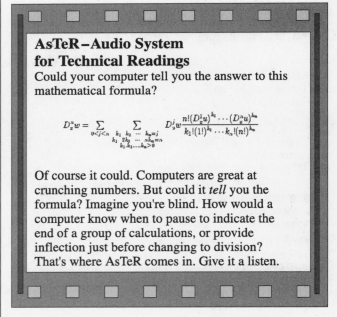

AsTeR – Audio System for Technical Readings

Could your computer tell you the answer to this mathematical formula?

$$D_z^n w = \sum_{0 < j < n} \sum_{\substack{k_1\ k_2\ \cdots\ k_n=j \\ k_1\ 2k_2\ \cdots\ nk_n=n \\ k_1, k_2, \ldots, k_n > 0}} D_z^j w \frac{n!(D_z^1 u)^{k_1} \cdots (D_z^n u)^{k_n}}{k_1!(1!)^{k_1} \cdots k_n!(n!)^{k_n}}$$

Of course it could. Computers are great at crunching numbers. But could it *tell* you the formula? Imagine you're blind. How would a computer know when to pause to indicate the end of a group of calculations, or provide inflection just before changing to division? That's where AsTeR comes in. Give it a listen.

Blackfoot, Old and New

From southern Alberta and northern Montana, this Algonquian language is undergoing a period of change. Review some of the differences between the old and new dialects. There are also a language sample and information on Blackfoot phonetics and syntax.

URL:

http://www.teleport.com/~napoleon/blackfoot/

Celtic Linguistics

Learn about a variety of ancient Anglo languages by joining, reading, and participating in the CELTLING mailing list. Topics revolve around anything relating to Celtic linguistics.

Mailing List:
    Address: **listserv@mitvma.mit.edu**
    Body of Message: **subscribe celtling** *<your name>*

Clay to Cuneiform

This page explores the history of the Cuneiform writing system. It starts by describing the clay tokens the ancients used to record their writings, and moves on to over 600 of the cuneiform signs and symbols of the ancient Akkadian language.

URL:
    **http://ruurq2.sron.ruu.nl/akkadian/
    cuneiform.html**

Computational Linguistics

If you're really into the science of linguistics, the Association for Computational Linguistics devotes this collection of pages to their quarterly publication which focuses on the research and psychology of language.

URL:
    **http://www-mitpress.mit.edu/jrnls-catalog/
    comp-ling.html**

Culture in Conversation

Conversational styles and behaviors reflect a person's culture. Explore how mismatches in conversational behavior creates misunderstanding between cultures. This abstract involves a comparison of Japanese and American men as the basis for the study.

URL:
    http://www.iware.com/~sss/abstract.html

Deafe and Dumbe

Deafe and Dumbe is an essay written in the 1600s on the subtle art of reading lips and sign language. The essay also includes illustrations of early signing and information on speechreading, and it is written in an enjoyable Olde English style.

URL:
    http://mambo.ucsc.edu/psl/bulwer.html

Electronic Mailing Lists

This page lists electronic mailing lists on various topics in linguistics, including links to other linguistics resources. Just what you always wanted—a pen pal who enjoys discussing Head-Driven Phrase Structure Grammar in Swahili.

URL:
    http://www.ling.rochester.edu/lists.html

Foreign Languages for Travelers

Taking a trip to Germany, Norway, or another European country? Don't leave home without checking the Foreign Languages for Travelers page. Learn important key phrases and common questions, such as "Where can we eat?," "How much does this cost?," and "Where is the bathroom?." These pages offer tips for more than a dozen European languages.

URL:
    http://www.travlang.com/languages/

Funknet

Funknet is a mailing list for people studying any of the various aspects of language and communication, including cognition, sociology, psychology, and other topics relating to cognitive and communicative behavior.

Mailing List:
    Address: **listserv@ricevm1.rice.edu**
    Body of Message: **subscribe funknet** *<your name>*

Gerlingl

A mailing list for the discussion of older Germanic languages (up to about 1500 B.C.), their linguistics, and philology.

Mailing List:
    Address: **listserv@vmd.cso.uiuc.edu**
    Body of Message: **subscribe gerlingl** *<your name>*

Human-Languages Page

Enter this electronic Tower of Babel, a concise organization of language information. Here you'll find dictionaries, tutorials, and audio samples of languages from around the world—everything from Aboriginal to Yiddish. This extensive resource also has pointers to linguistics labs and translation services. But if it's all Greek to you, it might just as well be Klingon.

URL:
    **http://www.willamette.edu/~tjones/
    Language-Page.html**

A
B
C
D
E
F
G
H
I
J
K
L
M
N
O
P
Q
R
S
T
U
V
W
X
Y
Z

Excerpt from the Net...

(from the Foreign Languages for Travelers page)

```
Suomea: Ostokset/Ruokailu
Finnish: (Shopping/Dining)

How much does this cost? = Paljonko tämä maksaa?
What is this? = Mikä tämä on?
I'll buy it. = Minä ostan sen.
I would like to buy ... = Haluaisin ostaa...
Do you have ... = Onko sinulla ...
Breakfast = Aamiainen
Lunch = Lounaas
Dinner = Päivällinen
Vegetarian = Kasvissyöjä
Kosher = Kosher, Puhdas ruoka
Cheers! = Kippis!
Please bring the bill. = Saisinko laskun.
Do you accept credit cards? = Hyväksyttekö luottokortin?


Perussanasto (Basic Words)

Yes = Kyllä
No = Ei
Thank you = Kiitos
Thank you very much = Kiitos oikein paljon
You're welcome = Tervetuloa
Please = Pyydän
Excuse me = Anteeksi
Hello = Hei
Goodbye = Näkemiin
So long = Moikka
I do not understand = En ymmärrä
How do you say this in English? = Kuinka tämä sanotaan suomeksi?
What is your name? = Mikä sinun nimesi on?
Nice to meet you. = Mukava tavata sinut.
Where is the bathroom? = Missä on WC?
```

Increase Your Vocabulary

Each week, you can learn ten new words on this page. The definitions, roots, synonyms, and an example of each of the words is given. Don't miss your weekly chance to enhance your use of the English language.

URL:
> http://users.aol.com/jomnet/words.html

Looking for software? Try Software.

Indo-European Languages

Discuss and exchange ideas related to the historical and comparative linguistics of the Indo-European languages. Any topic related to the diachronic linguistics of the Indo-European languages is suitable for discussion.

Mailing List:
> Address: **listproc@cornell.edu**
> Body of Message: **subscribe indoeuropean-l**
> *<your name>*

Feel the squeeze in Compression.

Latin-American

Become a Latin-lover by joining this Latin-American linguistics and languages discussion list.

Mailing List:
Address: **listserv@mitvma.mit.edu**
Body of Message: **subscribe latamlin** *<your name>*

Linguistics Archive

This archive includes software, papers, digests, syllabi, handouts, and fonts for anthropological linguistics. And if you're not into anthropological linguistics, the fonts make for pretty cool coded messages. Where's the Rosetta stone when you need it?

URL:
ftp://ftp.uu.net/pub/linguistics

Lip-reading Resources

This page offers links to a variety of essays and resources relating to lip-reading. Articles here range from the beginning of recorded history on the subject to a recent NATO advanced study. So if your family's always complaining because the TV is too loud, you might want to read these pages before it's too late.

URL:
http://mambo.ucsc.edu/psl/lipr.html

Louisiana Creole

The development of a language is an ongoing, complicated process. Many factors—such as environment, ethnic background, and social class—help to shape our language. Creole dialects, discussed on this page, have an especially diverse background, combining speech characteristics from the French, African slaves, and Spanish settlers.

URL:
http://www.teleport.com/~napoleon/louiscreole/intro.html

Metaphors and More

This gopher is a linguistic glossary that includes many conventional English terms, and a few you've probably never heard. A key describes the setup of the glossary, and the contents offer terms and words as starting points for your search.

URL:
gopher://sil.org/11/gopher_root/linguistics/glossary/contents

Online English Writing Lab

English is one of the hardest languages to learn, and learning to speak it correctly is difficult—even for people born and raised in English-speaking countries. Gerunds, participles, proper punctuation, and those awful dangling modifiers—find out what these and other terms mean. Remember, I before E except after C (and except for a whole long list of exceptions).

URL:
http://owl.trc.purdue.edu/prose.html

Sure, English is tough. But what are ya going to do? All the best books, magazines, and technical journals are in English. Even all of the computer programming languages anyone might care to use are based on English words. So jump in and learn the parts of speech, about punctuation, sentences, and spelling. After perusing Purdue's Online Writing Lab, you'll have a hand on the King's English. (Just don't tell the Queen.)

A
B
C
D
E
F
G
H
I
J
K
L
M
N
O
P
Q
R
S
T
U
V
W
X
Y
Z

Need to network? See Networks.

Resources for Studying Human Speech

Breaking down human speech into its smallest parts is just part of what researchers are doing at the University of Washington. The "Resources for Studying Human Speech" page offers a tutorial covering physics and visualization of speech sounds using waveform analysis and spectrograms. You'll also enjoy the animated red-and-blue talking lips.

URL:
> http://weber.u.washington.edu/~dillon/
> PhonResources.html

Can you say "S O C K S"? Yes, but what does it mean . . . besides those grungy fiber tubes on the ends of your lower extremities. Learn about the phones and phonemes of English, and other resources for studying human speech on the University of Washington's web.

Semiotics for Beginners

Semiotics is the study of signs in society. Signs may be words or images from which you can derive meaning. As of yet, semiology has not been recognized as a true science; but in the framework presented on this page, linguistics is a part of this general science. Semiotics can include "anything that can be seen as signifying something."

URL:
> http://www.aber.ac.uk/~dgc/semiotic.html

Sign Language Linguistics

Here's a mailing list that discusses all types of topics relating to sign language linguistics. What's your sign?

Mailing List:
> Address: **listserv@yalevm.cis.yale.edu**
> Body of Message: **subscribe slling** *<your name>*

Syntactic and Historical Linguistics

This is a college-level tutorial on the framework, theories, and history of syntax. Don your cap and gown, and keep a dictionary open while reading this extensive lecture.

URL:
> http://www.entmp.org/linguistics/synthinar/

Signs, signs, Everywhere are signs.
If you're having trouble reading the writing on the wall, sign on to this tutorial on semiology to have it spelled out for you. Ponder paradigms and syntagms, denotation and connotation, and metaphor and metonymy in our society, culture, literature, and world. *Semiotics for Beginners* will signal the way.

What's your sign?

MALACOLOGY

APLYSIA Hometank

While still under construction, APLYSIA is an information resource for the molluscan neuroscience community. At present, members of the molluscan and electrophysiological communities are invited to sample a prototype top-level database query page and register their comments. In the future, there will be a searchable molluscan database and a listing of news, events, and meetings.

URL:
http://www.med.cornell.edu/Aplysia/
Hometank.html

Aquatic Mollusks of North Dakota

North Dakotans are no slouches on the Internet. Smack dab in the middle of the North American continent, they're out to show the world that among many other things, they have shellfish in their state, too. Toward this end, they've provided molluscan database maps showing precisely where their freshwater mussels, pill clams, and snails have been found and collected.

URL:
http://pastel.npwrc.r8.fws.gov/resource/distr/
invert/mollusks/mollusks.htm

Bishop Museum Malacology Page

Hawaii is home to several species of brilliantly colored land and freshwater snails that are found nowhere else on Earth. Watch for the Bishop Museum to feature these snails and other native mollusca in the near future; but, in the meantime, check out the rest of their natural history pages.

URL:
http://bernice.bishop.hawaii.org/bishop/mala/
mala.html

Cephalopod Bibliography Online Service

The Smithsonian Institute Research Information System (SIRIS) hosts this searchable database of approximately 4,500 references to publications, books, and journal articles on cephalopods and cephalopod-related subjects located at the Division of Mollusks at the National Museum of Natural History.

URL:
telnet:siris.si.edu

Cephalopod Page

Explore the world of nautiluses, squids, cuttlefish, and octopi on this definitive page. Cephalopod means "head foot"; and from the standpoint of complexity and behavior, they are at the apex of invertebrate evolution. Members of this class include the largest invertebrate, the giant squid. You'll find lots of taxonomical information and photographs, and links to other cephalopod pages and mailing groups.

URL:
http://is.dal.ca/~wood/www.html

Clam dip from North Dakota? "Balderdash!" you say. "Not so fast," say we. Mosey your browser over to Aquatic Mollusks of North Dakota, and mussel in on the info.

Clams in the U.S.

What is the biggest clam caught and eaten in the United States? What are oyster borers? How do oysters produce pearls? If you've been clammering for the answers to these and other clammy questions, dig up this page for answers and beautiful illustrations.

URL:
http://www.wh.whoi.edu/faq/fishfaq5a.html

EuroSquid

EuroSquid is easier to remember than "Fishery Potential of North East Atlantic Squid Stocks," so that's what it's called. What it's about is a source established by the Department of Zoology at the University of Aberdeen for the dissemination of information on squids and related research projects. You'll find a bibliographic list of department publications, some interesting text and graphics on squids, and a squid-research bulletin board.

URL:
http://www.abdn.ac.uk/~nhi104/

A B C D E F G H I J K L M N O P Q R S T U V W X Y Z

Field Guide to Freshwater Mussels of the Midwest

Muscle your way over to this online field guide published by the Illinois Natural History Survey. It includes 78 species of freshwater mussels found in Midwest streams and lakes, over half of which are either rare, threatened, endangered, or extinct. Thorough text and excellent maps and graphics accompany each species description.

URL:
http://www.inhs.uiuc.edu/chf/pub/mussel_man/cover.html

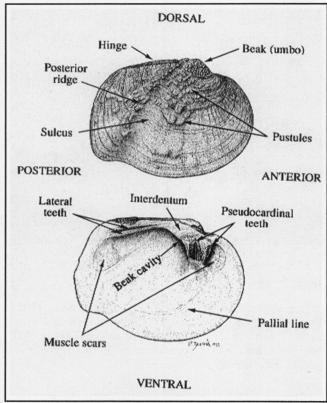

Learn about the anatomy of mussels with the *Field Guide to Freshwater Mussels of the Midwest*.

Introduction to the Brachiopoda

Explore the world of the brachiopods—marine animals that have a long and rich paleontological history. There are about 300 living species of brachiopods, and you can discover more about them here at this beautiful page with informative text and graphics.

URL:
http://ucmp1.berkeley.edu/brachiopoda/brachiopoda.html

Keys to the Invertebrates of the Woods Hole Region

If you live on the northeastern seaboard of the U.S., consult this online manual for the identification of the more common marine invertebrates around Woods Hole, Massachusetts. With your shell in hand, consult the database by clicking hypertext links describing your shell until you reach your particular species' identification.

URL:
http://www.mbl.edu/html/KEYS/INVERTS/14/14.0.html

Malacology Database at the Academy of Natural Science

The Academy of Natural Sciences in Philadelphia was founded in 1812, and is the oldest continually operating natural history museum in the U.S. Their malacology department boasts the second largest cataloged collection of marine invertebrates in the world. Their searchable online Gopher database currently has more than 430,000 cataloged lots containing about 12 million specimens.

URL:
gopher://erato.acnatsci.org

Mollusc DNA News

Read and join in discussions about research regarding molluscan DNA at this Usenet group.

URL:
news:bionet.molbio.molluscs

Mollusc Evolution Database

Search for information on your favorite molluscs or find online articles written by your favorite malacologist on this U.C. Berkeley database.

URL:
http://ucmp1.berkeley.edu/mollusca.html

Mollusc Evolution Mailing List

Get out of your shell by joining this forum for discussions of molluscan evolution, paleontology, taxonomy, and natural history. This list provides an interface between paleontological and neo-ontological molluscan workers, and also posts notices of meetings, symposia, literature, software, and other electronic happenings.

Mailing List:
Address: **listproc@ucmp1.berkeley.edu**
Body of Message: **subscribe mollusca** *<your name>*

Nudibranch Bibliography

To paraphrase Romeo's Juliet, "A sea-slug by any other name would still be a sea-slug." But thanks to this nudibranch database server, these beautiful creatures still get their credit. Nudibranchs are literally creatures that have naked, or exposed, gills, and do not possess a shell of any sort. Have compassion for them by checking out this cool bibliography database.

URL:

> gopher://gopher.ucsc.edu:70/7waissrc%3a/.WAIS/
> Nudibranch-bib.src

"Psssst . . . hey, buddy! Wanna read some racy stuff about nudibranchs? Yep, it's uncut and uncensored. And they're totally naked! What do you mean you're not into sea-slugs? Are you some kind of pervert, or something?" Expose yourself to the Nudibranch Bibliography before others do.

Shellfish Growing Area Classification Program

So many shellfish, so little time... Explore shortcuts to maritime coastal molluscan regional harvesting in French and English, as well as shellfish growing-area classification maps and detailed information regarding the Canadian shellfish sanitation program. You'll also be able to access the Atlantic Shellfish Classification Inventory.

URL:

> http://www.ns.doe.ca/epb/sfish/sfish.html

West Coast Marine Biology Field Guide

Here's a catchy online guide from CalPoly on common U.S. West Coast molluscs, gastropods, and sea hares. Pick a species' common name, and see a photo of it while learning its scientific name, habitat, and characteristics.

URL:

> http://www.calpoly.edu:8010/cgi-bin/db/db/
> marine-biology/Mollusks:/templates/index

Zebra Mussel Information

The zebra mussel *Dreissena polymorpha* is a pesky mollusc that inundates lakes and rivers. Zebra mussels have infested virtually every lock and dam in the upper Mississippi River, as well as several power generation plants. You can view status reports of the war on zebra mussels, see their current distribution, plot their progress with an online map, and find out when and where the next zebra mussel meetings will be held. You can also report any sightings online.

URL:

> http://www.nfrcg.gov/zebra.mussel/

MANUFACTURING

Consortium on Green Design and Manufacturing

The CGDM at U.C. Berkeley is a research organization dedicated to working with the manufacturing industry to facilitate the development of cleaner products and processes, and tools for management to implement in order to maintain these products and processes in a cost-effective way.

URL:

> http://euler.berkeley.edu/green/cgdm.html

Intelligent Manufacturing Systems

IMS is a small but international organization that promotes certain aspects of design and manufacturing processes to address product life cycles, process issues, and human and social issues.

URL:

> http://ksi.cpsc.ucalgary.ca/IMS/International/

A B C D E F G H I J K L M N O P Q R S T U V W X Y Z

Manufacturing Resources

Here at the mother of all manufacturing pages, you'll find dozens of links to great research material on manufacturing. A few of the topics are product introductions and disposals, manufacturing strategies, and engineering and technology.

URL:

http://www.warwick.ac.uk/~esrjf/manufact.html

MATHEMATICS

1000 Club

This Pi club is definitely not for home baking. Learn a thousand digits of Pi and you are in. If you are not that much of a number nerd, there is always the 100 Club. If you are a Pi rookie, just look into the Pi Club.

URL:

http://www.ts.umu.se/~olletg/pi/club_1000.html

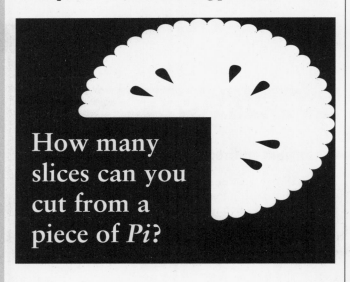

How many slices can you cut from a piece of *Pi*?

Babylonian and Egyptian Mathematics

Did you know that in 1700 B.C. the Babylonians had devised inverse tables that counted up to a billion? Separate reports from St. Andrews and Cambridge Universities are linked to this overview of similarities between the base 10 decimal system and the mathematical systems of the ancient Near East.

URL:

http://www.teleport.com/~ddonahue/
milo1st.html

Biographies of Women Mathematicians

In 1888, Sonya Kovaleskaya entered a French math competition with her paper, "On the Rotation of a Solid Body about a Fixed Point." The paper was so highly regarded that the prize money was increased from 3,000 to 5,000 francs. Similar stories and photos of dozens of women in the field of mathematics make up this ongoing project by students at Agnes Scott College.

URL:

http://www.scottlan.edu/lriddle/women/
women.htm

Blue Dog Can Count!!

Youngsters will enjoy filling in the formula and hearing their answers barked back at them by Blue Dog. Count the barks and they solve addition, subtraction, division, and multiplication problems.

URL:

http://kao.ini.cmu.edu:5550/bdf.html

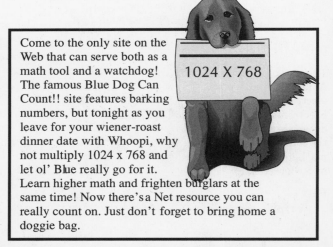

Come to the only site on the Web that can serve both as a math tool and a watchdog! The famous Blue Dog Can Count!! site features barking numbers, but tonight as you leave for your wiener-roast dinner date with Whoopi, why not multiply 1024 x 768 and let ol' Blue really go for it. Learn higher math and frighten burglars at the same time! Now there's a Net resource you can really count on. Just don't forget to bring home a doggie bag.

Center for Geometry Analysis Numerics and Graphics

The GANG gang at the University of Massachusetts, Amherst, maintains a modest but worthwhile site for interdisciplinary differential geometry research. Beautiful graphics accompany several MPEG animations with daunting titles, such as "Singly Perioidic Helicoid Deformation." There is additional information on the university's libraries, a share of software programs for workstations available through FTP, and links to other geometry resources.

URL:

http://www.gang.umass.edu

Clever Games for Clever People

This page sports pointers to 16 math games your child can play and learn about mathematics. Let them use their cleverness to develop new skills while they compete with friends. Each game explains what you need to play it and how. Parents, you'd better practice playing, too, or your child will outmaneuver you in no time.

URL:

http://www.cs.uidaho.edu/~casey931/conway/games.html

Conversion of Units

It's late at night, and you've got a hankering to know the relationship between Angstrom units and the Bohr radius. Look no further than this comprehensive calculator of conversion methods.

Note: Instructions in English and German.

URL:

http://www.chemie.fu-berlin.de/chemistry/general/units.html

e-MATH

There's some serious math going on here. e-MATH is the American Mathematical Society's resource for delivering electronic products and services to mathematicians. Electronic journals on e-MATH and the MathSci database are just a few items of interest available here to the mathematical community. e-MATH proves that a mathematician isn't just a machine for converting coffee into theorems.

URL:

http://e-math.ams.org

Explorer Outline for Mathematics

A page to keep kids counting and calculating forever, this hypertext database of K-12 mathematical resources includes lessons, curriculum notes, and software. The Explorer is a network database system for contributing, organizing, and delivering educational resources.

URL:

http://unite.ukans.edu/Browser/UNITEResource/MathematicsOutline.html

The Web will set you free.

The Fractal Microscope

The Latin word *fractus* means "to break up." When Benoit Mandelbrot discovered these beautifully bizarre mathematical images, he termed them "fractals" for their fragmented appearance. Now students in grades K through 12 can enjoy Mandelbrot's discoveries, where geometry mixes with art, as pupils master the science of mathematics. There is a stunning gallery of fractal images, good descriptions, and a bibliography of fractal reference books and papers.

URL:

http://www.ncsa.uiuc.edu/Edu/Fractal/Fractal_Home.html

Breaking up is hard to do, but if you're at the Fractal Microscope page, it's just part of the equation!

Gallery of Interactive Geometry

Math and art have long been intertwined. Now you can see this natural phenomenon for yourself using the University of Minnesota's Interactive Geometry page. Tweak a few variables and click the Send button to create custom Penrose tilings. In the process, you can select and visualize plane cross-sections of a lattice from 3 up to 13 dimensions! The eight interactive programs presented here are written using the W3Kit library, an object-oriented toolkit for building interactive World Wide Web applications.

URL:

http://www.geom.umn.edu/apps/gallery.html

The Geometry Center

Modern-day Euclids should make a bee-line for this page at the University of Minnesota. There are sections that cover all the angles on math and computer science research, geometrical application development, video animation production, and grades K through 16 math education. Don't miss the Gallery of Interactive Geometry, where you can square off with interactive programs such as QuasiTiler, Orbifold Pinball, and Unifweb.

URL:

http://www.geom.umn.edu

A B C D E F G H I J K L M N O P Q R S T U V W X Y Z

Googolplex

What is a googol? It's more than baby talk! If zero is your hero, you'll love this number. Add 100 zeros and you have a *googol*, but a *googolplex* stretches out even further. It's less than infinity, but almost as difficult to imagine. Have fun making something from a lot of nothings.

URL:

http://www.uni-frankfurt.de/~fp/Tools/ Googool.html

Graphics for the Calculus Classroom

Graphics and mathematics go together in real life, so why shouldn't they in the classroom, as well? Add some interesting animations, and you've got a first-rate tool for teaching a thorny branch of math, namely calculus. Fear not. This page will lead you into some excellent writings on the history and principles of each theorem advanced—from Archimedes' calculation of pi to how a ball bounces.

URL:

http://www.math.psu.edu/dna/graphics.html

Excerpt from the Net...

(from Largest Known Primes)

The Fundamental Theorem of Arithmetic shows that the primes are the building blocks of the positive integers: every positive integer is a product of prime numbers in one and only one way, except for the order of the factors.

The ancient Greeks proved (ca 300 BC) that there were infinitely many primes and that they were irregularly spaced (there can be arbitrarily large gaps between successive primes). On the other hand, in the nineteenth century it was shown that the number of primes less than or equal to n approaches n/(ln n) (as n gets very large); so a rough estimate for the nth prime is n ln n.

The Sieve of Eratosthenes is still the most efficient way of finding all very small primes (e.g., those less than 1,000,000).

Recently computers and cryptology have given a new emphasis to search for ever larger primes—at this site we store lists of thousands of these record breaking primes, all of which have over 1,000 digits!

The "Top Ten" Record Primes

On January 4, 1994 David Slowinski announced on the Internet that he and Paul Gage have found a new record prime: $2^{859433}-1$. The proof of this 258,716 digit number's primality (using the traditional Lucas-Lehmer test) took about 7.2 hours on a Cray C90 super computer. Richard Crandall independently verified the primality. (The complete decimal expansion of this prime is available here). The next largest primes and their discoverers are

```
2^756839-1 (227832 digits); Slowinski and Gage, 1992
391581*2^216193-1 (65087 digits); Noll and others, 1989
2^216091-1 (65050 digits); Slowinski, 1985
3*2^157169+1 (47314 digits); Jeffrey Young, 1995 (Fermat F157167 Factor)
9*2^149143+1 (44898 digits); Jeffrey Young, 1995
9*2^147073+1 (44275 digits); Jeffrey Young, 1995
9*2^145247+1 (43725 digits); Jeffrey Young, 1995
2^132049-1 (39751 digits); Slowinski, 1983
9*2^127003+1 (38233 digits); Jeffrey Young, 1995
```

The History of Numbers

We've got numbers by the trillions. The Babylonians started it all with their counting systems, eventually adopted by the Greeks. Then a famous mathematician (and song writer) named Paul Simon wrote, "When times are mysterious, serious numbers will speak to us always." This page also speaks to us with a historical overview of mathematics. A number of links leads to an enumeration of other numerical sources.

URL:
http://www-groups.dcs.st-and.ac.uk/~history/
HistTopics/History_overview.html

Inverse Symbolic Calculator

Deep Thought, the computer in Douglas Adam's *Hitchhiker's Guide to the Galaxy*, calculated that 42 is the answer to "life, the universe, and everything." According to the Inverse Symbolic Calculator, 42 is the answer to 67,610 functions and specialized tables of mathematical constants. Use the ISC to identify real numbers or produce identities with functions and real numbers. Some of the math gets pretty heady here, so you may want to cushion the shock of venturing in with a swig or two from a Pangalactic Gargleblaster.

URL:
http://www.cecm.sfu.ca/projects/ISC.html

Largest Known Primes

An integer greater than one is called a prime number if its only positive divisors are 1 and itself (like 2, 3, 5, 7, 11 and 13). Prime numbers are important in many math and computer science applications, such as cryptography and hashing. Learn all about primes (including the largest one known to date) and how they're used.

URL:
http://www.utm.edu/research/primes/
largest.html

Lessons, Tutorials, and Lecture Notes

Running out of math problems for your students? Run to the Mathematics Archives for dozens of new instructional ideas.

URL:
http://archives.math.utk.edu/tutorials.html

Mathematical Animation Gallery

Make those numbers dance right before your eyes! MAG shows you an image and the math formula that created it, and gives you the option to animate the whole kit-and-caboodle in either QuickTime or MPEG formats. You'll be rotating parametric plots around a Z axis in no time!

URL:
http://mathserv.math.sfu.ca/Animations/
animations.html

Mathematical MacTutor

Mathematical MacTutor is particularly strong in calculus, geometry, algebra, group theory, graph theory, number theory, and the history of mathematics. It has some interesting stacks on statistics, matrices, and complex analysis. In 1992 it won the Partnership Trust prize for innovation in mathematics teaching.

Note: Requires a Macintosh system.

URL:
http://www-groups.dcs.st-and.ac.uk/~history/
Mathematical_MacTutor.html

Mathematical Quotation Server

"Life is good for only two things, discovering mathematics and teaching mathematics" —Siméon Poisson. If you're looking for mathematical wisdom and witticisms, this page points to a collection of quotations culled from many sources. Conduct a keyword search to find a specific aphorism.

URL:
http://math.furman.edu/~mwoodard/mquot.html

MathMagic

Encouraging young mathematicians through telecommunications, MathMagic provides motivation and challenges through problem-solving strategies and communications skills. Any teacher with access to Internet electronic mail can participate.

URL:
http://forum.swarthmore.edu/mathmagic/
what.html

Anyone can browse the Web.

A B C D E F G H I J K L M N O P Q R S T U V W X Y Z

MATLAB Gallery

You can find movies and GIFs of math in action on this MathWorks-sponsored page. Several of the most notable to explore are Klein bottles, spherical harmonics, quiver plotting, the Sierpinski arrowhead, and an undulating Earth.

URL:

http://www.mathworks.com/gallery.html

Morphogenesis

Solve reaction-diffusion equations with Xmorphia, a tool derived from the Gray-Scott model illustrating the variety of patterns exhibited by a relatively simple parabolic partial differential equation. Did you know that such systems reconcile the complex organic look of the solutions with the simplicity of the equations being solved? Well, they do.

URL:

http://www.ccsf.caltech.edu/ismap/image.html

Multidimensional Analysis

Professor George W. Hart explains how traditional linear algebra causes one to miss the real mathematical properties of vectors and matrices. These structures contain dimensioned elements. Discover why linear algebra cannot be isomorphic to the algebra that scientists and engineers really need to use.

URL:

http://www.ctr.columbia.edu/~hart/
multanal.html

NA-Net

The NA-Net is a system developed to serve the community of numerical analysts and other researchers. It provides two independent databases and a weekly digest of articles on topics related to numerical analysis and those who practice it.

URL:

http://www.netlib.org/na-net/na_home.html

The Pavilion of Polyhedreality

Peek into this pavilion of practical polyhedra. All images are accompanied by short musical selections. Just click on any of the red notes before selecting a picture, and you can listen to a J.S. Bach piece as background music while each image is transferring.

URL:

http://www.li.net/~george/pavilion.html

Problem Solving

Today's math is complex, and half the battle is knowing how to go about solving problems. At this site, a variety of examples of word problems are given, as well as advice on how to explore and understand a problem.

URL:

http://www2.hawaii.edu/suremath/
solutionPaths.html

Excerpt from the Net...

```
(from Multidimensional Analysis)

Pop Quiz!

If you think that you know linear algebra and that there is nothing new and interesting
about matrices with scientific and engineering quantities like 1 meter, then take this
simple quiz, and you'll find out a few things:

1.Note that X above has no determinant, while Y and Z do have a determinant: the product
of the off-diagonal elements of X can not be subtracted from the product of the diagonal
elements of X, as would be necessary in the calculation of a 2-by-2 determinant. One might
now hypothesize a conjecture along the lines that "a square dimensioned  matrix has a
determinant if it can be squared." That conjecture is too strong however. The "if" part
holds but not the "only if." Find a 2-by-2 counterexample, i.e., a matrix with a
determinant but which can not be squared.

2.Find a square 2-by-2 matrix P such that P times P inverse gives a different result from P
inverse times P. Of course, in traditional linear algebra, the product of a matrix and its
inverse is the same regardless of order, (assuming an inverse exists,) but you're not in
Kansas anymore.

When you have solved these, check your answer here.
```

Shape Up

Geometry is just a bunch of shapes. Another creation by kids, this book covers shapes, forms, and concepts through children's unique observations and drawings. Fun to read.

URL:
http://192.152.5.115/room8geometry/ square_numbers.html

Slide Rule Home Page

Are you tired of LED displays and accuracy to 10 decimal places? Revive your old slide rule and visit the Slide Rule Home Page to reintroduce yourself to these mechanical wonders. For those of you who have no idea what a slide rule is, this page starts at the beginning and describes what they are, how they work, types of slide rules, and other interesting related topics.

URL:
http://photobooks.atdc.gatech.edu/~slipstick/ slipstik.html

Two Plus Two Equals Saxon Publishers

For the K-12 grades, this company publishes educational books and programs. This site offers information on book lists; authors; teachers; technical help; and many other resources of value to kids, teachers, parents, and anyone interested.

URL:
http://www.saxonpub.com

Uno, Dos, Tres...

With colorful pictures, even the youngest child can learn to count in Spanish. This enjoyable book was assembled by kindergartners.

URL:
http://buckman.pps.k12.or.us/spanish/ spanishbook.html

Web Spirograph

Who needs those chunky old store-bought Etch-A-Sketch toys when you can create your scribblings digitally on the Web!? Not only can you create cool patterns, but you learn about math and geometry at the same time.

URL:
http://juniper.tc.cornell.edu:8000/spiro/ spiro.html

Web Spirograph

With online Etch-A-Sketch and Lite Brite pages, Spirograph couldn't be far behind. You may have wasted hours jamming pens into those plastic gears and churned out monotonous pages as a kid, but this time you might actually learn something because it is all based on some simple mathematics. Simply enter the radii of the fixed and the rotating circles, as well as how far the pen is placed from the edge of the rotating circle. Positive values for the offset represent points outside the rotating circle and negative values are within. No assembly required. Batteries not included.

MEDICINE

Alzheimer Web

This is perhaps the best place to look for research information on Alzheimer's disease. From this site you can find out new information about diagnosis and treatment. Included also is a long list of research institutions that study the disease and offer home pages.

URL:
http://werple.mira.net.au/~dhs/ad.html

American Medical Association

One of the most influential lobbying groups in the medical profession, the American Medical Association offers this page to medical students and doctors. Among AMA resources and publications, there are directories, journal abstracts, and news about its current projects in Washington, D.C.

URL:
http://www.ama-assn.org/home/amahome.htm

A
B
C
D
E
F
G
H
I
J
K
L
M
N
O
P
Q
R
S
T
U
V
W
X
Y
Z

Archives of Dermatology

Read the most recent abstracts from the oldest dermatology journal in America. (How do we know it's the oldest? Because it has the most wrinkles, of course.) Provided by the American Medical Association, physicians and other health professionals are itching to explore this site to save valuable time. Select one of the last six issues of the journal to read abstracts of recent breakthroughs, techniques, and research.

URL:
> http://www.ama-assn.org/journals/standing/
> derm/dermhome.htm

Archives of Family Medicine

Compiled by the American Medical Association, this journal publishes clinical and research studies on family medicine. It includes articles about clinical topics, technological advances, and preventive medicine, to name just a few. The Archives of Family Medicine is a valuable resource for physicians and health-care professionals. On this page, you can peruse the table of contents for the current monthly issue, read up to six months of back issues online, and get directions for subscribing to the paper version of the publication.

URL:
> http://www.ama-assn.org/journals/standing/
> fami/famihome.htm

Archives of Internal Medicine

One of many AMA publications, the Archives of Internal Medicine includes original articles on daily patient care in general internal medicine and subspecialty areas.

URL:
> http://www.ama-assn.org/journals/standing/
> inte/intehome.htm

Archives of Neurology

Also published by the American Medical Association, the Archives of Neurology is online at the AMA's web site and is available for physicians, health-care professionals, and anyone else interested. You can see the table of contents for the current issue, read back issues, and search for abstracts of journal articles on the site. Subscription information for the paper versions of each of these AMA publications is included on each page.

URL:
> http://www.ama-assn.org/journals/standing/
> neur/neurhome.htm

> ## The Net is humanity's greatest achievement.

Archives of Ophthalmology

See your way clear to this page that was created especially for health professionals and physicians with an eye for information. Scan the past six months of issues for abstracts of articles on new instruments, socioeconomics, epidemiology, and biostatistics.

URL:
> http://www.ama-assn.org/journals/standing/
> opht/ophthome.htm

Archives of Otolaryngology

This AMA publication focuses on Otolaryngology—head and neck surgery. Like the other AMA publications on these pages, you can view the table of contents of the current issue (but not actually read the articles). Abstracts of articles in past issues are online and subscription information is given.

URL:
> http://www.ama-assn.org/journals/standing/
> otol/otolhome.htm

> Stand head and shoulders above the rest of your peers with knowledge gleaned from the Archives of Otolaryngology. Stick your neck out, but be mindful when you speak, or you might want to duck.

Archives of Psychiatry

This frequently cited journal in psychiatry is available online thanks to the American Medical Association. Physicians and other health professionals can browse the last six months of the journal. The articles in this journal deal with the biological origins of mental disorders and recent developments in drug research.

URL:
> http://www.ama-assn.org/journals/standing/psyc/psychome.htm

Archives of Surgery

This AMA publication focuses on surgery. Like the other AMA Archive journals listed above, the table of contents for the current journal and abstracts of articles from recent issues are available.

URL:
> http://www.ama-assn.org/journals/standing/surg/surghome.htm

Arthritis Foundation

The Arthritis Foundation runs this Web server to offer volumes of information on arthritis research and the search for a cure and ways to prevent this debilitating affliction. Of special interest to sufferers of arthritis is a question-and-answer section that contains brochures about each of the different types of arthritis and pain management. The foundation also offers links to many related resources on the Web and local chapters of the organization.

URL:
> http://www.arthritis.org

Breast Cancer Network

EduCare's thoughtful page includes valuable patient and clinical information on breast cancer. Patient resources include guides to help cope with the disease, self-diagnosis, and support groups. Clinical resources offer instructional material for health professionals and consultants.

URL:
> http://www.cancerhelp.com/ed/

> ## Kids, check out
> ## Kids and Amateurs.

CDC Home Travel Information

If you are thinking about leaving the country and would like the U.S. government to allow you to come back, you'd better visit this page to find out what vaccinations you will need to get before you take that trip. In addition, the Center for Disease Control provides a guide to the health risks and specific recommendations for your safety in different regions of the world.

URL:
> http://www.cdc.gov/travel/travel.html

CDC Prevention Guidelines

Every disease you can imagine, and many you can't, are listed in this A-Z guide to prevention from the Center for Disease Control. Topics range from AIDS to yellow fever. This excellent resource includes multiple guides for further reading on each disease.

URL:
> http://wwwonder.cdc.gov/wonder/prevguid/topics.htm

Center for Disease Control

The Center for Disease control tracks diseases around the country and the world. This will give you up-to-date information about outbreaks of disease and recommendations about how to prevent the spread of disease. You can get information about established diseases such as rabies, malaria, cholera, AIDS, and many others.

URL:
> http://www.cdc.gov/cdc.htm

Cure Paralysis Now

Cure Paralysis Now is a nonprofit organization dedicated to finding a cure for paralysis. Topics here include discussions of recent breakthroughs in research into restoring nerve functions, paralysis FAQs, graphics and nomenclature that describe the spinal column, and links to other related sites on the Internet. Information about spinal injuries is thoughtfully included, and you can make your own comments and read those of others.

URL:
> http://www.cureparalysis.org

A B C D E F G H I J K L M N O P Q R S T U V W X Y Z

Dental CyberWeb

Here's a place in cyberspace where bytes and bites go together. Dental CyberWeb is devoted to the exchange of dental information with links to public and professional areas, advertising and educational inquiries, and related dental resources on the Net. But what happens to a computer when it has a bad case of overbyte?

URL:
http://www.vv.com/dental-web/

Dental Trauma

If you're one of the 30 percent of people who have suffered dental trauma from an accident, here's a page you can sink your, uh, dentures into. This technical introduction to dental trauma research includes information about causes, treatments, statistics, and graphs on permanent and deciduous dentition.

URL:
http://www.unige.ch/smd/ortdent.html

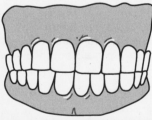

Be true to your teeth, and they'll never be false to you. If you're chomping at the bit for Dental Trauma information, better brush up here first.

Diabetes Documents

People suffering from diabetes have a place to come for detailed information about their affliction. Sponsored by the National Institute of Health, this page has links for information about the types of diabetes, including insulin- and non-insulin-dependent diabetes; diabetic eye disease; hypoglycemia; and overviews and statistics on diabetes in general.

URL:
http://www.niddk.nih.gov/DiabetesDocs.html

Dialysis and Chronic Renal Failure

Here's a collection of informative documents about dialysis and renal failure. These articles are organized by topic (such as Basic Hemodialysis, Peritoneal Dialysis, and Chronic Renal Failure); by Problem Area; and by Organ System. The documents focus on a wide variety of focused topics under these categories, including nutrition, lipids, infections, urological, and gynecological aspects.

URL:
http://www.medtext.com/fdh.htm

Digestive Diseases

If you've been aching for someone to explain the numerous digestion-related maladies, here's your page. Clear explanations of heartburn, hiatal hernias, hemorrhoids, lactose intolerance, gallstones, Crohn's Disease, constipation, and diarrhea are let loose for your perusal. A lengthy list of organizations for patients and health professionals is included.

URL:
http://www.niddk.nih.gov/DigestiveDocs.html

Emergency Services Web Site List

This fantastic server connects you with emergency services throughout the world. Besides providing a good list of links for emergency medical services, there are other connections to law enforcement agencies; fire departments; disaster management and hazardous materials teams; and many other interesting links and tidbits of information.

URL:
http://gilligan.uafadm.alaska.edu/www_911.htm

Emotional Trauma

You can experience emotional trauma when a loved one has died, when you've survived a near-death experience, and when it's April 15th. Seek a good counselor (or accountant) when you need comforting, but check out this page to understand the biological roots and symptoms of emotional trauma.

URL:
http://gladstone.uoregon.edu/~dvb/trauma.htm

Food and Drug Administration

For information on new medications, foods, cosmetics, biologics, or toxicology, visit the home page for the U.S. Food and Drug Administration.

URL:
http://www.fda.gov

Hodgkin's Disease

A pretty, 29-year-old woman named Amanda who lives on the south coast of England has created what may be the world's foremost resource for sufferers of Hodgkin's Disease. Amanda describes her own personal battle with this disease from before the time she was diagnosed to present. She explains in great detail her thoughts, fears, and the ongoing treatment she is enduring. On her pages, Amanda also provides links to a great deal of technical information about HD and other forms of cancer and its treatments.

URL:

http://www.dircon.co.uk/adastra/amanda/
amanda1.html

HospitalWeb

HospitalWeb provides you with links to every hospital that is on the World Wide Web. Although this list is far from complete, many of the major medical centers in the country are listed here. If you are looking for more information about departments and specialties of particular hospitals or medical centers, this is a good starting place. The hospitals in this list are categorized by state.

URL:

http://dem0nmac.mgh.harvard.edu/
hospitalweb.html

Lupus Home Page

Brought to you by Hamline University, the Lupus Home Page is the quintessential body of knowledge on lupus. Read and learn about the symptoms of lupus, how to live well with the disease, and just exactly what it is. These pages offer a ton of helpful information on lupus and other similar diseases, such as kidney disease, skin disease, heart disease, and blood disorders.

URL:

http://www.hamline.edu/~lupus/

Medical Newsletters and Journals

Medical journals and newsletters abound on the Internet. Although some journals are available in their entirety only by subscription, many of the newsletters are complete and informative.

URL:

http://www.dsg.ki.se/journal/medweb.html

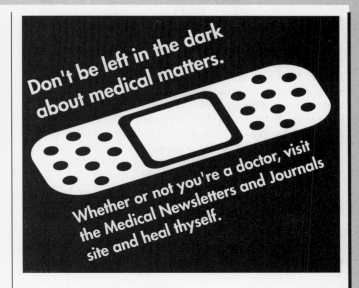

Don't be left in the dark about medical matters. Whether or not you're a doctor, visit the Medical Newsletters and Journals site and heal thyself.

Medical School and the MCAT

So, you want to be a doctor, eh? The drama of the trauma center. The prestige of being a foremost surgeon. The thrill of driving your $90,000 European sports car. The joy of medical school, endless exams, tests, and internships. Nobody said it would be easy! If you still want to be a physician, check out this page for information on the MCAT exam, test dates, AMCAS, applications, interviews, and critical school rankings.

URL:

http://www.kaplan.com/mcat/

Medical Schools in the United States

Created and maintained by the Vanderbilt University Medical Center, this page will connect you directly to virtually every medical school in the U.S. Depending on the school, various learning resources are available for doctors, medical students, and anyone with an interest in the field of medicine.

URL:

http://vumclib.mc.vanderbilt.edu/~aubrey/
medstu/medical_schools.html

Looking for a science project? Look in Kids and Amateurs.

A
B
C
D
E
F
G
H
I
J
K
L
M
N
O
P
Q
R
S
T
U
V
W
X
Y
Z

Medicare and Medicaid Information

Like the Medicaid system itself, this is a complicated site from the Health Care Financing Administration of the federal Medicare and Medicaid programs. Available are up-to-date guides, handbooks, and pamphlets of interest to Americans who are concerned about how their tax dollars are being used (to pay for simple bandages and rolls of gauze that cost dozens of times less in local drug stores, for example). Billions of taxpayer dollars could be saved every year by reforming the laws that prohibit Medicare from bidding competitively for medical supplies and services.

URL:
http://www.ssa.gov/hcfa/hcfahp2.html

MedWeb: Cardiology

This is a comprehensive list of sights relating to cardiology on the Internet. Included are many national societies, organizations, and cardiology centers throughout the country.

URL:
http://www.cc.emory.edu/WHSCL/
medweb.cardiology.html

Mike's Lymphoma Resource Page

Mike, a cancer survivor, compiles this page that describes the diagnosis and treatment of Hodgkin's and non-Hodgkin's lymphoma. Mike has also included a list of related cancer sites, and he gets down to the bone with details on marrow transplants.

URL:
http://users.aol.com/kittyba/lymphoma.html

Multiple Sclerosis

Aapo Halko, a 33-year-old mathematician, researcher, and student in Helsinki, Finland, has created the consummate Multiple Sclerosis resource. On his MS page, Aapo provides an abundance of information about MS, its symptoms, treatment, and ongoing research. He also offers links to other MS web pages and resources on the Net.

URL:
http://www.helsinki.fi/~ahalko/ms.html

National Cancer Institute

The National Cancer Institute is the principal government agency devoted to cancer research and training. Connect to public data sources, intramural and extramural research projects, and the International Cancer Information Center.

URL:
http://www.nci.nih.gov

National Institute of Allergies and Infectious Disease

This federally funded institution has created research and education projects to aid in the study of infectious diseases—especially those that relate to the human immune system. Other topics include organ transplantation, asthma, allergies, genetics, minority concerns, and AIDS.

URL:
http://web.fie.com/web/fed/nih/

National Institute on Aging

Contributed by the National Institute of Health, this web page contains information about the effects of aging—including some graphic images of human brains. For neurological questions, this is a fine place to begin your research. If you already have a background in the neurological sciences, abstracts of recent articles are included for your convenience.

URL:
http://adobe.nia.nih.gov

National Institutes of Health

The NIH is the central government distributor of research funds for medical research in the U.S. This is a great place to look for information on research labs and other scientific resources prepared by the government. Here, you can get many general health questions answered, and keep tabs on how your tax dollars are being spent at the same time!

URL:
http://www.nih.gov

A B C D E F G H I J K L **M** N O P Q R S T U V W X Y Z

Excerpt from the Net...

(from National Institute of Allergies and Infectious Disease)

Tropical Medicine

NIAID pays special attention to the six diseases (filariasis, leishmaniasis, leprosy, malaria, schistosomiasis, and trypanosomiasis) selected for emphasis by the World Health Organization/World Bank/United Nations Development Programme Special Program for Research and Training in Tropical Diseases (TDR/WHO) because of their global importance and the fact that U.S. citizens may be exposed to these conditions. For example, approximately 1,000 civilian cases of malaria are reported each year resulting from foreign travel, immigration, or transfusions or associated with an endemic focus in the San Diego area. In addition, efficient insect vectors for malaria and other tropical diseases are present in many areas and may be responsible for reintroducing the diseases

In November 1991, troops who served in Operation Desert Storm were excluded from blood donations for 2 years because of the discovery of largely asymptomatic disseminated leishmaniasis in several veterans. Blood transmission of American trypanosomiasis (Chagas' disease) in the United States also was recently documented; an epidemic of mosquito-transmitted encephalitis has occurred in Florida; and epidemic cholera, which appeared in the Americas for the first time this century in 1991, has spread to all Central American and South American countries except Uruguay.

Fortunately, the ongoing revolution in immunology and molecular biology has produced powerful research tools that are being used to obtain a better understanding of disease agents and the immune response they provoke in animal models and humans and to broaden the science base to develop new or improved diagnostic tests, vaccines, therapeutic agents, and prevention or control efforts.

National Physician Assistant Page

A physician may be told to "heal thyself," but the doctor may need some help healing others. That's where you come into this growing medical field, the Physician Assistant profession. General information about the profession includes a complete listing of the training programs available, job opportunities, and links to related resources.

URL:
http://www.halcyon.com/physasst/

Neuroscience on the Internet

You don't have to be a brain surgeon to use this search engine. Just type in any word relating to the neurosciences and you are sure to find what you're looking for. In addition to the searchable database, this page offers links and directions for using other Internet indices, as well.

URL:
http://ivory.lm.com/~nab/cusi.html

Pediatrics and Adolescent Medicine

Commentary, case quizzes, articles on sports medicine, editorials, and practice management advice can be found here in the oldest journal in U.S. pediatric literature, the *Archives of Pediatrics and Adolescent Medicine*.

URL:
http://www.ama-assn.org/journals/standing/ajdc/ajdchome.htm

Planned Parenthood Online

Planned Parenthood provides information about contraception, abortion services, women's health, STDs, and pointers to other locations on the Web with more related information. Also among these pages is a list of all the health center locations in the U.S.

URL:
http://www.ppca.org

Prostate Cancer InfoLink

Find out if you are at risk for prostate cancer at this site created by the CoMed Communications Internet Health Forum. The Prostate Cancer InfoLink includes information about the diagnosis and treatment of prostate cancer, as well as where to turn for help and support. Especially if you're male and the years are adding up, you'll want to check in here and read up on prostate cancer.

URL:
http://www.comed.com/Prostate/

Skin Cancer: An Introduction

Before you leave the house, learn the facts about skin cancer and how you can improve your odds at avoiding it. These pages bring together information from a number of government sources and physician groups to inform you of the dangers of skin cancer and preventative measures you can take.

URL:
http://www.maui.net/~southsky/introto.html

They all laughed when I sat down to play at the organ, but they didn't know it was a transplant. If you've got the heart to stomach the exchange of vital body part information at the Transplantation Resources on the Internet page, then we surely see eye to eye.

Transplantation Resources on the Internet

You left your heart in San Francisco. Your kidney in Kalamazoo. And your lung in Ft. Lauderdale. At least you kept your brain so you could buy this wonderful book and learn about transplantation resources on the Internet. Visit transplant centers, organ registries, and online donor networks to locate valuable body parts.

URL:
http://www.med.umich.edu/trans/transweb/resources_index.html

TraumaNET

Created and maintained by Louisiana State University, TraumaNET is a good jumping off point to many other trauma and critical care sites and organizations on the Internet. A few of the links here lead to the Flightweb for Aeromedical Resources on the Net, Mothers Against Drunk Driving, the TraumaNET Gopher, and the Emergency Services Guide.

URL:
http://www.trauma.lsumc.edu/htmls/homepage2.html

Urologic Disease Documents

Here's a collection of basic learning guides on the subject of urologic diseases, including prostate enlargement, urinary tract infections, Peyronie's Disease, and interstitial cystitis. Even more valuable than these informative guides is the directory of kidney and urologic disease organizations—a comprehensive list of all the major institutions around the country that study these diseases, complete with their addresses and a brief synopsis of the activities and research at each institution.

URL:
http://www.niddk.nih.gov/UrologicDocs.html

Virtual Nursing Center

This is a useful web sight for nurses and other hospital staff. This site contains medical dictionaries, glossaries, online nursing journals, information about educational courses, disease and poison databases, and pharmacy information.

URL:
http://www-sci.lib.uci.edu/~martindale/Nursing.html

Vision Science

Provided by NASA (although we're not sure why), this page is the Web's premier launching pad for research into vision and ophthalmic sciences. There's no actual research here, but there must be hundreds of links to vision and ophthalmic research groups, institutes, university departments, organizations, conferences and symposia. You'll also find software, journals, bibliographies, commercial products, tutorials, FAQs, and links to other Internet resources, including Usenet groups and mailing lists.

URL:

http://vision.arc.nasa.gov/VisionScience/
VisionScience.html

World Health Organization

The World Health Organization produces reports about global radiation, eradication of diseases, and statistics about world health and medical problems. On their home page, they make these reports and findings available, along with press releases, newsletters, and other forms of general information about the organization and world health issues.

URL:

http://www.who.org

You're Going to Die (Eventually)

Find out just how long you have left to live by playing The Longevity Game. The online game is the folly of statisticians from Northwestern Mutual Life, an insurance company. (Who else gets paid to predict when you're going to die?) Answer questions about your lifestyle and health, then click the Evaluate button to see if you win the game. The high score is held by someone who is 120 years old now and lives a healthy lifestyle. If that's you, look forward to your 141st birthday. Otherwise, you lose.

URL:

http://www.northwesternmutual.com/
longevit/longevit.htm

METEOROLOGY

Atmospheric Technology

Dedicated to providing scientists with atmospheric observing systems, this page includes information on technology projects. Topics include the Winter Icing and Storms Project, operation of a research aircraft, atmospheric sounding systems, remote sensing systems, and analysis software.

URL:

http://www.atd.ucar.edu

Center of Ocean-Land-Atmosphere Studies

COLA is a part of the Institute of Global Environment and Society (IGES), and it is dedicated to researching and predicting climate changes and interactions of the atmosphere, land, and the oceans. This page offers information on the projects, activities, and faculty of the center, as well as the text of lectures and seminars, current weather forecasts, and much more.

URL:

http://grads.iges.org/home.html

Climate Model Diagnosis and Intercomparison

The Program for Climate Model Diagnosis and Intercomparison is an atmospheric research program that develops tools and methods to improve the study of global climate models (GCMs). This page provides papers, data, software, and information relating to the program's projects and research.

URL:

http://www-pcmdi.llnl.gov

Climate Research Programs

Here's a good-sized list of climate research programs underway worldwide. Included are climate norms for the U.S., nineteenth century weather data recorded at military posts, and global climate norms. A general description of the objectives, background, resources, and impact of each program is given.

URL:

http://www.ncdc.noaa.gov/homepg/climres.html

The Web is too cool.

A B C D E F G H I J K L M N O P Q R S T U V W X Y Z

Current Weather Images

Nearly real-time images, movies, and interactive reports of current weather conditions in the United States. You can find many current satellite photos, video, and infrared images here.

URL:

http://wxweb.msu.edu/weather/

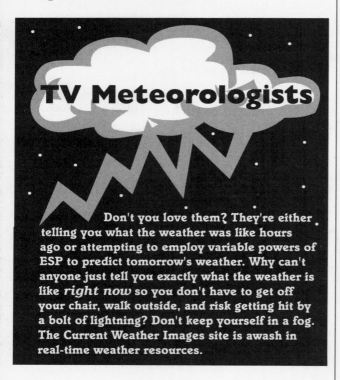

TV Meteorologists

Don't you love them? They're either telling you what the weather was like hours ago or attempting to employ variable powers of ESP to predict tomorrow's weather. Why can't anyone just tell you exactly what the weather is like *right now* so you don't have to get off your chair, walk outside, and risk getting hit by a bolt of lightning? Don't keep yourself in a fog. The Current Weather Images site is awash in real-time weather resources.

Daily Planet

Great Caesar's ghost! With Jimmy Olson, Lois Lane, and Clark Kent constantly involved in some wacky adventure, the Daily Planet was left with only its weather staff and was forced to refocus its editorial content. Our superhero and friends may be confined to a life of late night cable TV re-runs, but we can watch the *Daily Planet* web page for current weather maps, satellite images, multimedia learning modules, forecasts, and lists of popular weather resources.

URL:

http://www.atmos.uiuc.edu

Everyone should be on the Web.

If you're not on the Web, you don't exist.

Earth Space Research Group

"Donde está the little boy" in Spanish means "Where is El Niño?" You can find out on the Earth Space Research Group page from the University of Santa Barbara. This site includes information on Indian Ocean monsoons, the ozone layer, and El Niño reports. Data comes from satellite pictures, mappings, and other advanced imaging technologies.

URL:

http://www.crseo.ucsb.edu/esrg.html

Global Energy and Water Cycle Experiment

GEWEX is a research program that investigates the hydraulic cycle and energy fluxes in the atmosphere, on the surface of the land, and in the top layers of the ocean, to predict changes in the climate. This page includes information on GEWIX projects, news, data, reports, and documents, and offers more specific information on projects such as the ones related to radiative fluxes.

URL:

http://www.cais.com/gewex/gewex.html

Gold Coast Weather

This site is designed for quick access to Internet weather and ocean information with emphasis on items of interest to residents of California's Gold Coast (San Luis Obispo, Santa Barbara, and Ventura counties). The information is updated daily with satellite images and charts.

URL:

http://www.vcnet.com/goldcoastwx/
cscalwx.html

Hurricane Watch

Board up the doors and windows, the wind is a-blowin'! This page offers information on hurricanes around the world. With maps, historical information, images, and news about tropical storms, hurricanes, and severe weather, you'll be prepared in the event Mother Nature gets long winded.

URL:

http://www.sims.net/links/hurricane.html

Excerpt from the Net...

(from The Storm Chaser Homepage)

DISCLAIMER: Storm chasing exposes one to many weather hazards, such as lightning, dangerous roads, damaging winds, hail, and flying debris which puts the chaser's life at risk, particularly those who have little or no experience and/or storm structure education. Learning to deal with these is best done by understanding supercells and thunderstorms, and driving with an experienced chase partner. The author(s) of these pages and the contents therein is (are) not responsible for any of your actions as a result of what you see here!

LDEO Climate Data Catalog

The Climate Data Catalog is both a catalog and a library of datasets. It assists in locating the climate data you want and works with it once you've found it. The interface allows you to make plots, tables, and files from any dataset, its subsets, or processed versions thereof.

URL:
http://rainbow.ldgo.columbia.edu/datacatalog.html

Monthly Temperature Anomalies

Create a contour map or a time series plot of temperature anomalies derived from the merged Global Historic Climate Network and the Monthly Climatic Data of the World. The station data are quality controlled in time and space and then mapped to a 5x5-degree grid. The color scale represents anomalies in degrees Celsius. The generated image is suitable for display on a color monitor or a PostScript device.

URL:
http://www.ncdc.noaa.gov/onlineprod/ghcnmcdwmonth/form.html

National Center for Atmospheric Research

Explore NCAR's many activities, programs, and research efforts to better understand the Earth's climate systems.

URL:
http://www.ucar.edu

NJ Online Weather

Get a five-day forecast of the weather outlook in your city, complete with pictures. Enter your ZIP code, and you'll know at a glance what the week has in store. Entries from *The Old Farmer's Almanac* are also presented.

URL:
http://www.nj.com

Storm Chaser Home Page

You'd have to be crazy to spend endless hours on the road in treacherous weather conditions, plotting midcourse corrections whenever the wind changes, eating on the run, and all for what? To risk your life hunting tornadoes and hurricanes! If you're a crazy storm chaser, too, hunt down this page to exchange the latest weather information with the National Weather Service (NWS) and other chasers.

URL:
http://taiga.geog.niu.edu/chaser/chaser.html

Storm Chasing

This meteorologist has been chasing tornadoes since 1972. Although this sounds like it might be a fraternity prank, it actually requires good forecasting methods and extensive safety precautions. It's interesting reading.

URL:
http://www.nssl.uoknor.edu/personal/Doswell/stchs.html

TOGA/COARE

The Tropical Ocean Global Atmosphere (TOGA) and Coupled Ocean Atmosphere Response Experiment (COARE) are parts of the Center for Ocean Atmospheric Prediction Studies Department at Florida State University. This page offers information on oceanographical and meterological studies the department conducts, including high-resolution soundings, radiation experiments, and sea and land data processing.

URL:
http://www.coaps.fsu.edu/COARE/

Like lizards? Check out Herpetology.

A B C D E F G H I J K L **M** N O P Q R S T U V W X Y Z

Tropical Storm Watch Information

FEMA, the Federal Emergency Management Agency, built this page to inform you of current storm watches and to distribute hurricane preparedness material, maps, situation reports, news releases, weather advisories, and even insurance advice. It may be too late to build your house out of brick, but at least with this page you'll know when to lash down the cattle and Aunt Sally, too.

URL:

http://www.fema.gov/fema/trop.html

> *Federal Emergency Management Agency*
> **The last time a hurricane came whipping through and tossed the Winnebago across the county, you promised you'd hitch up the old '74 Gremlin and take the family to some place safe, like New Madrid or central California. Since life in a trailer suits you, perhaps the best thing to do is check the Tropical Storm Watch page from the Federal Emergency Management Agency.**

Ultimate Weather Source

This page offers extensive information on weather, road conditions, and earthquakes, as well as access to other statistical data and resources. Detailed maps and discussions are included. The data provided here is very specific and updated daily.

URL:

http://www.geopages.com/RodeoDrive/1091/

Universal Weather and Aviation

Universal Weather and Aviation provides aviation conditions, aviation communication and management information, maps, and NotiFax weather warnings. Weather graphics and textual weather reports are included. Weather reports are easy to get with Universal's clickable maps.

URL:

http://www.univ-wea.com

Weather and Climate Images

This page offers access to information on weather and precipitation forecasts worldwide, including daily forecasts on weather patterns and characteristics such as El Niño and other climactic anomalies. It also provides links to other NOAA resources and services.

URL:

http://grads.iges.org/pix/head.html

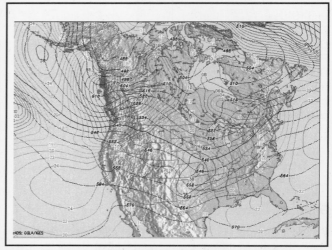

Pressure maps, precipitation maps, temperature maps, winds and divergence maps, radar images, and others are as close as a click of your mouse. Wade through the web at the Center for Ocean-Land-Atmosphere Studies (see Weather and Climate Images).

Weather Maps To Go

Using your X Window terminal, you can specify your own ready-to-order weather maps. Plug in your display address; specify the station, image size, and many other interactive parameters; and soon you'll have a custom weather report on your screen.

URL:

http://rs560.cl.msu.edu/weather/getmegif.html

Weather Predictions

Today, we take for granted meteorologists' abilities to fairly accurately predict weather over a short period of time. But what about accurate predictions for next year, or even ten—or a hundred—years from now? CLIVAR (Climate Variability and Predictability) is part of the World Climate Research Program that would like to do just that. This page gives an overview of the program and its goals, and projects underway for long-term weather prediction.

URL:

http://www.dkrz.de/clivar/hp.html

The Web will really launch you.

WeatherNet

WeatherNet offers weather forecasts that are up-to-date, and its page brings together weather information from resources such as the Tropical Weather page, Travel Cities Weather page, Radar & Satellite page, and Purdue's Hurricane page. WeatherNet also provides a link to CNN's weather page featuring worldwide forecasts and weather news. This page also includes an extensive list of links to various climate centers, organizations, and radio and TV stations,

URL:
http://cirrus.sprl.umich.edu/wxnet/

You're Not in Kansas Anymore!

What could be more terrifying than a tornado bearing down on your home. What causes these storms, and where are they most common? Find out why some tornadoes spin clockwise, and others counterclockwise. Twist yourself into this page for interesting reading about tornadoes.

URL:
http://cc.usu.edu/~kforsyth/Tornado.html

MOBILE COMPUTING

Cambridge Mobile Special Interest Group

The Mobile Special Interest Group promotes mobile awareness and allows fellow researchers to exchange ideas related to mobile computing. It offers a collection of publications relating to mobile computing and cellular telecommunications. Current areas of interest include high-speed multimedia wireless LANs, a mobile application framework for GSM, systems supporting location-aware applications, support for application mobility, distributed mobile computing, and wireless office and wireless applications. There is also a list of current conferences.

URL:
**http://www.cl.cam.ac.uk/users/ct10000/
mobile.html**

It's a Wireless World

Dedicated to making information about the wireless and mobile data industry available to anyone interested, this site provides an alphabetically ordered vendor index, the latest wireless industry news, articles, book and magazine lists, details of trade shows, employment opportunities, and hardware vendors. There's also an excellent list of acronyms that relate to wireless and mobile computing.

URL:
http://www.iadfw.net/tcocklin/

Mobile and Wireless Computing

This World Wide Web Virtual Library for mobile and wireless computing is an excellent place to locate information about the next revolution in telecommunications. Included are lists of papers and universities currently doing research in this area. There is information about upcoming conferences and projects, as well as links to many other sites. This virtual library offers a large collection of resources relating to mobile and wireless computing. There are calls for papers, conference details, links to many project and research groups working in the area, online journals, newsletters, many product descriptions, standards, a bibliography, and much more.

URL:
**http://snapple.cs.washington.edu:600/mobile/
mobile_www.html**

Mobile Computing Bibliography

A searchable bibliography of over 160 papers, books, and reports concerning mobile and wireless computing. The full bibliography can also be downloaded in BibTeX, Refer, and PostScript formats.

URL:
**http://liinwww.ira.uka.de/bibliography/
Distributed/mobile.html**

Mobile Office

The Mobile Office web is a comprehensive source of information about portable computers and communications. It offers news, reviews, a survey on portable computers and products, and links to other web sites offering material about mobile and portable computing.

URL:
http://www.mobileoffice.com

A
B
C
D
E
F
G
H
I
J
K
L
M
N
O
P
Q
R
S
T
U
V
W
X
Y
Z

Mobile Satellite Telecommunications

This index of mobile satellite telecommunications resources on the Web has sections for mobile satellite system operators, service providers, equipment, related research, conferences, periodicals, newsletters, and other satellite-related links. A very interesting site for anyone interested in the key to wireless communication and satellite technology.

URL:
http://www.wp.com/mcintosh_page_o_stuff/ tcomm.html

Mobilis

Mobilis, the mobile lifestyle magazine, is an online web publication featuring interviews, tutorials, reviews, and opinions concerning all aspects of PDAs, wireless communication, and mobile peripherals. It is well presented with plenty of graphics and explanatory diagrams. All previous issues are also available at this site.

URL:
http://www.volksware.com/mobilis/

Motorola Wireless Data Group

The site offers a slide show about wireless communications (2.5 MB in MOV format), a FAQ, an online magazine about wireless data, press releases, product specifications, technology developments, and much more about wireless telecommunications and mobile computing.

URL:
http://www.mot.com/MIMS/WDG/

Pen-Based Computing

Maintained by the publisher of the industry newsletter *Pen-Based Computing: The Journal of Stylus Systems*, this page includes a growing number of articles from the newsletter, details of how to get a free issue, and an index of all back issues.

URL:
http://www.volksware.com/pbc/

It had to happen sooner or later –computing for the typing impaired. But are pen-based computers really pen-based because people would rather write freehand? Or is it because they're just too damn small to put a keyboard on? Also, isn't it ironic that the only people who can afford them (doctors), have handwriting so bad that it takes   a pharmacist to decipher it?

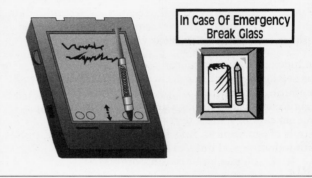

In Case Of Emergency Break Glass

UMTS

The Universal Mobile Telecommunications System (UMTS) is the realization of a new generation of mobile communications technology. With UMTS, personal services will be based on a combination of fixed and wireless/mobile services to form a seamless end-to-end service for the user. This paper introduces the UMTS concept, gives a user perspective, describes the general features of UMTS, examines network and radio interface perspectives, and looks at the evolution of the system.

URL:
http://www.vtt.fi/tte/nh/UMTS/umts.html

Wireless Dealers Web

Offered by the Wireless Dealers Association—an organization of wireless dealers, agents, retailers, and resellers, the WDW site provides news and information on the wireless industry, some industry white papers, and selected articles from their wireless magazine.

URL:
http://wirelessdealers.com

Going shopping on the Web? Remember to pick up some ecash.

Wireless Libraries

Here's a dandy bibliography of wireless data communications publications and articles. There are a few online articles, plus links to wireless, mobile, and pen-based computing Internet resources.

URL:

http://www.duc.auburn.edu/~fostecd/docs/
wireless.html

Yahoo's Mobile Computing Index

An index of resources relating to mobile computing. It has sections for conferences, wireless companies, portable computing, packet radio, PDAs, institutes researching mobile computing, online journals and magazines, product catalogs, links to research projects and systems, details of conferences, information on the wireless mobile data industry, and links to other mobility sites.

URL:

http://www.yahoo.com/
Computers_and_Internet/Mobile_Computing/

MUSIC

All Music

This list is dedicated to the discussion of all types and forms of music. Any music aspects from appreciation to performance are welcome. Since the topic of ALLMUSIC is quite broad, this tends to be quite a busy forum.

Mailing List:
Address: **listserv@american.edu**
Body of Message: **subscribe allmusic** *<your name>*

Ambient Music

The Ambient Music mailing list is intended for information and discussion pertaining to all forms of ambient music—from Brian Eno's Ambient series to environmental sound effects recordings; from The Orb's ambient dub to Jim O'Rourke's ambient industrial drones. Originally conceived by Eno as "sonic wallpaper," the word *ambient* as a musical term has in recent years come to embrace a very diverse range of artists and influences.

Mailing List:
Address: **majordomo@hyperreal.com**
Body of Message: **subscribe ambient**

Asian Contemporary Music

The Asian Contemporary Music Discussion Group (ACTMUS-L) is an electronic forum for the exchange of information and ideas on music written by contemporary Asian composers. The list is unmoderated, and composers, performers, theorists, musicologists, and scholars with interests in this area are welcome to subscribe and participate in the discussion.

Mailing List:
Address: **listserv@ubvm.cc.buffalo.edu**
Body of Message: **subscribe actmus-l** *<your name>*

Bluegrass

Here's a twangy forum for the discussion of bluegrass and Old-Time music in general—including, but not limited to, recordings, bands, individual performers, live performances, publications, business aspects, venues, history, and ethical issues. Early commercial country music is also an acceptable topic.

Mailing List:
Address: **listserv@ukcc.uky.edu**
Body of Message: **subscribe bgrass-l** *<your name>*

Blues Music Discussion

If you've been down so long that it looks like up to you, consider joining the mailing list for the discussion of blues music and performers. Charley Patton, Bessie Smith, Robert Cray, and Stevie Ray Vaughan are all fair topics.

Mailing List:
Address: **listserv@brownvm.brown.edu**
Body of Message: **subscribe blues-l** *<your name>*

Brahms at the Piano

In 1889, Johannes Brahms recorded a segment of the first of his *Ungarische Tanze* in an arrangement for solo piano. Using denoising procedures currently under development at Yale University's Center for Studies in Music Technology, researchers analyze and reconstruct this historic recording in hopes of contributing to the understanding of performance practices in the late nineteenth century.

URL:

http://www.music.yale.edu/research/brahms/
brahms1.html

A B C D E F G H I J K L M N O P Q R S T U V W X Y Z

Brahms at the Piano

Before CDs, LPs, cassettes, and, yes, even 8-track tapes, there were waxed cylinders to play your favorite hits. Invented before the turn of the century by Thomas Edison (what didn't he invent, besides the Internet?), these cylinders were used to record the pop stars of the day—Enrico Caruso, Sarah Bernhardt, and Chaliapin, to name a few. The great composer Johannes Brahms became a recording artist as well in 1889 by playing two segments of his famous "Hungarian Dances" for a representative of Edison. Over a hundred years later, a team of thoughtful scientists are attempting to make that recording sound a little bit better through a sophisticated electronic denoising process that consists of signal analysis with orthogonal trigonometric and wavelet bases. Read more and hear the fruits of their labors at *Brahms at the Piano*.

CAIRSS for Music Database of Music Research Literature

The Computer-Assisted Information Retrieval System is a bibliographic database of music research literature that contains information from articles on music that appear in 15 primary journals. The CAIRSS database also contains information from over a thousand other selected articles and journals.

Note: At the UTSA screen, type **library**
     then type **LOCAL**
     finally, type **CMUS**

URLs:
  telnet:utsaibm.utsa.edu

  http://galaxy.einet.net/hytelnet/FUL064.html

**Need some great music?
Lend your ear to the Music:
MIDI section.**

Canadian Association of Music Libraries (CAML)

CAML-L is the electronic mail distribution list for the Canadian Association of Music Libraries, Archives and Documentation Centres. The list can be used for reference queries and discussions of all musical topics of interest to music information professionals. It will also provide electronic access to documents of interest to the musical community by acting as a clearinghouse and distribution center.

Mailing List:
  Address: **listserv@listserv.ucalgary.ca**
  Body of Message: **subscribe caml-l** *<your name>*

Chinese Music

This mailing list, run by the association for Chinese Music Research Network, was set up to distribute information and facilitate discussion regarding Chinese music.

Mailing List:
  Address: **listserv@uhccvm.its.hawaii.edu**
  Body of Message: **subscribe acmr-l** *<your name>*

Classical Composer Biographies

If you're fascinated with classical composers and their lives, look no further than this semi-informal page documenting the major biggies. Rock me, Amadeus.

URL:
  http://www.cl.cam.ac.uk/users/mn200/music/composers.html

Classical Music Forum

If you don't know a mazurka from a minuet, you'd be better off joining this forum of discussion on all aspects of classical music. That way, next time you'll know your Handel from your Haydn.

Mailing List:
  Address: **listserv@brownvm.brown.edu**
  Body of Message: **subscribe classm-l** *<your name>*

Classical Music of India

Sit down with your sitar at this forum to exchange information on the classical music of India.

Mailing List:
  Address: **listserv@vm1.nodak.edu**
  Body of Message: **subscribe sangeet** *<your name>*

Collection of Historic Musical Instruments

This page is the electronic entryway to the museum at Edinburgh University, which boasts over a thousand musical instruments from the past four centuries. The collection includes many artfully crafted stringed, woodwind, brass, and percussion instruments from around the world. These pages offer graphic images of the instruments, and a section on the instrumental history of the orchestra, theater, dance, and popular music. You can also obtain drawings and publications relating to these historical instruments.

URL:
http://www.music.ed.ac.uk/euchmi

Film and Television Music

Join in an open, unmoderated list devoted to discussion of dramatic music for films and television. Discussion topics include current and past film scores, film music composers, technical and aesthetic aspects of film scoring, film music history and criticism, and recordings of film music.

Mailing List:
Address: **listserv@iubvm.ucs.indiana.edu**
Body of Message: **subscribe filmus-l** *<your name>*

Folk Music

The Folk Music page lists a wide array of online information about folk music, including information about artists, albums, categories by location and instrument, concert schedules, email lists, FTP sites, and Usenet groups. There are also scads of links to other folk music-related pages.

URL:
http://www.eit.com/web/folk/folkhome.html

Chantses are, you'll want to titillate your taste for medieval musicology at the Gregorian Chant page.

Since, after all, we're all just folks, there's a music that speaks to us folks called, appropriately enough, *folk*. If your melodic yet methodic research sways you in this direction, we suggest that the only possible thing left for you to do is to folk-off at the Folk Music Page.

Gregorian Chant

Thanks to some monks in Spain who had a top-ten CD, Gregorian chants are back in style again, at least among the cognoscente. This page contains a lengthy and excellent hypertext treatise on Gregorian chant, from its foundations and history to its influence and performance. Included are related musicology sites, links to medieval history and literature sites, libraries and microfilm collections, chant conferences and travel resources, and other Gregorian chant pages.

URL:
http://www.music.princeton.edu/chant_html/

The Guitar Pre-1650

If you want to pick up your guitar and play "Yesterday," you should know that the guitar was born more than 300 years before Chet Atkins, Jim Hall, or Chuck Berry began their careers. You'll find an instructive treatise here on the history of this marvelous stringed instrument and information about its forerunners, including the lute and the oud.

URL:
http://www4.ncsu.edu/eos/users/s/sfcallic/SCA/
ARTS.CRAFTS/Fret.html

Want to be a doctor? Study up in Medicine.

Harmony Central: Guitar Resources

The Guitar Forum at Harmony Central features a wide spectrum of guitar-related subjects including instruction, tablature, how effects work, newsgroups, commercial guitar sites, and other items of merit. You'll never have to play an air-chord again.

URL:
> http://harmony-central.mit.edu/Guitar/

Jewish Music

The Jewish Music mailing list is for discussion of all varieties of world Jewish music including Sephardic, Klezmer, Cantorial, folk, and Israeli.

Mailing List:
> Address: **listserv@shamash.nysernet.org**
> Body of Message: **subscribe jewish-music** *<your name>*

Just a Few Midi Files for You

This page features a collection of downloadable MIDI files in zip format. This list is continually updated, and there is no particular style or organization—what is here is always an eclectic mix.

URL:
> http://www.warwick.ac.uk/~phulm/midis.html

The Just Intonation Network

All musicians have played the ancient Chinese song *Tu Ning*, and if you're a musician, now's your chance to learn how to play it differently. Just Intonation is a musical philosophy devoted to systems of tuning where each interval can be represented by ratios of whole numbers, with a preference for the smallest numbers adherent to a given musical purpose. If this doesn't make any sense, you can learn more about the JIN and request a free sample hard-copy issue of their magazine at this page.

URL:
> http://www.dnai.com/~jinetwk/

Köchel's Catalog of Mozart's Works

It's a hot topic of debate among football buddies during half-time. Which Köchel number is Mozart's *Serenade for String Quartet and Bass in G Major* (otherwise known as *Eine Kleine Nachtmusik*)? A case of Becks is yours to collect from Hans and Franz by simply punching in this URL. Now you can prove, once and for all, that it's #525. Until, of course, next week's game.

URL:
> http://www.webcom.com/~music/composer/
> works/mozart/top.html

Lamc-l

This is the mailing list for the academic discussion of Latin American music.

Mailing List:
> Address: **listserv@iubvm.ucs.indiana.edu**
> Body of Message: **subscribe lamc-l** *<your name>*

Latin American Folk Music

This mailing list is dedicated to the discussion and promotion of Andean and other Latin American folk music. There is a FAQ introduction that contains guidelines regarding the purpose of the list.

Mailing List:
> Address: **majordomo@discpro.org**
> Body of Message: **subscribe andino**

Latin Music Theory

Carpe diem! Seize the day, or at least the music, with the Thesaurus Musicarum Latinarum Database for Latin Music Theory.

Mailing List:
> Address: **listserv@iubvm.ucs.indiana.edu**
> Body of Message: **subscribe tml-l** *<your name>*

The Mammoth Music Meta-List

Researching music from around the world? As its name implies, the Mammoth Music Meta-List is a resource of Cenozoic proportions. Worldwide styles and genres of music are its focus. There is information, as well as links to lyrics, discographies, reviews, radio stations, performances, sounds, MIDI, music festivals, schools, libraries, research, and online magazines.

URL:
> http://www.pathfinder.com/vibe/mmm/
> music.html

Music and Brain Information Database

The word "psychomusicology" may make you think of an Alfred Hitchcock soundtrack, but it really isn't as scary as it sounds. With the MBI database, you can drain your brain reading up on lots of references and abstracts on music as they relate to behavior, neurobiology, perception, and biophysics. Maybe the neighbor's collection of Barry Manilow really *is* driving you crazy.

URLs:
> telnet:mila.ps.uci.edu

> http://galaxy.einet.net/hytelnet/FUL063.html

Excerpt from the Net...

(from the Music and Brain Information Database (MBI))

Welcome to the Music and Brain Information Database (MBI) MBI is funded by a start-up grant from the National Association of Music Merchants. Our goal is to establish a comprehensive data base of scientific research (references and abstracts) on music as related to behavior, the brain and allied fields, in order to foster interdisciplinary knowledge. Topics included: auditory system, human and animal behavior, creativity, human brain / neuropsychology of music, effects of music on behavior and physiology, music education / medicine / performance / and therapy, neurobiology, perception and psychophysics. All citations and abstracts from the following journals will be included: Bulletin of the Council for Research in Music Education, Journal of Research in Music Education, Music Perception, Psychology of Music, Psychomusicology.

Music Education

Do you play an instrument and want to teach others how to play music as well? This mailing list features discussion on all aspects of music education. There are monthly notebooks available.

Mailing List:
Address: **listserv@vm1.spcs.umn.edu**
Body of Message: **subscribe music-ed** *<your name>*

Music Technology

You and your synthesizer can groove with this mailing list for the discussion of new media and information technologies in relation to musical experience.

Mailing List:
Address: **listproc@bgu.edu**
Body of Message: **subscribe techno-l** *<your name>*

Music Theory

Read or disseminate information regarding re-theorizing music with this discussion of music theory.

Mailing List:
Address: **listserv@uci.edu**
Body of Message: **subscribe polyphony** *<your name>*

Musico-Textual Studies

Got some lyrics and a tune in your head? Head for this discussion list on Musico-Textual Studies.

Mailing List:
Address: **listserv@msu.edu**
Body of Message: **subscribe h-mustex** *<your name>*

Oi!

In the course of your music research, you won't want to forget about Oi!, the melodic type of punk rock that originated in Britain around 1980 and spread around the world. Sociologically, Oi! is a coalition of street punks and skinheads who emphasize working-class concerns. Musically, it is distinguished by choral chanting. The Oi! list is a way of exchanging information and opinions about Oi! music.

Mailing List:
Address: **listserv@nizkor.almanac.bc.ca**
Body of Message: **subscribe oi-list** *<your name>*

Opera Schedule Server

No opera enthusiast on the Web who knows his Figaro from her Fidelio should be without a bookmark to this wonderful web page and database maintained by some very thoughtful folks in Hungary. You can search opera schedules two years in advance for cities from Aberswyth to Zwolle, and approximately 170 others around the world, and you can conduct your search with either a text interface or through a clickable database. There is also information on, and pictures of, some of the world's major opera houses, as well as links to operatic companies and other opera-related sources on the Web.

URL:
http://www.fsz.bme.hu/opera/main.html

Nobody sends you email?
Join a mailing list.

A B C D E F G H I J K L **M** N O P Q R S T U V W X Y Z

It ain't over 'til the fat lady sings Fidelio. Find out where she'll stage her next performance at the Opera Schedule Server.

Princeton University Department of Music

Princeton offers a complete listing of its music department's curriculum, faculty and staff, graduate students, researchers, composers, and alumni. These pages offer up-to-date information on events such as concerts, recordings, and seminars. There is a section on Princeton's Web resources with links to the Winham Laboratory for computer music and the CMIX Home Page of C programming for music mixing and filtering. Also included are a section on filter design and the Medieval Latin Theory Database.

URL:
   http://www.music.princeton.edu

Rice University Music Gopher

Here's a marvelous cornucopia of music-related material pulled in from various corners of the Internet. You'll find topics ranging from ethnomusicology to techno music reviews and MIDI bibliographies on this web site.

URL:
   gopher://riceinfo.rice.edu:70/11/Subject/Music

Having trouble reading Usenet news? Type "news:" into your Web browser.

RISM-US

The mission of Répertoire International des Sources Musicales—RISM (International Inventory of Musical Sources) is to identify and describe music sources and writings from ancient times through about the year 1825. There are a number of online searchable databases, including the U.S. RISM Music Manuscripts Database, the U.S. RISM Libretto Project, and the RISM Names Authority File. In addition, RISM publishes works on a variety of subjects in musicology. The page also contains telnet, gopher, and Web links to HOLLIS, the Harvard University Library Online Information System. More than 30,000 names of composers, librettists, and copyists are listed in the RISM-L News Server, a database of composer names.

URL:
   http://rism.harvard.edu/./RISM/

Rocklist

How does Bonnie Raitt? Is Elton John? If there is such a thing as academic discussion of rock and popular music, you'll find it here on this mailing list.

Mailing List:
   Address: **listserv@kentvm.kent.edu**
   Body of Message: **subscribe rocklist** *<your name>*

The Soundroom

Percussive sound samples and resources for electronic composers and musicians are here in the Soundroom. There are downloadable sections of 8-bit/22kHz rhythm loops.

Note: Most samples are in Macintosh AIFF format.

URL:
   http://snhungar.kings.edu/Sndroom.html

Southern Gospel Music

Welcome to the forum that discusses anything related to Southern gospel music. The **singing_news-talk** list is a service of *The Singing News Magazine*, a premier Southern Gospel Music publication.

Mailing List:
   Address: **majordomo@world.std.com**
   Body of Message: **subscribe singing_news-talk**

Thesaurus Musicarum Latinarum Database for Latin Music

"Nihil humanum ab musicarum alienum est. " (Nothing concerning music is alien to humanity.) Latin lovers everywhere need look no further than this great collection of treatises on secular and liturgical music written between the fourth and seventeenth centuries. Be prepared to have some decent knowledge of that dead language, and in no time at all you'll be reading the Rolling Stone of history. *Carpe diem* and *Caveat emptor!*

URL:
> gopher://iubvm.ucs.indiana.edu/11/tml

Trax Weekly

TraxWeekly is a newsletter dedicated to tracker musicians (MOD makers). It is released on a weekly basis, and contains articles on how to track. There are reviews of music packs and other big music releases. One or two people are also interviewed each week.

Mailing List:
> Address: **listserv@unseen.aztec.co.za**
> Body of Message: **subscribe trax-weekly** *<your name>*

Trombone

Trombone pictures, music, mouthpiece specifications, discographies, reviews, archives, related Web links, and everything else about the trombone can be found right here.

URL:
> http://www.missouri.edu/~cceric/

> If seventy-six trombones were in the big parade, was that the start of heavy metal? Toot your own horn at the Trombone Home Page!

U.C. San Diego Music Library

Here you'll find information regarding the University of California at San Diego's Music Library and Department, a calendar of events, a photo gallery of some of the library's collections, a jazz database, and many other links to music-related sounds and resources.

URL:
> http://orpheus.ucsd.edu/music/

UCSB Department of Music

The University of California at Santa Barbara Music Department's page provides a list of complete curriculum for music composition, theory, and history programs. The Center for Computer Music Research and Composition houses a multimedia center with a real-time digital recording and monitoring studio and a direct digital synthesis studio. The page offers complete information regarding all aspects of the University's programs, faculty, and admissions.

URL:
> http://boethius.music.ucsb.edu/music_dept.html

University of Maryland at College Park inforM system

If ethnomusicology is your bag, check out *The EthnoMusicology/Digest*. A subdirectory contains back issues of *Ethnomusicology Research Digest* (ERD), and there is a searchable gopher database of relevant terms.

URL:
> http://www.inform.umd.edu:8080

UvA/Alfa-Informatica Musicology

This University of Amsterdam page is devoted to several computational musicology projects. One of the most ambitious is the Amsterdam Catalogue of Csound Computer Instruments (ACCCI), which contains a large library of standardized computer instrument samples using Software Sound Synthesis (SWSS) available for DOS, Mac, NeXT, Sun, and SGI platforms. Also included are studies on Music Cognition Research, a musicological analysis of Stockhausen's "Plus-Minus" 1963 composition, and a catalog of algorithmic tools that can be used to modify or generate musical structures.

URL:
> http://mars.let.uva.nl/musicology.html

Web Music Database

Over 6,000 record albums and CDs are organized in this online database, searchable by title, artist, track, style, language, and country.

URL:
> http://www.roadkill.com/~burnett/MDB/

World Music

Music is the universal language, and you can harmonize here at this discussion list for the study of world music.

Mailing List:
Address: **listserv@kentvm.kent.edu**
Body of Message: **subscribe wmusic-l** *<your name>*

You say gypsy, and I say reggae,
You say disco, and I say—
"Give me a break!"

Klezmer, flamenco,
The samba, the tango,
Let's call the whole thing
World Music.

MUSIC: MIDI

The Animal Studio MIDI Page

A group of hobbyists in the Netherlands have put up this all-purpose MIDI site for fun, to help others working with complex MIDI setups, and to get answers to some of their own questions about MIDI. They offer a collection of MIDI documents, some music files, a question-and-answer section, and links to other MIDI-related resources.

URL:
http://www.xs4all.nl/~epzwiers/astudio/midi/

The Gopher's time is short.

CERL Sound Group

If digital audio signal processing and state-of-the-art developments in sound computation are just your cup of java, don't miss the Computer Engineering Research Lab Sound Group page. Download the latest Mac programs for Lemur sinusoidal analysis/synthesis and the latest demo of Lime music notation software. Here you can also read about upcoming conferences, or catch up on the latest news in the Sound and Computing Special Interest Group.

URL:
http://datura.cerl.uiuc.edu

Classical MIDI Archives

Over 20MB and 1,200+ files of mostly classical music for MIDI arranged by composer. Certain major composers such as Bach and Beethoven have their own pages, due to the volume of musical material. Textual information accompanies many compositions.

URL:
http://www.prs.net/midi.html

Emusic

Electronic Music and all attendant topics are the welcome conversations of this mailing list for emusic composition, criticism, technology, and technique. Timbral research, MIDI programming, sly tricks for old machines, and reviews of new toys are encouraged.

Mailing List:
Address: **listserv@american.edu**
Body of Message: **subscribe emusic-l** *<your name>*

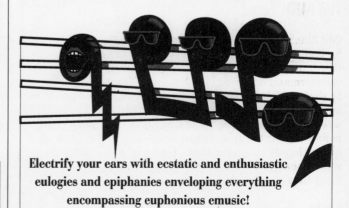

Electrify your ears with ecstatic and enthusiastic eulogies and epiphanies enveloping everything encompassing euphonious emusic!

Harmony Central: MIDI Tools and Resources

The Musical Instrument Digital Interface communications protocol has revolutionized music during its short existence. You'll find a plethora of MIDI-related topics on this page including documentation, forums, ways to create interfaces, and platform-specific software and software-development tools.

URL:

http://harmony-central.mit.edu/MIDI/

Kurzweil K2000 Samples and Programs Patches

This directory has more binary Kurzweil sound files than you can shake a drumstick at. This site also features K2000 synthesizer programming tutorials; C language header (.h) files; sequenced songs; DOS, Atari, and Mac synth software; and general information about the K2000.

URL:

ftp://bach.nevada.edu/pub/k2000

Mac MIDI Utility Archives

Strictly for Macs, this archive contains a plethora of worthwhile software programs, patches, patterns, Hypercard stacks, QuickTime converters, MIDI players, editors, toolkits, and demos.

URL:

gopher://gopher.archive.merit.edu:7055/11/ mac/sound/midi

The MIDI Farm

Old MacDonald had an electronic farm. And on this farm he had a MIDI. If you're a moosical animal, check out this innovative page chock-full of MIDI computer software and hardware information. You can milk a bushel of downloadable files from the Silo. You can cluck over that new song "E-I-E-I-O!" by The Love Donkeys in a live chat forum called The Barn. Then stick a hayseed between your teeth and read *Down On the Farm*, a bimonthly newsletter about MIDI and musical instruments. How ya gonna keep 'em off of The Farm after they've seen this page?

URL:

http://www.midifarm.com

MIDI FTP Berkeley NetJam

Scads of downloadable files on everything from musical composition to software programs to MIDI patches make this a popular site on the Internet for musicians.

URL:

ftp://ftp.uwp.edu/pub/music

MIDIGATE for Windows

MIDIGATE is a stand-alone, Web browser-helper application for Windows containing a number of features designed to facilitate MIDI sequence queuing and playback.

URL:

http://www.prs.net/midigate.html

Forget about listening to the radio while you work. Forget playing audio CDs on your computer's CD-ROM drive while you surf the Web. Tune in some digital classics with MIDIGATE.

MIDIGATE is a great little program for Windows that will queue and play MIDI music files while you work! When you run across a great classic on the Web—Tchaikovsky, for example—just click on the link to it, and MIDIGATE will download it and queue it up for playing. We don't know how many songs you can queue, but we couldn't find the limit.

A
B
C
D
E
F
G
H
I
J
K
L
M
N
O
P
Q
R
S
T
U
V
W
X
Y
Z

MIDIWerks MIDIStuff Page

MIDI isn't a type of skirt, it's a communications standard for electronic musical instruments and computers. And as this page describes, it's not just something you go out and buy, plug in, and bang around. This page offers several good articles on MIDI ranging from basic to in-depth, and also a list of other MIDI resources and links.

URL:

http://www.io.com/~midiwrks/midi.html

Music Machines

So, you've just been invited to play keyboard in that hot new folk-funk-ska-metal band, but you don't have a clue where to begin. The Music Machines page is devoted to information about music equipment and technology: synthesizers, drum machines, effects, and the like. There are lots of images, descriptions, reviews, schematics, and do-it-yourself tips to help get you on the road toward tinkling those ivories.

URL:

http://www.hyperreal.com/music/machines/

Synthesizer Patches

This FTP site features patches, sounds, and files for over 25 different synthesizer brands, makes, and models. Pick your synth and download to your heart's content.

URL:

ftp://wozzeck.tfo.arizona.edu/pub/
patch_editing/patches

USA New Gear Price List

The USA New Gear Price List is a compilation of market prices for electronic music and professional audio equipment. Use this list to eliminate the guesswork when buying MIDI and related computer-based music gear.

URL:

http://www.pitt.edu/~cjp/newgear.html

**Fascinated with fishes?
School yourself in Ichthyology.**

A Brief History of Electronic Instruments

Did you know that the first electronic music instrument was created over 120 years ago? Elisha Gray's musical telegraph was invented two years before his nemesis, Alexander Graham Bell, patented the telephone. Gray reproduced an experiment that his nephew had tried, and developed the principle of self-vibrating sonic electromagnetic devices. This story, and other histories of electromagnetic, triode lamp, and transistor instruments await you at this page that also offers a version in French.

URL:

http://maury.ief-paris-sud.fr:8001/~thierry/
history/history.html

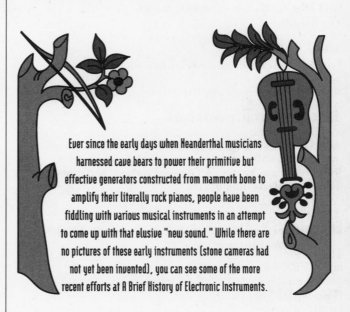

Ever since the early days when Neanderthal musicians harnessed cave bears to power their primitive but effective generators constructed from mammoth bone to amplify their literally rock pianos, people have been fiddling with various musical instruments in an attempt to come up with that elusive "new sound." While there are no pictures of these early instruments (stone cameras had not yet been invented), you can see some of the more recent efforts at A Brief History of Electronic Instruments.

Akai Sampler

Akai has its enthusiasts, and this page is dedicated to any resource even remotely related to their line of samplers. You'll find technical specifications, info on Akai's newest products, MIDI sysex implementation charts, Akai-related software, a how-to guide on upgrading the S900 operating system, and a FAQ list. There is a list of pointers to other Akai-related sources.

URL:

http://www.cs.ruu.nl/~jules/Akai/

CZ-101/patches

The Casio CZ-101 continues to impress musicians for its affordability and clarity of sound. There's a whole array of patches that you can download here for this "little synthesizer that could, and still does."

URL:
http://hyperreal.com/music/machines/
manufacturers/Casio/CZ-101/patches/

E-mu Samplers and Modules Archive

E-mu looms large in the synthesizer world, and this site offers information on how to join the E-mu sampler mailing list plus archives of sounds and utilities for their range of samplers and sound modules. There are pointers to other synth-related sites, as well as E-mu's company page.

URL:
http://www.spies.com/Emu/

Ensoniq

Ensoniq produces an excellent line of synthesizers and synth-related products. This page provides several customer services, including direct contact with the company via email. You can download the latest Ensoniq drivers and hear audio demos of the company's products.

URL:
http://www.ensoniq.com

Ensoniq VFX Home Page

The Ensoniq keyboard page has lots of useful information on its VFX line, plus a mailing list, a patch archive, and details on the *Transoniq Hacker*, the newsletter for all Ensoniq owners. You'll also find a FAQ guide to the VFX family of synths.

URL:
http://www.cs.colorado.edu/~mccreary/vfx/

Korg Editor Pages

If you play a Korg synthesizer, you'll want to be-bop over to this page and grab the nifty editor/librarian shareware program for Windows that's available. Features include sound and graphical envelope editing, infinite sound storage in a single file, and group sound storage.

URL:
http://elektron.et.tudelft.nl/~joostn/korg.html

Kurzweil Music Systems Online

What do musicians Stevie Wonder, Pink Floyd, Rick Wakeman, and Kenny Rogers have in common? They use Kurzweil keyboards. This site includes information on Kurzweil's keyboard products and many links to related resources.

Mail:
Address: **kurzweil@aol.com**

URL:
http://www.youngchang.com/kurzweil/

MIDI Guitar

Yes, Virginia, there is a MIDI guitar synthesizer. How it sounds, of course, is another matter; but if you're interested in owning or playing one, riff your browser over to this page of MIDI and non-MIDI guitar links. Soon to be available is a MIDI guitar FAQ and how to do MIDI guitar on a budget.

URL:
http://www.epix.net/~joelc/midi_git.html

No, you don't have to wear a midi-skirt to play a MIDI guitar, but depending upon your musical ability, it just might help...

Oberheim Xpander/Matrix 12 Patches

If music be the food of love, play on...and if you play an Oberheim, here are some lovely sound patches. All are freely downloadable from this FTP site. Grab 'em and gig!

URL:
ftp://ftp.synthcom.com/pub/synth/patches/
xp-m12/

A B C D E F G H I J K L **M** N O P Q R S T U V W X Y Z

Roland D50 Patches

Lots of different free patches for the Roland are available at this nice FTP site. Grab a mix of sound files from bass, bells, brass, strings, and percussion. Most are around 14KB in length.

URL:
    ftp://ftp.synthcom.com/pub/synth/patches/d50/

SID Home Page

Providing technical information about Commodore's SID 6581 synthesizer chip is the focus of this site. Here you can find detailed technical data about the SID and vote for your favorite SID tunes. The page also features general info on famous SID composers; hardware projects that use the SID; samples; a history of SID music; and links to other SID and Commodore resources.

URL:
    http://stud1.tuwien.ac.at/~e9426444/

Sovtek

Sovtek is an interesting company that produces synth and synth-related products. You can read an engrossing history of its parent company, Electro-Harmonix, and get the latest information on product lines and catalog updates.

URL:
    http://www.turnstyle.com/sovtek/

SY-List Netserver

Yamaha is one of the leading brands of synths and related components, and this page provides access to lots of information, sounds, and programs for the Yamaha SY series. You'll find a related mailing list as well.

URL:
    http://www.neuroinformatik.ruhr-uni-bochum.de/
    ini/PEOPLE/heja/sy-list.html

Synth-l

This list is dedicated to the discussion of the less-esoteric aspects of synthesizers. It concentrates on the discussion of both availability and capability of music software and hardware.

Mailing List:
    Address: **listserv@american.edu**
    Body of Message: **subscribe synth-l** *<your name>*

Synthcom Systems

This page has a little of everything you might want regarding synthesizers and musical equipment. There are downloadable patches for popular synth brands, interactive used-gear price lists, current product updates, and links to other excellent MIDI and keyboard-related sites.

URL:
    http://www.synthcom.com

Synthesizer Discussion Group

Make magic with your MIDI synthesizer with help from this Usenet group that focuses on the practical and lofty aspirations of computer music. Harmonize with like-minded artists like yourself to discuss techniques, tips, tricks, styles, equipment, and more.

URL:
    news:rec.music.makers.synth

The Theremin

If you enjoy waving your hands in the air to produce unearthly sounds, then playing a theremin is right up your alley. This musical device and synth precursor was invented by and named after Leon Theremin in 1918. The instrument is still popular today in experimental music circles. Two metal antennae control pitch and volume by varying the frequency between two oscillators. You'll find all kinds of theremin-related information here.

URL:
    http://www.vuse.vanderbilt.edu/~jbbarile/
        wwwfiles/theremin.html

Virtual Keyboard Museum

Relive those salad days of yore when a high-tech keyboard meant owning a Wurlitzer with a Leslie. Three decades of electric keys, makes, and models that changed the sonic backdrop of pop music are featured at this very hip museum. Here are excellent graphics and text on the instruments and trivia on what famous songs or albums featured them. Learn about such pre-synth and synth wonders as the Mellotron, Hammond organ, Fender Rhodes electric piano, the famous Moog, the Yamaha DX-7, and sampling equipment.

URL:
    http://users.aol.com/KeyMuseum/

NANOTECHNOLOGY

Introduction to Nanotechnology

Ye of such little minds (and little eyes, and little hands) are perfect for a career in nanotechnology. You have to think extremely small to visualize this anticipated manufacturing technology. With this overview of molecular nanotechnology, learn how any stable chemical structure can be built in miniature proportions by assembling individual atoms. Various approaches are reviewed and compared.

URL:
http://nanotech.rutgers.edu/nanotech/intro.html

Let's Get Small

It's not just wild and crazy, it's infinitesimally minuscule. It's nanotechnology, and it'll turn all your big problems into insignificant and easily discarded bits, like the tiny balls of lint collecting in the bottom of your trouser pockets. So don't let those problems reduce you to tears. Just shrink your world.

Nano Link

This is a generous list of nanotechnology sites throughout the Web. This list includes many universities and other research centers that have information about nanotechnology on the Internet. There are also many government agencies and private corporations worth exploring.

URL:
http://sunsite.nus.sg/MEMEX/nanolink.html

NanoFab

The Stanford NanoFabrication Research Facility (or NanoFab) is a state-of-the-art fabrication facility on the grounds of Stanford University. Stop by and read about some of the intriguing projects underway here.

URL:
http://www-nanonet.stanford.edu/NanoFab/

NanoNet

This nanotechnology home page was created for scientists and engineers who have interesting research ideas for machines built on an atomic level. To increase the speed at which this new field becomes mainstream, five research campuses have volunteered sophisticated equipment and share information via NanoNet.

URL:
http://snf.stanford.edu/NNUN/

Nanostructure Devices

Offered by the Fasol Laboratory at the University of Tokyo, this page describes the research into nanotechnology underway at this lab, and elsewhere in Japan. Here, you can find information about research projects, lab staff, publications, conferences, events, and seminars.

URL:
http://kappa.iis.u-tokyo.ac.jp/~fasol/homex.html

NanoStructure Lab

This lab at the University of Minnesota develops new nanotechnologies for fabricating structures and devices substantially smaller than current technology permits, and they're breaking new ground with nanoscale electronic, optoelectronic, and magnetic storage devices. Read about the NanoStructure Lab's current projects, their staff, staff member's publications, and the facilities.

URL:
http://www.umn.edu/nlhome/m017/nanolab/

Nanotechnology at XEROX

Computer chips made from the atoms in ordinary beach sand and diamonds created from pencil lead atoms. This is what nanotechnology might do for the future of manufacturing. If you can wield a tiny pair of atomic tweezers, you might be able to arrange atoms inexpensively, easily, and in all the right places.

URL:
http://nano.xerox.com/nano

Like birds? Flutter into Ornithology.

A B C D E F G H I J K L M N O P Q R S T U V W X Y Z

Nanotechnology in Space

This draft article envisions some of the new capabilities molecular nanotechnology will provide for space operations. Methodology and goals are expressed with the main focus on space. Alternative methods of launching payloads into space with chemical rockets and sky hooks seem like something out of *Star Trek*. The article even examines sample architectures for large space colonies.

URL:

> http://nano.xerox.com/nanotech/nano4/
> mckendreePaper.html

Nanotechnology on the Web

Collecting the best of nanotechnology resources on the Web, this page has pointers to everything to do with microscopic engineering and construction techniques at molecular and atomic levels. Topics include molecular modeling projects, nanotechnology transfer services, lists of electronic magazines and journals on nanotechnology, images, and much more.

URL:

> http://www.lucifer.com/~sean/Nano.html#NEW

Excerpt from the Net...

```
(from Nanotechnology at XEROX)
```

Manufactured products are made from atoms. The properties of those products depend on how those atoms are arranged. If we rearrange the atoms in graphite (as in a pencil lead) we can make diamond. If we rearrange the atoms in sand (and add a few other trace elements) we can make computer chips. If we rearrange the atoms in dirt, water and air we can make potatoes.

Today's manufacturing methods are very crude at the molecular level. Casting, grinding, milling and even lithography move atoms in great thundering statistical herds. It's like trying to make things out of LEGO blocks with boxing gloves on your hands. Yes, you can push the LEGO blocks into great heaps and pile them up, but you can't really snap them together the way you'd like.

In the future, nanotechnology will let us take off the boxing gloves. We'll be able to snap together the fundamental building blocks of nature easily, inexpensively and in almost any arrangement that we desire. This will be essential if we are to continue the revolution in computer hardware beyond about the next decade, and will also let us build a broad range of manufactured products more cleanly, more precisely, more flexibly, and at lower cost.

It's worth pointing out that the word "nanotechnology" has become very popular and is used to describe a broad range of research where the characteristic dimensions are less than about 1,000 nanometers. For example, continued improvements in lithography have resulted in line widths that are less than one micron: this work is often called "nanotechnology." Sub-micron lithography is clearly very valuable (ask anyone who uses a computer!) but it is equally clear that lithography will not let us build semiconductor devices in which individual dopant atoms are located at specific lattice sites. Many of the exponentially improving trends in computer hardware capability have remained steady for the last 50 years. There is fairly widespread confidence that these trends are likely to continue for at least another ten years, but then lithography starts to reach its fundamental limits.

If we are to continue these trends we will have to develop a new "post-lithographic" manufacturing technology which will let us inexpensively build computer systems with mole quantities of logic elements that are molecular in both size and precision and are interconnected in complex and highly idiosyncratic patterns. Nanotechnology will let us do this.

The next time you're feeling sluggish from too much good food, don't bother going to the doctor to have your cholesterol checked—call Cardio-router. That's right! We'll release a thousand tiny Vasco-Borer 2,000s into your bloodstream, and you'll be bouncing off the walls again in no time. (Okay, we're not there yet, but just check out "Nanotechnology on the Web", and you'll be amazed at the stuff we may someday do with tiny machines.)

Nanoworld Australia

All things small and smaller is the focus of the NanoWorld page maintained by the Centre for Microscopy and Microanalysis. The center is dedicated to understanding the structures and composition of all materials at atomic, molecular, cellular, and macromolecular scales. Its Net-acclaimed gallery includes photomicroscopic pictures of blood cells, plankton, pollen, insects, and other diminutive subjects.

URL:
    http://www.uq.oz.au/nanoworld/nanohome.html

Nanoworld USA

This is a great introductory page for anyone interested in nanotechnology. Discussed here are the basic ideas and terms used to describe nanotechnology, and the different ways nanotechnology might change our lives. Links to other sites on the Net are offered for you to explore.

URL:
    http://www.nanothinc.com/NanoWorld/
    nwtoc.html

Notre Dame Microelectronics Lab

Notre Dame's microelectronics lab is conducting extensive research into microelectronic devices on a very small scale that can function at extremely low temperatures. Check out their facilities, and into some of their amazing research projects.

URL:
    http://www.nd.edu/~micro/

Small Is Beautiful

Small Is Beautiful is a NASA page that serves as an excellent index to many nanotechnology and nanorobotics resources on the Web—including many of the labs and other resources listed in this section. Besides university labs, this page lists journals, conferences, DNA nanotechnology, nanotechnology for K-12 education, images and visualization resources, and media accounts and essays.

URL:
    http://www.nas.nasa.gov/NAS/Education/
    nanotech/nanotech.html

There Is Plenty of Room at the Bottom

In his classic 1959 lecture, Richard Feynman discusses miniaturization and how small we can go. He describes nanotechnology as a field in which much can be done in principle but little has been done as yet. See how far we have gone in order to reach for the bottom.

URL:
    http://nano.xerox.com/nanotech/feynman.html

NAVAL ARCHITECTURE

America's Cup Online

Despite the fact that this year's races are over for now, the America's Cup Online site remains an interesting place to explore sailboat design and competition racing. Fine pictures, race results, and information about future races are maintained in this popular media center.

URL:
    http://www.ac95.org

A B C D E F G H I J K L M N O P Q R S T U V W X Y Z

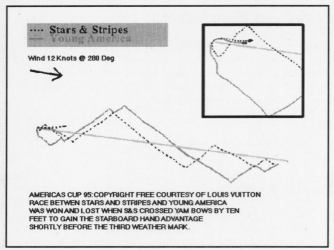

Check out the great photographs and detailed analysis of each round robin and the finals of America's Cup '95 at America's Cup Online.

American Society of Naval Engineers

Ahoy, mates. ASNE advances the knowledge and practice of naval engineering. To learn more about naval engineering as a career, drop an anchor on this page.

URL:

http://www.jhuapl.edu/ASNE/

Boat Builder's Place

Boat plans are free on the Internet! This web page provides information on books, supplies, kits, and that well-designed boat plan that you've been searching for. Although most of the boats are less than 18 feet in length, the plans are well-drawn and easy to download.

URL:

http://www.bateau.com

Boat Plans Online

Most boat-building sites focus primarily on wooden boats, but Boat Plans Online features building plans for boats constructed from modern materials—fiberglass, steel, aluminum, and cold-molded wood. This site provides technical information, building plans, and links to other resources of interest.

URL:

http://www.mpcs.com/boats/

National Shipbuilding Research and Documentation Center

This site is for serious ship builders and contains a tremendous amount of information on naval architecture and shipbuilding. Set up by the University of Michigan Transportation Research Institute, this site provides information about technical and non-technical shipbuilding—including surface preparation, design/production integration, human resource innovations, marine industry standards, welding, and industrial engineering.

URL:

http://www.nsnet.com/doccenter.html

Naval Research Laboratory

The NRL is the research and development laboratory for the Navy, and it provides this page to offer information about the programs, facilities, and personnel of this researchers community. This page also allows access to some of the divisions of the NRL, such as the X-ray Astromony Lab.

URL:

http://www.cmf.nrl.navy.mil/home.html

The Sailing Site

Hoist your sails for the Sailing Site, an island link in the information sea on everything from sailing to boat construction. A long list of sites and home pages keeps you afloat for hours while exploring interesting Naval ports and major regattas around the world.

URL:

http://www.gosailing.com/websites.html

Sources of Information for Recreational Boat Building

This interesting and large document discusses how to get useful boat and ship designs. You'll find information on helpful books and computer software, listings of boat-building schools, and a variety of Internet addresses to help you get even more information. If you are serious about building a seaworthy vessel, begin with this document.

URL:

http://www.efn.org/~jkohnen/nautical/
sources.lst

NETWORKS

ATM Cluster Networking

Does your old 100 Mbit Ethernet network have you tapping your fingers while you wait for your email to download? Well then, it's time to upgrade to ATM. See what they're doing at Cornell University to do away with those pesky latency delays and minuscule bandwidths.

URL:

http://www.cs.cornell.edu/Info/Projects/U-Net

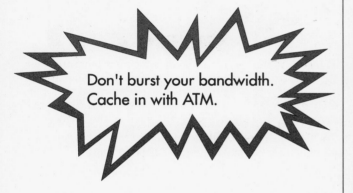

Don't burst your bandwidth. Cache in with ATM.

ATM Dictionary

From AAL to UNI and everything in between, this gopher defines the terms and acronyms associated with ATM (Asynchronous Transfer Mode) networking. This gopher is also searchable by keyword.

URL:

gopher://cell-relay.indiana.edu/11/FAQ/
dictionary

ATM Forum

Asynchronous Transfer Mode (ATM) is fast becoming the base technology for the next generation of global communications. ATM lends itself to a diverse range of applications and different interface speeds and distances. On their pages, the ATM Forum offers a tutorial on ATM basics, white papers on ATM technology, FAQs, a large acronym handbook, press releases, and other links and information relating to ATM technology.

URL:

http://www.atmforum.com

It's easy to use an ATM. You put your card in the slot, type in your password, and . . . huh? Oh. Wrong ATM. Unlike the kind of machine you get money out of, **Asynchronous Transfer Mode** (ATM) equipment is something you sink a lot of money into. Asynchronous Transfer Mode and Broadband ISDN are wickedly fast networking solutions, and these pages tell you all about them in pictures, words, and sounds.

ATM Research at the U.S. Naval Research Labs

Dedicated to research and development of ATM technology, this page offers volumes of background and introductory materials and articles on ATM, including draft standards, newsgroups, ATM test projects and results, hardware research, testbed descriptions, and links to research projects about advancing ATM technology.

URL:

http://netlab.itd.nrl.navy.mil/ATM.html

Cell Relay

A new party game? No! It's everything you ever wanted to know about ATM networking, including the archives to the cell-relay, IPATM, and ROLC mailing lists. Also provided here is an excellent "Getting Started" section that offers an ATM FAQ and ATM tutorials. Other resources include ATM documents and bibliographies, schedules of upcoming events, calls for papers, and a collection of ATM software toolkits and simulation software.

URL:

http://frame-relay.indiana.edu/cell-relay/

A B C D E F G H I J K L M N O P Q R S T U V W X Y Z

Computer Communications Tutorials

Here's a first-class resource—a very comprehensive educational reference point for the computer communications industry. It offers in-depth tutorials, articles, and presentations on many areas of networks and communications, including Asynchronous Transfer Mode (ATM), Ethernet, Fiber Distributed Data Interface (FDDI), Fast Ethernet, Fibre Channel technology, Internet Protocol (IP) and TCP/IP, network management, programming tutorials, system administration, Token Ring technology, and much more. Here's a resource you shouldn't miss if you work with, or study, networking technology.

URL:

http://www.iol.unh.edu/training/
Training_Homepage.html

Ethernet FAQ

Ethernet got your mind in the ether? Here's a frequently asked question list that will answer all your questions about Ethernet, the network cabling and signaling specification developed in the late 1970s. This FAQ has a lot of good general information about Ethernet and the Ethernet standards, plus Ethernet cabling information; specs on Ethernet devices and components; and information on error handling, testing, and troubleshooting Ethernet networks.

URL:

http://smurfland.cit.buffalo.edu/NetMan/
FAQs/ethernet.faq

Ethernet Page

The proverbial drawing on the coffee shop napkin may sound like a cliché, but here you can see the actual genesis of the Ethernet topology and protocol. (This drawing makes it pretty obvious where the term "Ethernet" came from.) This page offers a wealth of information about Ethernet and its associated IEEE standards. Read about the original 10 Mbps system and the new 100 Mbps Fast Ethernet system. There are also quick reference guides, introductory descriptions, Ethernet software for PCs, FAQs, technical papers, reports, troubleshooting numbers, newsgroup lists, and links to vendor pages. The software includes packet drivers and documentation, a list of packet driver access methods, and packet driver applications.

URL:

http://wwwhost.ots.utexas.edu/ethernet/
ethernet-home.html

Hey, Mac, address your Ethernet troubles by colliding with the *Ethernet Page*.

Fast Ethernet

Fast Ethernet is a networking standard that calls for a 100 Megabit-per-second data transfer rate. This tutorial offers pointers to papers, guides, descriptions, comparisons, and reviews on Fast Ethernet resources on the Web. It also gives a list of commercial sites that supply Fast Ethernet products with a short description of each company's products.

Note: This page requires an HTML 3 compliant browser.

URL:

http://ganges.cs.tcd.ie/msc-course/tut1/
grp1/start.html

Fast Ethernet Consortium

The Fast Ethernet Consortium tests Fast Ethernet (IEEE 802.3u) products and software for interoperability and conformance to standards. On their web pages, the consortium offers details about their test suites, equipment lists, product listings, newsletters, announcements, membership information, articles, an excellent Fast Ethernet reference guide, and an introduction to auto-negotiation.

URL:

http://www.iol.unh.edu/consortiums/fe/
fast_ethernet_consortium.html

The Net is the new medium.

FDDI

No, this is not a new federal bureaucracy. FDDI is yet another networking acronym. This one stands for Fiber Distributed Data Interface. This informative FAQ describes and explains FDDI technology; lists the various FDDI standards; details the differences between FDDI and FDDI-II; and explains dual homing, DAS, SAS, wrapped ring, the advantages and disadvantages of concentrators, beaconing rings, and port connection rules.

URL:
> gopher://cell-relay.indiana.edu/00/FAQ/
> FDDI-FAQ.txt

Fibre Channel Standard

The Fibre Channel Standard (FCS) defines a high-speed data transfer interface that can be used to connect workstations, mainframes, supercomputers, storage devices, and displays. Provided by the CERN organization, this page offers news, specifications, and guides to the Fibre Channel Standard and related products. The products are classified as testers, fabrics, interfaces, storage devices, routers, concentrators, converters, components, and software. You can also peruse the FCS products by manufacturer name.

URL:
> http://www.cern.ch/HSI/fcs/fcs.html

High Speed Interconnect

High Speed Interconnect (HSI) is a project of part of the CERN organization to provide high-speed links and interfaces for Data Acquisition systems. Read about the European take on ATM, HIPPI, Fibre Channel, SCI, and other optical systems. There's also some examples of how you can use these technologies in different parts of a data acquisition system.

URL:
> http://www.cern.ch/HSI/

IBM Networking

Details, announcements, and product specifications of IBM's networking activities and products. The page offers a catalog of networking products, technical white papers, monthly technical perspectives, and links to online networking zines and other sites.

URL:
> http://www.raleigh.ibm.com/nethome.html

Looking for Earth Science resources? Check Earth Science, Ecology, Geography, and Geology.

Internet Management Navigator

The Internet Management Navigator is a page created by Eric van Hengstum of the Netherlands that provides a good introduction to the Simple Network Management Protocol (SNMP), the Common Management Information Protocol (CMIP), and other network management protocols. Eric's page also has links to many articles, specs, standards lists, references, FAQs, mailing lists, software, and newsgroups relating to network management on the Internet.

URL:
> http://wwwsnmp.cs.utwente.nl/int/

IP Next Generation

IP Next Generation (or IPng) is a new version of the Internet Protocol (IP) that is designed to be an evolutionary step from IPv4. IPng is designed to run well on high-performance networks such as ATM, and at the same time be efficient for low-bandwidth networks—wireless, for example. The IPng home page offers a complete overview of IPng and the motivations behind it; details and discussions of the IPng working group; and IPng specs, presentations, software implementations, and news of proposed standards.

URL:
> http://playground.sun.com/pub/ipng/html/
> ipng-main.html

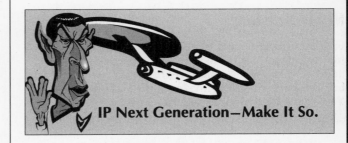

IP Next Generation—Make It So.

A B C D E F G H I J K L M N O P Q R S T U V W X Y Z

ISDN

Integrated Services Digital Network (ISDN) is a completely digital telecommunications network designed to carry voice, data, and video signals. Provided by Dan Kegel, this page is an excellent resource for information on ISDN, with links to hundreds of ISDN resources on the Internet. There's an excellent introduction to ISDN concepts and systems. There are also links for relevant news, magazine and periodical articles, provider listings, discussion archives, software archives, and descriptions and reviews of ISDN hardware.

URL:
> http://alumni.caltech.edu/~dank/isdn/

LAN Technology Scorecard

Here's a huge list of networking technologies including all the types of Ethernet, FDDI, SONET, ATM, and more. For each technology, the cable type, transmission speeds, packet type, state of standard, and product availability are given. A short description of each is also given, along with a list of vendors supporting each technology.

URL:
> http://web.syr.edu/~jmwobus/comfaqs/
> lan-technology

Modems

This collection of information on modems gives an introduction to modems, and explains modulation protocols, error-control protocols, data-compression protocols, cellular protocols, fax modems, and future directions. It offers advice on buying a high-speed modem and software, and correctly configuring them. Also available are a modem dictionary, modem standards, a technical reference for Hayes modems, FAQs, and links to some modem manufacturer's sites.

URL:
> http://www.slac.stanford.edu/winters/pub/
> www/net/modem.html

Mrinfo Gateway

Find out what's up and what's down with this tool for interrogating the state of multicast routers. Type the IP address or hostname of a server running mrouted in your browser's search dialog, and learn the status of that host. You'll need to be mrouted in order for this service to work.

URL:
> http://www.cl.cam.ac.uk/htbin/mrinfo

Network Reference

This collection of network references covers the network layer, network services, security, privacy, and wide area networking. There are FAQs and tutorials on ATM, Ethernet, SCI, and the Dynamic Host Configuration Protocol. The network services cover network management, the X Window system, X.500, X.400, video conferencing, the Andrew File System, DCE, and Netware. The wide area networking section covers ISDN, modems, and a telecommunications library.

URL:
> http://www.slac.stanford.edu/winters/pub/
> www/net/reference.html

OSI and Network Services Tutorial

Here's an informative article covering all seven of the layers of the Open System Interconnection (OSI) reference model. There are separate sections for each of the physical, datalink, network, transport, session, presentation, and application layers. Applications described include X.400 Mail, X.500 Directory, X.700 Management, and EDI. There are also sections covering IP Multicast, IP Next Generation, and ATM, although these are not part of the OSI model. Each section is well illustrated with graphic diagrams and explanations, and all sections are highly hyperlinked. This is an excellent resource for learning about the workings and messages of the OSI reference model.

URL:
> http://ganges.cs.tcd.ie/4ba2/

OSI Reference Model FAQ

This FAQ explains OSI and the OSI reference model. It introduces ASN.1—the Abstract Syntax Notation number 1—as a precise and parseable specification of how to structure protocol data units (PDUs). It lists the OSI standards, the relationship between OSI and TCP/IP, offers some constructive criticisms of OSI, lists free OSI implementations, discusses new and planned OSI standards, and finally provides book and journal references.

URL:
> http://www.cis.ohio-state.edu/hypertext/
> faq/usenet/osi-protocols/faq.html

Take flight in Aerospace and Space Technology.

> **Reach for the stars.
> Gaze into Astronomy and
> Physics: Astrophysics.**

The Simple Web

Another informative page offered to the world by Eric van Hengstum, this one provides links and strategies for general network management. It's structured into the three main architectures—Open Systems Interconnection management, Telecommunications Management Network, and Internet management. For each section, there are links to articles, book lists, theses, standards lists, newsgroups, newsletters, and links to working groups.

URL:
   http://snmp.cs.utwente.nl/

SNMP FAQ

The Simple Network Management Protocol (SNMP) is a protocol for Internet network management services. This FAQ introduces SNMP and related RFCs, and provides references and links to many related web resources. There are also pointers to SNMP software, and an explanation of MIBs—the objects that describe an SNMP manageable entity.

URL:
   http://wwwsnmp.cs.utwente.nl/int/DIRfaq/

TCP/IP: An Introduction

You've probably heard or read a hundred times that TCP/IP is the "glue" that holds the Internet together. But you probably still wonder what TCP and IP are, and what they actually do. Wonder not. This page tears the veil of mystery from IP and TCP, explaining exactly what they are, what they do, and how they do it. This short illustrated tutorial also gives a short history of the development of these protocols and covers addressing, subnets, and common problems people encounter with TCP/IP.

URL:
   http://pclt.cis.yale.edu/pclt/comm/tcpip.htm

Token Ring Guides

ASTRAL (Alliance for Strategic Token Ring Advancement and Leadership—is this a made-up acronym or what?) is an organization of member companies that make and sell Token Ring networking products. On their site, ASTRAL provides vendor-independent guides and tutorials about the Token Ring topology and protocol. They provide both press releases and white papers, the latter of which are an excellent introduction to Token Ring. They also provide a guide for network managers, a value assessment study, and tips for extending the life of a Token Ring network.

URL:
   http://www.astral.org

VG AnyLAN Consortium

The VG-AnyLAN Consortium tests 100VG-AnyLAN (IEEE 802.12) products and software for interoperability and conformance to standards. On their pages, they offer newsletters, product listings, information about their test suites, tutorials, and pointers to related sites. The tutorials include in-depth discussions of VG-AnyLAN concepts, a detailed study of the operations of VG-AnyLAN, a visual overview of 100VG-AnyLAN networks and components, and technical references.

URL:
   http://www.iol.unh.edu/consortiums/vganylan/
   vg_consortium.html

VG AnyLan FAQ

VG is a new 100 Mbps Ethernet and Token Ring networking technology. This HTML FAQ describes VG, the state of the standard, the components of a VG network, details about how packets move around on a VG network, security features, a vendor list, references, and pointers to related sites with VG information.

URL:
   http://www.io.com/~richardr/vg/

X Window FAQs

This FTP server houses a collection of FAQs covering many aspects of the X-Window window manager system, including a bibliography, conference announcements, using X in day-to-day life, configuring the system, where to get source code, binaries, or patches for the popular platforms, using tools and applications with X, programming, and much more.

URL:
   ftp://rtfm.mit.edu/pub/usenet/news.answers/
   x-faq

A
B
C
D
E
F
G
H
I
J
K
L
M
N
O
P
Q
R
S
T
U
V
W
X
Y
Z

OCEANOGRAPHY

About Octopi...

This page will sucker you into learning more about those squishy cousins of squid, cuttlefish, and the chambered nautilus. Find out more fascinating facts about octopi here at this succulent Mote Marine Laboratory page in Sarasota, Florida.

URL:

    http://www.marinelab.sarasota.fl.us/
    OCTOPI.HTM

About Red Tide...

Red tide is an ocean phenomenon caused by several species of marine phytoplankton—microscopic plants that produce potent chemical toxins. Since shellfish consume these plants for nourishment, they in turn build up high toxicity levels. People who eat these shellfish often contract paralytic shellfish poisoning, and can die as a result. Read this document on red tide before you down that next oyster.

URL:

    http://www.marinelab.sarasota.fl.us/~mhenry/
    WREDTIDE.HTM

AquaNIC

This network information center offers links to many updated sources related to oceanography. This server contains everything from newsletters to job announcements and images.

URL:

    gopher://thorplus.lib.purdue.edu:70/h/
    databases/AquaNIC/home.html

El Niño

The El Niño affects weather conditions around the world. By disrupting the Pacific Ocean atmosphere, certain consequences result, such as increased rainfall. Predicting El Niño conditions and the resulting climate variations is essential. This page describes non-El Niño conditions as compared to the El Niño, and you'll find detailed graphs, an animation, and additional links and references.

URL:

    http://www.pmel.noaa.gov/toga-tao/
    el-nino-story.html

El Niño may be just a little boy to you, but this youngster is responsible for dramatic climate conditions in the Pacific. Find out if El Niño has been malo or simpatico this year.

Global Ocean Ecosystems

U.S. GLOBEC (GLOBal ocean ECosystems dynamics) is a research program organized by oceanographers and fisheries scientists to study the effect of global climate changes on the abundance and production of animals in the sea.

URL:

    http://www.usglobec.berkeley.edu/usglobec/
    globec.homepage.html

Hydrographic Atlas of the Southern Ocean

The next time you quest for a southern oceanic hydrographic atlas of hydrographic parameters on zonal and meridional sections, chill out here. A team of researchers from several Antarctic research institutions has collected oceanic temperature, salinity, and oxygen data south of 30 degrees latitude. You'll find the atlas maps in GIF format. Now see if you can find the penguins.

URL:

    http://www.awi-bremerhaven.de/Atlas/SO/
    Deckblatt.html

Interactive Marine Observations

When you're bobbing along the bottom of the beautiful, briny sea, you'll never need to surface with this page on your waterproof PC. Click on the image map and zoom into temperature, pressure, wind, and water conditions reported by buoys and CMAN stations.

URL:

    http://thunder.met.fsu.edu/nws/
    public_html/buoy/

JAM

Discuss your pet theories about oceanography and earth sciences. This list will assist you with finding references and establishing contacts. You'll find job advertisements and conference and workshop announcements.

Mailing List:
  Address: **majordomo@ccsdec1.ufsia.ac.be**
  Body of Message: **subscribe jam**

Marine Mammal Science

The field of marine mammal science has a growing appeal for young people. Yet many students do not clearly understand what the field involves. At this site you will find suggestions and information on how to plan education and work experience in order to develop a career in this field. You'll learn about marine mammal science, like how many anchovies it takes to get a killer whale to do back flips but not bite your leg off at the knee.

URL:
  http://www.rtis.com/nat/user/elsberry/marspec/ mmstrat.html

MIT Sea Grant AUV Lab

Move over Robo-Cop and make room for Robo-Lobster! MIT's Sea Grant AUV Lab is dedicated to the development of fully autonomous underwater vehicles (AUVs). These vehicles are capable of accomplishing missions without tethers, cables, or remote control. (But can they remove the barnacles from their own hulls?)

URL:
  http://web.mit.edu/afs/athena/org/s/seagrant/ www/auv.htm

National Oceanographic Data Center

The NODC provides ocean data management and ocean data services to researchers. A substantial amount of information on ocean buoy systems, salinity, navigation, and more, can be accessed from online databases, maps, graphs, pictures, and CD-ROMs.

URL:
  http://www.nodc.noaa.gov

Hug some trees in the Forestry section.

Domestic Science? Of course it's a real science!

Ocean Awareness

Find out fun facts about what oceans really accomplish for our planet and the creatures who live there. You can read status reports that discuss the importance of ocean resources and their current problems from pollution. There is a list of pointers to ocean-related organizations.

URL:
  http://www.cs.fsu.edu/projects/sp95ug/group1.7/ ocean1.html

Aren't oceans just wonderful bodies of water? You can sail, swim, surf, snorkel, scuba, and even sink in them! Before you attempt any of these activities, however, we suggest that you slosh your wave over to the Ocean Awareness page that literally features current news and resources for sea horses and you. Wade through the section on ocean pollution. Even if you're affluent, you won't want to frolic in the effluent!

Ocean Planet

Ocean Planet is an exhibition at the Smithsonian's National Museum of Natural History. This is an actual traveling exhibit that will tour the country over a period of four years in an effort to study and understand environmental issues that affect the health of the world's oceans. On this, the home page of the Ocean Planet exhibit, you can read an overview about the exhibition, read a message from the curator, and keep tabs on its progress throughout the nation.

URL:
  http://seawifs.gsfc.nasa.gov/ocean_planet.html

Ocean Resources

The oceans are abundant with resources of all kinds—food for people and animals, a wealth of biodiversity, and mineral resources that are just being discovered. There's even gold in those waters, as well as magnesium, bromine, and salt. Hydrothermal vents help to create new ecosystems with even more resources. Read about the oceans' bounty here.

URL:
http://www.cs.fsu.edu/projects/sp95ug/
group1.7/resource.html

OCEANIC

If you dropped a bottle with a message into the ocean, where would it end up? You have no idea unless you check out the Ocean Information Center (OCEANIC) page. Their World Ocean Circulation Experiment is vital for pinpointing the proper drop point to get your message to land in front of Bill Gates' beach house. He'll be so impressed that he may just accept your request to be a Windows 99 beta tester.

URL:
http://diu.cms.udel.edu

Oceanography Society

The Oceanography Society was founded in 1988 to disseminate knowledge of oceanography through research and education and to facilitate communication among oceanographers. You'll find a complete index of issues of *Oceanography* magazine online, plus a Web-based BBS.

URLs:
ftp://ftp.tos.org

http://www.tos.org

Oceanography Virtual Library

Wade through this large collection of oceanography resources that are organized mainly by geographical location and Internet service type (e.g., web, FTP, and Gopher). You'll find electronic publications, email discussion groups, and conference announcements. On this page you'll probably find links to virtually every oceanography department, group, or organization with a site on the Internet!

URL:
http://www.mth.uea.ac.uk/ocean/
oceanography.html

There's an ocean of information on the Internet. Set your sails for the Oceanography Virtual Library.

Pacific Marine Environmental Laboratory (PMEL)

PMEL conducts studies to improve our understanding of the complex physical and geochemical processes of the world's oceans. The Research Projects page includes studies on atmospheric climate and chemistry, carbon dioxide, chlorofluorocarbon tracing, fisheries, thermal modeling and analysis, tsunamis, and more.

URL:
http://www.pmel.noaa.gov/research.html

Red Tide Update

Surf the Web before you surf the water. If you live along the southwest Florida coast, you'll want to check the local red tide conditions from your computer via the Net. There are localized daily updates to let you know if and where there are problems.

URL:
http://www.marinelab.sarasota.fl.us/~mhenry/
rtupdate.html

Satellite Oceanography Lab - University of Hawaii

A tremendous collection of oceanographic images and videos from satellites and the space shuttle. Be sure to see the *spectacular* directory!

URL:
ftp://satftp.soest.hawaii.edu/pub

Find out about your relatives
in Primatology . . . uh . . .
Genealogy.

Scripps Institution of Oceanography

This San Diego-based research center offers extensive information on current projects and research at Scripps, as well as access to many other oceanography resources on the Internet.

URL:

http://sio.ucsd.edu

Scripps Institution of Oceanography Library

Here's an excellent collection of bibliographies, electronic periodicals, and articles that relate to oceanography. There is in-depth information regarding San Diego and Southern California wave, wind, and ocean temperature data, as well as current conditions. In addition, you'll find oceanographic datasets, satellite images, tide predictions, and a directory of oceanographic and earth science institutions.

URL:

http://orpheus.ucsd.edu/sio/

Sea Grant Mid-Atlantic Region

Set sail for a better understanding of our Eastern coastal, estuarine, and ocean resources. This page contains items of interest related to marine ecology, environmental science, and biotechnology. Maps of Chesapeake Bay and the Mid-Atlantic region are among the many environmental resources.

URL:

http://www.mdsg.umd.edu

Sea Surface Temperature

View an archive of sea surface temperature satellite images. These Advanced Very High Resolution Radiometer (AVHRR) images are processed from data collected by NOAA satellites. The archive contains in excess of 20,000 images from April 1979 to the present, and new images are added daily. These images are 1,024 by 1,024 pixels, and can be viewed directly from the web page. You'll find a classroom lesson plan describing some of the uses of these images.

URL:

http://dcz.gso.uri.edu/avhrr-archive/
archive.html

SeaWiFS

The Sea-viewing Wide Field-of-view Sensor (SeaWiFS) Project uses an Earth-orbiting ocean color-sensor to quantify the ocean's role in the global carbon cycle and other biogeochemical cycles. Part of NASA's Mission to Planet Earth, the SeaWiFS Project is designed to look at our planet from space in order to understand it better. It's like monitoring the color changes of a global mood ring.

URL:

http://seawifs.gsfc.nasa.gov/scripts/SEAWIFS.html

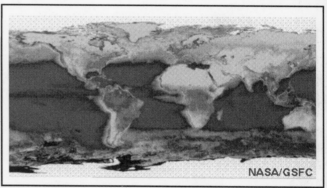

NASA's SeaWiFS project provides dozens of amazing images on their web pages. This is a global biosphere image showing relative temperatures on the surface of the Earth. Other images show ocean temperatures, vegetation density, and pinpoint manmade environmental problems.

SelectSite Ocean Technology

Check out this site for pointers to businesses, conferences, and publications specializing in ocean technology. Newsgroups and mailing list references, as well as other marine technology listings are found here.

URL:

http://www.selectsite.com/oceantech/

Sources of Oceanography and Climate Information

Make a splash with this comprehensive directory of oceanography and climate information on the Web! There are sections on conference announcements, ocean models, oceanographic data, and climate. You'll also find global change information, newsgroups, and online oceanographic and fisheries institutions.

URL:

http://www-ocean.ml.csiro.au/index/
www-resources.html

A B C D E F G H I J K L M N O P Q R S T U V W X Y Z

Tide and Current Predictor

Select your area and predict your tide and current in the U.S. and throughout the world. This page will calculate basic and customized predictions of harmonic tidal activity. The results can be generated as a tabular list, text plot, graphic plot, or a one-month calendar forecast.

URL:

http://tbone.biol.sc.edu/tide/sitesel.html

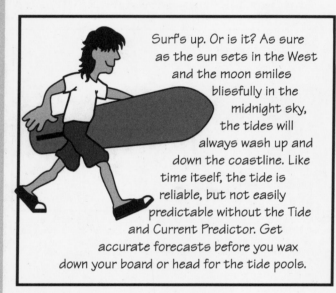

Surf's up. Or is it? As sure as the sun sets in the West and the moon smiles blissfully in the midnight sky, the tides will always wash up and down the coastline. Like time itself, the tide is reliable, but not easily predictable without the Tide and Current Predictor. Get accurate forecasts before you wax down your board or head for the tide pools.

Underwater Farming

Today, more than 60 species of aquatic plants and animals are grown commercially. Not only does aquaculture increase the amount of seafood we can eat, but it also enables us to learn more about fish and aid mother nature in replenishing diminishing fish stocks. This page focuses mostly on aquaculture in and around Australia.

URL:

http://www.dpie.gov.au/resources.energy/
fisheries/fishfacts/ff8.html

Button down the hatches with a firewall.

Woods Hole Oceanographic Institution

WHOI is the largest independent marine science research facility in the United States. It is dedicated to the study of all aspects of marine science and to the education of marine scientists. WHOI's page features pointers to multidisciplinary projects, applied ocean physics and engineering, marine chemistry and geochemistry, and physical oceanography.

URLs:

ftp://ftp.whoi.edu

http://www.whoi.edu/html/whoi-by-dept.html

Yahoo Oceanography Index

Here's a large collection of oceanography resources, including aquariums, companies, fisheries, institutes, ships, weather, and marine biology. There are links to several ongoing projects, information networks, data collections, reports, and tools—such as a tide predictor.

URL:

http://www.yahoo.com/Science/Oceanography/

OPERATING SYSTEMS

A/UX

Perhaps the weirdest implementation of Unix ever created, A/UX is Apple's version of Unix that runs on a Macintosh. Fortunately, this odd bird can sustain flight thanks to Jagubox's A/UX directory with loads of neat, useful, and critical A/UX stuff.

URL:

http://jagubox.gsfc.nasa.gov/aux/

CP/M

About a million years ago it seems, Gary Kildall invented CP/M, a Control Program for Microcomputers. CP/M heralded a major event in personal computing and laid the foundation for MS-DOS, which reigned for over a decade (and continues to haunt Microsoft's Windows operating system). Find out more about the venerable CP/M and its derivatives and emulators in this Usenet group.

URL:

news:comp.os.cpm

FreeBSD

FreeBSD is a state-of-the-art operating system for personal computers with 386, 486, and Pentium processors. FreeBSD provides many advanced features previously available only on much more expensive computers, yet FreeBSD is available from the Net without charge. Among its many impressive features, FreeBSD is "binary compatible" with programs built for SCO, BSDI, NetBSD, 386BSD, and Linux. It can even run Windows applications using WINE, a Windows emulator.

URL:

> http://www.freebsd.org

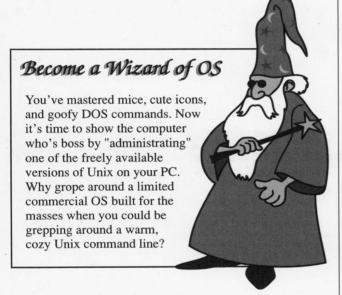

Become a Wizard of OS

You've mastered mice, cute icons, and goofy DOS commands. Now it's time to show the computer who's boss by "administrating" one of the freely available versions of Unix on your PC. Why grope around a limited commercial OS built for the masses when you could be grepping around a warm, cozy Unix command line?

Greg Carter's Windows 95 Help Page

It's probably a good guess that Greg Carter likes Windows 95. Why else would anyone put up a collection of web pages about an operating system? Whatever his reasons, Greg offers a lot of great information, tips, and tricks about Windows 95, especially using third-party Internet applications with Windows 95.

URL:

> http://www.neosoft.com/~gregcms/

Publish it on the Web.

Inside Microsoft Windows 95

The Cobb Group is a commercial publisher of computer- and software-related magazines. This page is a commercial solicitation to subscribe to the Cobb Group's *Inside Microsoft Windows 95*; however, they do offer useful and interesting articles online from past issues. Topics include running problematic DOS programs under Windows 95, copying files in the same folder, and using SLIP to connect to an Internet service provider.

URL:

> http://www.zdnet.com/~cobb/win95/

IRIX

Silicon Graphics makes available their Advantage Online support page for the latest information, patches, and common Q&A about the IRIX operating system.

Note: You must be a Silicon Graphics software support customer to access some parts of this site.

URL:

> http://www.sgi.com/Support/advantage.html

Linux Operating System

Linux, originally developed by Linus Torvalds of Helsinki, is a freely distributable implementation of Unix for 386, 486, and Pentium machines. The Linux project is one of the most active today, involving teams of programmers around the world who hack away at this popular and constantly changing operating system.

URL:

> http://www.linux.org

Mac OS

Until recently, Apple wouldn't admit that it had an "operating system." Now, with a dedicated web page, the Mac OS is clearly an important technological asset for the company that brought icons and mice to our desktops. You'll find information on Apple's strategy for Mac OS, overviews of key product concepts, licensing details, and plans for new directions. Programmers have access to developer notes, release notes, detailed technical discussions, and more. Plus, you can download software updates, examples, demos, documentation, tools, and other items to help you use the Mac OS more effectively.

URL:

> http://www.austin.apple.com/macos/
> macosmain.html

Microsoft Windows

The official page for news, information, and product updates for Microsoft Windows. Microsoft includes a fully searchable technical information and software library for "help yourself" customer and developer support.

URL:
http://www.microsoft.com/windows/

Solaris

The sun also rises over Solaris, Sun Microsystem's robust and scaleable operating environment. Sun provides news, information on Solaris products and technologies, patches, and customer-support resources.

URL:
http://www.sun.com/cgi-bin/show?sunsoft/
solaris/index.body

Unix Reference Desk

This one-stop hypertext help desk for Unix users contains references to material culled from a number of different sources on computing in the Unix environment. Resource sections include General Information, Texinfo Pages, Applications, Programming, IBM AIX Systems, HP-UX Systems, Unix for PCs, Sun Systems, X Windows, Networking, Security, and Humor.

URL:
http://www.eecs.nwu.edu/unix.html

Windows95.com

Bill Gates built Windows 95, but Steve Jenkins built Windows95.com. Here you'll find extensive information about Microsoft's Windows 95, including a vast collection of software and online tutorials. The tutorials are organized in a walk-through manner; simply choose hyperlinks for tasks you are trying to complete, and specific instructions with pictures of dialog boxes are presented to help you out.

URL:
http://www.windows95.com

Be a star-gazer. Scope out Astronomy.

Can you say S O C K S? Find out what it means in Linguistics.

XFree86

The XFree86 Project, Inc. is a nonprofit organization that produces an X Window server for PC-based Unix and Unix-like systems. XFree86 offers a high-performance graphical user interface that is compatible with popular operating systems like FreeBSD, NetBSD, and Linux. Find the latest versions, documentation, and support here.

URL:
http://www.xfree86.org

ORNITHOLOGY

Best Birding Sites

Fly to Europe and discover some of the best bird sites in all of Europe—indeed, perhaps the world. These sites range from England and Ireland all the way up to Holland, and the author has chosen what he feels are the most spectacular areas to view a wide range of species. The author describes each site by country, and includes the types of birds you can spot at each.

URL:
http://bchannel.avonibp.co.uk/productioncos/
greenumbrella/birds.html

Bird Books

The Natural History Book Service (NHBS) is a mail-order bookstore in Devon, England. Yes, it's a commercial enterprise, but it may just be the best source in the world for bird books, sounds, and video. From this page, you can follow links to birds of Africa; Australasia; the Antarctic; North, Central, and South America; and other locations in the world.

URL:
http://www.nhbs.co.uk/booknet/su14.html

Excerpt from the Net...

(from Bird Facts)

Moves to Counteract Conditions That Adversely Affect Birds

Wilderness Society senior ecologist David Wilcove has suggested several steps to counteract the conditions that adversely affect birds:

Reduce logging in the national forests in the East

Encourage the protection of large forest tracts

Protect key migrations points such as the Delware Peninsula, Cape May, N.J., and Point Pelee, Ontario.

A California winery owner is creating 90 acres of artificial wetlands on his Sonoma land to provide a habitat for migratory songbirds and waterfowl.

A conservancy on Hilton Head Island, S.C., is using reclaimed/treated water from a tertiary water treatment plant owned by the Hilton Head Plantation to restore the swamp area and recharge the aquifer. The conservancy is home to wading birds such as heron, egret, and ibis, which require standing water around the black gum trees they use as nesting sites. Before the program was initiated, the local water table was lowered by the increased industrial and development demand and a drought; the birds had started to leave. With the new program, birds are returning.

In 1991, in the June issue of the Urban Audubon (a monthly newsletter of the NYC Audubon Society), a program called "Captains for Conservation" was launched. Its goal is to inspire "captains of industry" who own tall skyscrapers to turn off the lights on their buildings during migration season. In bad weather, birds must fly at lower altitudes and the rain/mist/fog can combine with lights to disorient birds, causing them to crash into skyscrapers. For more information, write to Sarah Elliot at 333 East 34 St., New York, NY, 10016.

California wind farms are taking the following steps to protect birds that fly into the blades of the wind turbines:

Retrofitting power lines to avoid electrocution

Experimenting with different color blads and high frequency sounds that might deter birds

Paying more attention to bird migration patterns when choosing new development sites.

The lead pellets left behind by hunters are sometimes swallowed by geese and ducks and cause death by lead poisoning. A ban on the use of lead shot was enacted by the U.S. Fish and Wildlife Service during the 1987-88 season; usage was phased out by 1991.

Many birds and small animals have suffocated when they have gotten caught in plastic "six-pack" rings that have been carelessly discarded by people. Forma-Pak, a company in Stockton, Calif., has developed a "clean top carrier," which snaps over the top of a can, is made of recycled plastic, and should not pose a threat to birds and other small animals. It is being tested by Miller Lite in Tennessee.

> ## How to use a Gopher: type the URL into your Web browser.

Bird Facts

Beginning with an explanation of the food chain and how birds fit into it, this page discusses the role of birds as environmental indicators. The basic needs of birds are detailed, and good information on endangered species and success stories are part of this easy-to-understand page that is geared to the junior high school level.

URL:
    http://www.nceet.snre.umich.edu/Curriculum/
    birdfacts.html

Bird Menus

Been looking for the consummate authority on the diet and nesting habits of British birds? This article provides information on the nesting habits and diets for a variety of birds common in the U.K. It gives suggestions for assisting birds with their nesting and also some helpful hints on how to attract food insects to the bird's areas.

URL:
    http://rs306.ccs.bbk.ac.uk/flora/birdmenu.htm

Birds of Prey

This page offers detailed information on a substantial list of raptors from throughout the world. Each species has a fact page with information on its characteristics, habitat, ranges, and much more. Take a close-up look at Leanne, a horned owl—one of many birds featured.

URL:
    http://www.raptor.cvm.umn.edu/raptor/rfacts/
    rfacts.html

> ## Ready for an uplifting experience? Read about Aviation.

Extinct Birds

Since the year 1600, more than 90 species of birds have vanished into extinction. This page lists nearly a dozen species now extinct—including the Crested Caracara, the Carolina Parakeet, the Cuban Macaw, and the Passenger Pigeon. There is information on each species, accompanied by a photo for those that became extinct in recent years and paintings of those that have been extinct longer.

URL:
    http://straylight.tamu.edu/bene/lg/
    birds_now_extinct.html

Forests, Flames, and Feathers

The destruction of fire is far-reaching—not only for humans, but also for wildlife. Not all birds can fly safely away from the flames of a bushfire. Those that do survive may be injured or end up homeless. This page addresses some ways to help these birds in need, including tips for first aid, feeding, and providing shelter.

URL:
    http://www.vicnet.net.au/vicnet/RAOU/
    firebird.html

The Mute Swan

The Mute Swan is a favorite friend in Britain. If you frequent public parks or the countryside in much of England, you know these large and graceful birds well. This page offers interesting details about the Mute Swan, including its history, customs, and life cycle.

URL:
    http://www.airtime.co.uk/users/cygnus/
    muteswan.htm

Ostriches

Time to get your head out of the sand and understand these birds. They are the largest bird in the world, flightless, and very aggressive. Ostriches are now valuable sources of low-cholesterol meat, decorative feathers, a fine leather, and very large eggs.

URL:
    http://netvet.wustl.edu/species/birds/ostrich.txt

Parrots

This is a well-written introduction to psittacines. These beautiful, talkative birds have many unique characteristics, and their natural need to socialize has made them very popular pets. With a larger brain than other species of birds, they have a great ability to learn.

URL:
> http://www.ub.tu-clausthal.de/PAhtml/
> intr000.html

Penguin Page

This site features one of the best collections of human knowledge on penguins north of the Antarctic. Links here include information on the species of penguins; their behaviors, reproductive habits, and relatives; and many more topics relating to these cute diving birds. So put on your tux and dive into the Penguin Page.

URL:
> http://www.vni.net/~kwelch/penguin.html

Penguins on the Net

They're there in black and white (and 254 other indexed colors). Waddle over to the Penguin Page.

Peregrine Falcons

You can come to know a couple of peregrines on an almost personal basis with the photos and text on these pages. Two falcons chose the University of Calgary for nesting and were studied during their stay. Also available here are many other links of interest relating to peregrines and other birds of prey.

URL:
> http://ksi.cpsc.ucalgary.ca/falcon/

Pheasant

This article goes into great detail to describe the conservation status of the 50 species of pheasant. There's an extensive table listing the categories of pheasants, and a five-year action plan including projects to provide basic information on the distribution, habitats, and conditions of pheasants.

URL:
> http://www.open.ac.uk/OU/Academic/Biology/
> P_McGowa/TRAGOPAN/TRAGOPAN_2-6.html

Raptor Center

Funded by private donations and the University of Minnesota, the Raptor Center is an international emergency medical facility for raptors (birds of prey). The Raptor Center claims to have treated over a tenth of the population of free bald eagles in the lower 48 states. Check out the center's state-of-the-art veterinary medical facility and the expert staff, and find out about membership, volunteer programs, and internships.

URL:
> http://www.raptor.cvm.umn.edu

Shorebird Migration Maps

Where did the Willet go? Find out on this page of current information on the sandpiper and plover migrations within the lower 48 states. Click on a species and view a color-coded migration map.

URL:
> http://www.utm.edu/~phertzel/migration.htm

Tragopans

Study the mating system and breeding ecology of 11 of the species of pheasant called Cabot's Tragopan. This is actually the formal report of a study that took place in the Zhejiang Province of China from 1990 to 1992.

URL:
> http://www.open.ac.uk/OU/Academic/Biology/
> P_McGowa/TRAGOPAN/TRAGOPAN_2-7.html

A
B
C
D
E
F
G
H
I
J
K
L
M
N
O
P
Q
R
S
T
U
V
W
X
Y
Z

PALEONTOLOGY

Chicago Field Museum

Among many excellent interactive exhibits, the Chicago Field Museum includes an uncomplicated tour of *Life Over Time,* beginning at least 3.8 billion years ago, and progressing through the eons to the Age of Dinosaurs. Thrill to several animations of prehistoric critters eating and running! Test your knowledge with its online brain games, and more.

URL:
> http://www.bvis.uic.edu/museum/
> Dna_To_Dinosaurs.html

Continental Scientific Drilling Program

Drilling, observing, and sampling the Earth's continental crust, DOSECC, Inc. represents the scientific drilling interests of 44 U.S. universities and 3 national laboratories. Current and past projects are documented online for your Web drilling and sampling pleasure.

URL:
> http://www-odp.tamu.edu/~csd/

Earthnet

Earthnet, a mammoth undertaking, is a collection of hundreds of links to geoscience and paleontology information from museums, academic institutions, and other sources found on the Internet.

URL:
> http://jacobson.isgs.uiuc.edu

Electronic Prehistoric Shark Museum

Peter Benchley wrote a book (which later became a movie series) about a guppy, the Great White shark. That's how the aquatic antagonist in *Jaws* would compare to its ancient ancestors that cruised the currents 5 to 25 million years ago. Since these sharks have no bones about them, only their seven-inch-long pearly whites remain as fossils today, giving us an impression of their 40- to 100-foot size that dwarfs today's Great White.

URL:
> http://turnpike.net/emporium/C/celestial/
> epsm.htm

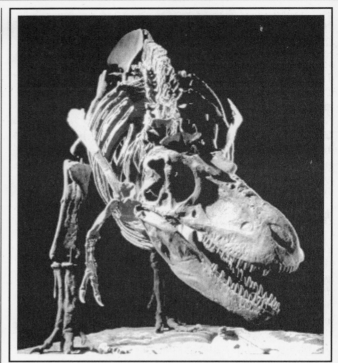
Tyrannosaurid—Albertosaurus, one of the images on Dino Russ's Earthnet page that you can really sink your teeth into.

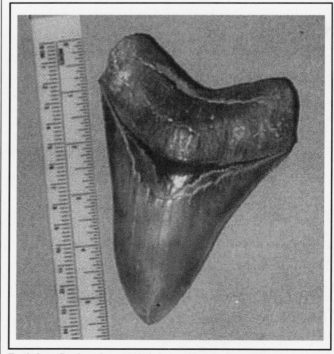
Tooth from Carcharodon megalodon, a prehistoric Great White, on display at the Electronic Prehistoric Shark Museum.

The Field Museum of Natural History

Check out this easy-to-read overview of dinosaurs and the periods in which they lived. Take a tour through the Jurassic period or view the skeleton of a Stegosaurus. These pages include some stunning drawings and graphics.

URL:

http://web66.coled.umn.edu/hillside/franklin/
Dinosaurs/Project.html

Hall of Dinosaurs

You are standing in an antechamber now. Ahead stretches a series of long hallways where you can see various amazing skeletons. In front of you is a map. It is not your usual map. Called a *cladogram*, it is a map that paleontologists use all the time to understand how animals are related to each other through time. Conveniently explore and savor each page of the University of California's Museum of Paleontology.

URL:

http://ucmp1.berkeley.edu/exhibittext/
cladecham.html

International Organization of Paleobotany

The IOP home page satisfies your interest in paleobiology, paleobotany, and past environments. Search the Plant Fossil Record Database to reveal records of fossil plants, pollen, seeds, wood, flowers, and leaves. Cool stuff, if you're into it.

URL:

http://www.uel.ac.uk/palaeo/

Introduction to Dilophosaurus

Enjoy a guided tour of Dilophosaurus! This tour is narrated by the discoverer of Dilophosaurus, Sam Welles, Professor Emeritus at the Museum of Paleontology. Thus begins the prof, "In the summer of 1942, Dr. Camp and I were on a joint expedition into the Navajo country . . ." You can hear the rest for yourself.

URL:

http://ucmp1.berkeley.edu/dilophosaur/
intro.html

Listen and watch as Dr. Welles introduces you to the Dilophosaurus. This virtual story includes images and audio narration of his fantastic discovery over 50 years ago.

New Mexico Friends of Paleontology

The Friends of Paleontology at the New Mexico Museum of Natural History are benevolent folks who like collecting fossils, under the supervision of the museum's curatorial staff. They also prepare and curate fossils; educate the public about paleontology; organize field trips; provide scholarships to paleontology students; award grants for support of fieldwork, preparation equipment, and research publications; plus promote world peace and good will.

URL:

http://www.unm.edu/~greywolf/test/nmfp.html

Paleontological Association Newsletter

A comprehensive newsletter on the world of paleontology, including association meetings, programs, conferences and conference reports, announcements, book reviews, and commentaries.

URL:

http://www.nhm.ac.uk/paleonet/PalAss/
Newsl26.html

Like birds? Flutter into Ornithology.

A
B
C
D
E
F
G
H
I
J
K
L
M
N
O
P
Q
R
S
T
U
V
W
X
Y
Z

Paleontological Research Institution

PRI houses one of the ten largest American collections of fossil invertebrates, including mollusks, corals, sponges, sea urchins, and others. The collection is particularly strong in Cenozoic fossils, microfossils, Caribbean corals, and Paleozoic fossils from upstate New York. PRI also publishes the oldest paleontological journal in the Western Hemisphere, *Bulletins of American Paleontology*.

URL:

    http://www.englib.cornell.edu/pri/pri1.html

Paleontological Society

The Paleontological Society reports on current activities; grants-in-aid; its international research program; publications; an online version of *Priscum,* the society's newsletter; the *Code for Fossil Collecting,* plus links to other paleontology resources.

URL:

    http://www.uic.edu/orgs/paleo/homepage.html

Palynology and Palaeoclimatology

Australian National University's Bioinformatics Hypermedia Service will be of interest to pollen analysts, anthropologists, archaeologists, paleontologists, geophysicists, and others interested in palaeoclimates and palaeoenvironments. Includes a wide variety of resources and Web pointers.

URL:

    http://life.anu.edu.au/landscape_ecology/
      pollen.html

Royal Tyrrell Museum of Palaeontology

Located near Calgary, the Tyrrell Museum is one of the premier paleontology museums in the world. This informative web page focuses on the study of prehistoric plants and animals in Alberta, Canada. Be sure to check out the Devonian Reef exhibit.

URL:

    http://www.freenet.calgary.ab.ca/science/tyrrell/

Excerpt from the Net...

(from World's First Dinosaur Skeleton)

Hadrosaurus: A Duckbill that Roamed the
    Coast of what is Now Pennsylvania

Hadrosaurus foulkii

In life, Hadrosaurus foulkii , the world's first dinosaur skeleton found in 1858, would dwarf a six-foot man. Nearly as tall as a two-story building, the animal weighed up to four tons and was a member of the dinosaur family that later became known as "duckbills" because of the bird-like nature of their jaws and frontal skull structure. Hadrosaurus foulkii was up to 30 feet long from the tip of its nose to the tip of its tail, but despite such bulk, the animal was not particularly ferocious. Much the same as a large, placid cow, it was a plant eater that browsed leaves and branches along the marshes and shrub lands of the Atlantic coast. It was also a good swimmer and could have regularly lolled in the water a substantial distance from shore.

Duckbills lived and traveled in herds, not unlike ponderous flocks of birds. They laid eggs in nests. Some paleontologists believe they protected the eggs until hatched and then continued to nurture the brood for a substantial period of time, just as birds do.

It is not all that difficult to imagine how the edges of conifer forests along the Cretaceous coast of what is now Pennsylvania were once alive with teeming rookeries of such duckbill dinosaurs. One of those creatures ultimately got into trouble near fast moving water and died. Its body floated out to sea, where it sank and was quickly covered in mud and sediment. After the flesh decayed, the bones absorbed minerals, surviving intact as a skeleton until about 70 million years later when a Haddonfield workman wrenched one from sticky marl, hefted it aloft into harsh sunlight and wondered aloud what it could possibly be.

The Hadrosaurus foulkii showing relative size to a man.

World's First Dinosaur Skeleton

These days, kids know more about dinosaurs than you do. But if you have access to the Web, you can get the edge before *you* become extinct. Start by taking a grand tour of the first dinosaur ever excavated. In the summer of 1858, William Parker Foulke, a Victorian gentleman and fossil hobbyist, discovered the world's first nearly complete skeleton of a dinosaur at Haddonfield, New Jersey—an event that would rock the scientific world and forever change our view of natural history. Excavate categories such as Finding the Bones in 1858, The Meaning of the Find, The Creature, Re-establishing the Site, and National Historic Site Designation, 1994.

URL:

http://tigger.levins.com/xdinosaur.html

PDAs

Hewlett Packard PDAs

While not quite as popular as Apple's Newton, HP's model 100 and 200 nevertheless have their fans, and this page lists a bounty of resources on the Internet that relate to HP PDAs. Included are Usenet newsgroups, web pages, FAQs, and FTP sites. All of these resources are hot links to the resources themselves, so this is a great starting point.

URL:

http://www.mt.cs.keio.ac.jp/person/itojun/HPLX/

Looking for Earth Science resources? Check Earth Science, Ecology, Geography, and Geology.

Reach for the stars. Gaze into Astronomy and Physics: Astrophysics.

HP 200LX Page

This page is the proverbial "everything you ever wanted to know about the HP 200LX." The page starts out with a good outline describing exactly what the HP 200XL is and what it can and can't do. Then it goes on to point to FAQs and information about accessories, software, and competing devices.

URL:

http://www.io.com/~rob/hp200lx/

International Newton Software

Tired of playing hangman in English? Try your hand with it in German or French! This page offers a collection of free international software for Apple's Newton PDA. Applications include *flashCard*, a self-quiz program; *shopList*, for creating and organizing your trips to the grocery store; and *transForm*, for designing forms. The languages these programs support are French, German, and Spanish.

URL:

http://www.eecs.wsu.edu/~schlimme/newton/index.shtml

PDA:
Personal Digital Assistant
Printing Decoder Aborts
Probably Doesn't Add
Press Down Accurately
Puny Display Anemic
Phrases Distort Absolutely
Pretty Damn Aggravating

A
B
C
D
E
F
G
H
I
J
K
L
M
N
O
P
Q
R
S
T
U
V
W
X
Y
Z

Mobile Computing and Personal Digital Assistants

Product reviews and specifications, conference announcements, links to research projects, sections on wireless networks and multimedia capabilities, and more related to PDAs. This index is also searchable by keyword.

URL:
> http://splat.baker.com/grand-unification-theory/
> mobile-pda/

Newton

Apple Computer's Newton is indeed a lesson in physics. It's been a long time since a computer has attracted such an avid following of users. Read about this popular Personal Digital Assistant from the folks who created it—Apple. The Newton pages offer detailed information and specs about the Newton, products you can use with your Newton, and a direct link to Apple's customer service department.

URL:
> http://www.apple.com/documents/newton.html

Newton Discussion Groups

Just can't get enough of the Newton from static web pages and FTP archives? Drop in on the Newton Usenet groups to converse with other Newton users. Get your questions answered, or help others by answering their questions.

URLs:
> news:comp.binaries.newton
> news:comp.sys.newton.announce
> news:comp.sys.newton.misc
> news:comp.sys.newton.programmer

PDA Uses and Wishlists

Tristan Bonstone wanted to know what people use their PDAs for, so he created a survey form and posted it in PDA-related Usenet groups. Read about the results and Tristan's findings on this page. Besides responses summarized by platform, Tristan also offers respondent's wishes for new PDAs and the answers to questions like "Describe the perfect PDA," and "What capabilities are you missing?"

URL:
> http://www.clever.net/tristan/survey.html

What's a PDA Good For? The screens are too small, they're hard to read. Your nine year old can probably do a better job of reading your scratches than they do. You can't run Photoshop on them—not to mention Dark Forces. And yet people seem to love their PDAs. Find out why, and what they use them for, at the PDA Survey (PDA Uses and Wishlists).

Tech Notes on the HP 100/200 LX

Albert Nurick and Tech.Net provide this excellent page with technical details about the HP 100/200 LX line of Personal Digital Assistants. It includes links to relevant Usenet newsgroups, software applications, electronic texts about the LX, development tools, and links to related resources on the Net.

URL:
> http://www.tech.net/technotes/hplx/

The Ultimate Newton

If you're a fan of Apple's Newton, check out this page for more information on Newton hardware, software, and leatherware than you can shake your stylus at. Included are links to a wealth of information about the Newton and using it; Newton FTP sites; software companies; and Newton products vendors, publications, and user groups.

URL:
> http://rainbow.rmii.com/~rbruce/

WriteWare

WriteWare is a company that makes beautiful replacement styluses for Apple's Newton and other Personal Digital Assistants. This page is a catalog of their products—most of which look like very expensive pens rather than PDA pointers.

URL:
> http://bigkahuna.sbusiness.com/writeware/

Yahoo's PDA Index

A collection of product descriptions, reviews, surveys, FAQs, magazines, newsgroups, business articles, and more about Personal Digital Assistants. There are special sections for the Apple Newton, Hewlett-Packard 100, Psion, and Sony Magic Link products—complete with reviews and specifications.

URL:
> http://www.yahoo.com/
> Computers_and_Internet/Hardware/PDAs/

PHOTOGRAPHY

Amateur General Photography

Other than remembering to remove the lens cap, what important things should you know about photography? Find out in these Usenet groups. Overexpose yourself to information on amateur photography—including tips on film, lighting, equipment, darkroom assistance, and more. Look into it and see what develops.

URLs:
> news:rec.photo
> news:rec.photo.help
> news:rec.photo.marketplace
> news:rec.photo.misc

Bengt's Photo Page

Focus in on Bengt's Photo Page for technology, current research, and suggestions for taking and making better photographs. Experts answer basic questions on darkroom design, polarization filters, special effects, experiments, alternative photography techniques, digital imaging, photo CD, and other related topics.

URL:
> http://www.math.liu.se/~behal/photo/

> Whether you just got those prints back from the store or finished drying them in your own darkroom, if you're not happy with your work, you'll find Bengt's Photo Page to be a photographer's delight.

Center for Creative Photography

The University of Arizona at Tucson is home to a creative collection of over 70,000 photographs from Ansel Adams to Edward Weston. Visit the collection online and explore their library, educational programs, and museum shop.

URL:
> http://www.ccp.arizona.edu/ccp.html

The Center Hall Foundation

If photos of early American life interest you, check out this well-designed page devoted to pre-Civil War, Civil War, and World War I. Early daguerreotypes and black-and-whites document the American experience in war and peace. Excellent graphics of the research building give a sense that you are actually using the facility to explore the collections.

URL:
> http://soho.ios.com/~arcodd/CHFmain.html

Depth of Field Plotter

This is a tremendously innovative and useful page that calculates and generates a graph in GIF format illustrating depth of field calculations for given focal lengths and apertures (f-stops). This resource is provided to the Web by Nick Sushkin—and if you don't trust his computations, you can even view the calculation to check the plot for yourself.

URL:
> http://oh114.wpi.edu/cgi-bin/htdof

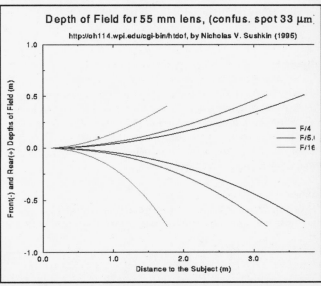

If you know what to do with this depth of field plot, you're a better photographer than we are. But whether or not you know what you're doing, you can play with focal lengths and apertures on Nick Sushkin's Depth of Field Plotter to create your own interesting graphics.

A
B
C
D
E
F
G
H
I
J
K
L
M
N
O
P
Q
R
S
T
U
V
W
X
Y
Z

Digital Camera Guide

If you're tired of dealing with messy film cameras and unreliable film developers (not to mention paying for film and development), maybe it's time to take your photography digital. The Digital Camera Guide is an amazingly useful guide to all of the digital cameras on the market—from Agfa to Kodak and Sony. There's an alphabetic listing of each manufacturer and camera model, and separate tables for handhelds/portables, camera packs and systems, and cameras that output still video images. Of course, references to manufacturers and camera models are links to those companies or spec sheets and ads for the cameras. The summary listings here include information about the maximum resolutions for each camera, the color and sensor depths, storage methods, and manufacturer's suggested prices. If you're thinking about buying a digital camera, don't miss this guide.

URL:

http://rainbow.rmii.com/~plugin/
DigitalCameraGuide.html

Digital Photography '95

A very popular site on the Internet, *digital photography '95* is a gallery of beautiful and artistic digital photographs. If you need inspiration or you would like to be in next year's gallery, pay it a visit.

URL:

http://www.bradley.edu/exhibit95/

Glossary of Photographic Terms

Having trouble keeping your hot shoes off your unipods? Check in with the glossary of photographic terms to keep your photo-nomenclature correct. This glossary sheds some light onto all of the important terms of photography. From adjustable-focus to zoom lenses—they're all exposed here.

URL:

http://www.kodak.com/aboutKodak/bu/ci/
photography/bPictures/glossary/termsA.shtml

Take flight in Aerospace and Space Technology.

KODAK Digital Images Offering

Everyone likes something for nothing, that's why everyone likes this page of free stock photography (for noncommercial use). Take your pick from 7 categories with 12 sample images each of flowers, famous city monuments, night scenes, landscapes, man-made and natural patterns, and a variety selection. The images are in both JPEG and Image Pac format using Kodak's Photo-CD process.

URL:

http://www.kodak.com/digitalImaging/
digitalImaging.shtml

Lighting and Action Tips for the Kodak DC40

Learn how to use a Kodak DC40 digital camera to optimize the lighting for outdoor and fast-action shots. Some of the information here about lighting is specific to the Kodak camera, but there are also great tips on capturing the peak of action, panning, and hints on directional shooting that apply to any camera—digital or film.

URL:

http://wwww.kodak.com/daiHome/
dc40Samples/dc40ActionLighting.shtml

DIGITAL PHOTOGRAPHY

It's about time someone invented a camera that doesn't need film. Film is so messy, not to mention hard to load. And what about the cost? Film companies and camera shops have had us by the shutter buttons for a century. Digital cameras are great. You can take pictures all day long, then download them into your computer and put mustaches on your co-workers. The funny part? Most camera shops won't carry them because they don't want people to realize they can buy a camera that doesn't need film or developing! No worries though. Slide into Kodak's web for a searchable database of dealers.

Minolta Users' Group

The Minolta Users' Group helps you learn how to work Minolta's Maxxum line of autofocus cameras. Listed topics include lenses, cameras, frequently asked questions, and suggested repair service. Lots of technical specifications—but, unfortunately, very few pictures of the cameras themselves. How about that. Camera-shy cameras.

URL:
http://tronic.rit.edu/Minolta/

Mysteries of the Darkroom

Mysterious things can happen in dark rooms, especially when you've got bizarre chemicals lying about in shallow pans. The **rec.photo.darkroom** newsgroup exposes you to techniques, tools, and equipment. Broad, engaging discussions address the latest *developments* in darkroom procedures and technology.

URL:
news:rec.photo.darkroom

National Press Photographers Association

This Usenet newsgroup offers photographers moderated access to a wide variety of postings from professional press photographers and articles from the NPPA. The NPPA issues feature information about free-lance work, fluorescent light filters, and reviving the f/64 group.

URL:
news:bit.listserv.nppa-l

Photo CD Transfer Sites

Search Kodak's online database for a site near you that can transfer your pictures to a Kodak Photo CD. This database is searchable by lab name, address, city, state, zip code, and area code. In addition, you can choose to narrow your search to owner sites (those that have the hardware in-house to actually do the conversions) and transfer sites that send the work out to other sites.

URL:
http://www.kodak.com/digitalImages/
piwSites/piwSites.shtml

Remember the Photo CD? Don't worry about it. We forgot, too. Relive the nostalgic marketing hoopla on the Internet.

Photo Information Service

The Photo Information Service is a collection of pages created by an unknown photographer to help budding photographers with any questions about photography. Offered on these pages are an online photo gallery; a web chat page (where you can ask your questions and get answers); information about photographers and models; and links to photography-related newsgroups.

URL:
http://www.southwind.net/~janet3/

photo.net

photo.net is filled with information on cameras, tutorials on traditional and digital photography techniques, and equipment reviews of everything from 35mm cameras to color printers.

URL:
http://swissnet.ai.mit.edu/photo/

PHOTO>Electronic Imaging Magazine

PHOTO>Electronic Imaging brings readers in-depth articles and tutorials on new imaging trends and technologies. Included is information on the Lightspace language, digital facsimiles, and other technologies in photography and digital imagery.

URL:
http://www.novalink.com/pei.html

A
B
C
D
E
F
G
H
I
J
K
L
M
N
O
P
Q
R
S
T
U
V
W
X
Y
Z

Excerpt from the Net...

(from photo.net)

Scanning Photos for the Web

by Philip Greenspun for photo.net

So you want to do it right? (or how I did Costa Rica)

Because of the pathetic lack of standards on the Web there is no way to produce an image to look good on your screen and also have it look good on anyone else's screen. For one thing, because there is no standard for monitor gamma, nor any way of including the target gamma in an HTML document, various remote computer systems may turn all the semi-dark areas of your image pitch black (or wash them out for that matter).

Currently, here is what I've decided produces the best results:

    Power Macintosh with a Trinitron monitor (17" or larger) with gamma set to 2.2 (this makes your Macintosh screen darker, but closer to the average Unix box and PC). You will need at least 32 MB of RAM to work with medium-size images; 64 MB would be better.

    Open the IMG00n.PCD files on the PhotoCD from Adobe PhotoShop using the Kodak Color Management System (included with PhotoShop), transforming from Photo YCC space to Adobe RGB.

    Edit the image, paying particular attention to the "Levels" command (try out Auto Levels if you are a beginner).

    When you have it exactly right, save the image as a PhotoShop 3.0 format file.

    Use the Unsharp Mask filter once with the default settings.

    "Save a copy" of your image as JPEG, medium quality.

    Use the Revert command to reload the unsharpened image from disk. You're doing this because you only want to sharpen a picture once and you'll have to run Unsharp Mask again after resizing.

    Use the Image Size command to shrink the picture, then use Unsharp Mask again.

    Save the small image as JPEG, low quality (for a thumbnail that Netscape and other modern Web browsers can read).

    Bash the image down to 6 bits Indexed Color and then save as a CompuServe GIF (for a thumbnail that archaic browsers can read; note that NCSA Mosaic for X Window can only display a maximum of 50 colors from each GIF so using 8-bit GIFs is pointless).

Photography Home Page

Richard Jacobs—photographer extraordinaire—has put up a web page with links to all of his favorite photography, camera, and graphics resources on the Web. Included are links to camera clubs; nature and garden photographic societies; Jacob's own photo gallery; and information about equipment for sale, software, and photography publications.

URL:
    http://www.mcs.net/~rjacobs/home.html

Photon Magazine

An excellent monthly online magazine, Photon publishes articles, news, technical Q&A, and maintains an extensive online photo library featuring the work of professional photographers. Find advice on advanced imaging techniques from expert scientists and photographers.

URL:
    http://www.scotborders.co.uk/photon/

The Net is the new medium.

PhotoServe

The PhotoServe pages offer an opportunity for professional photographers to display and study each other's works, read essays, and swap tips and techniques for better photography. The gallery includes works from some new talented artists, a bibliography, an updated essay section, and a list of useful photographic resources on the Internet.

URL:

http://www.photoserve.com/

PhotoSight

PhotoSight is a photography resource center on the Web offering photographic exhibitions, information about photographic products, and tips and techniques for taking better pictures.

URL:

http://www.webcom.com/~zume/PhotoSight/

Réseau Art and Photographie

The Réseau Art and Photographie Web is a resource for anyone interested in photography as an artform. Instead of focusing on the technical, this site is dedicated to the aesthetics of photography, and follows the activities of several photographic art organizations including the Centre National de l'Audiovisuel, the Galerie Nei Liicht Dudelange, and Café Crème.

URL:

http://www.restena.lu/cna/cna.html

Tips for Better Digital Pictures

Even if you're a professional photographer, it never hurts to bone up on the basics. This page reviews them all by presenting ten tips for taking better pictures. Although this page is specific to digital photography, the rules here apply also to film photography.

URL:

http://www.kodak.com/daiHome/dc40Samples/
dc40TenTips.shtml

Tips for Shutterbugs

Are the complexities of shutter speeds, f-stops, lenses, and filtering keeping you from experiencing the joy of photography? Perhaps a guide that takes you by the hand and starts at the very beginning is just what you need. Eastman Kodak presents this guide to better pictures for you. The guide covers a number of important topics including exposure, lighting, flash photography, lenses, filters and attachments, composition, close-ups, camera care and handling—and, of course—choosing the best film.

URL:

http://www.kodak.com/ciHome/photography/
bPictures/pictureTaking/pictureTaking.shtml

Top 10 Techniques for Better Pictures

"Keep your camera ready" is tip number one. You never know when a UFO might buzz by, or when a famous movie star might knock at your door looking for directions! Kodak, the world's premier film and photographic materials manufacturer, offers this page of common sense tips for beginners and experts alike.

URL:

http://www.kodak.com/ciHome/photography/
bPictures/pictureTaking/top10/
10TipsContents.shtml

If you want to know the truth of it, several of our favorite **Top 10 Techniques for Better Pictures** are remembering to put film in the camera and taking the lens cap off before madly clicking away. Of course, actually remembering to develop the film is right up there, too! Last tip—don't try to shoot group portraits of killer bees. . .

A B C D E F G H I J K L M N O P Q R S T U V W X Y Z

UCR Museum of Photography

The ambitious folks at the University of California-Riverside Museum of Photography feature a number of excellent online photographic exhibits. New works by contemporary photographers, a children's video and photograph section, and a great section of zoomable photos are waiting for you to inspect under the Virtual Magnifying Glass. Then, leave your comments on the Wall. Kilroy was here.

URL:
http://www.cmp.ucr.edu/netscape.html

PHYSICS

E-Print Theory Archives

Search for papers on many disciplines in the physical sciences. Database sections include high-energy physics; matter theory; astrophysics; general relativity; nuclear theory; superconductivity; plus chemical, accelerator, nuclear, quantum, plasma, atomic, molecular, and optical physics. My head hurts already.

URL:
http://xxx.lanl.gov

Fundamental Physical Constants

Incredibly useful for those in the physical sciences, this list is a thorough list of constants that are used to solve physics problems. This means you don't have to flip through your old physics books to find the equation that you need. Everything will be at your fingertips at this web site.

URL:
http://physics.nist.gov/PhysRefData/codata86/codata86.html

High Energy Physics Links

Become highly physical with this collection of resources relating to high-energy physics. There are job and conference listings, online journals, software, news, and links to physics research institutions.

URL:
http://www-hep.phys.cmu.edu:8001/~brahm/physics.html

Movies

View a collection of MPEG physics movies, including a flight around a rotating black hole; relativistic flight through a crystal lattice; large-object flight; and a nonrelativistic high-speed flight of large objects. See the real Einstein ring, and get out the popcorn as you watch the embedding of a black hole.

URL:
http://www.tat.physik.uni-tuebingen.de/movies.html

Museum of the University of Naples Physics Department

Get physical with this online exhibit of early instruments maintained by the Department of Physics. The 400-piece collection spans the time period from 1645 to 1900, and contains many unique or rare scientific apparati. Included are antique optics, lenses, and microscopes, and early devices that measure heat and electromagnetism.

URL:
http://hpl33.na.infn.it/Museum/Museum.html

Physics Around the World

Physics Around the World is a chain reaction of science resources and information available on the Web. Included is a terrific search engine and the Market Place for Physicists, where scientists can buy or sell used instruments.

URL:
http://www.physics.mcgill.ca/physics-services/physics_services2.html

Planet Earth Physics Page

Check out this large collection of physics resources divided into sections covering specialized fields, physics institutes, laws, periodic tables, and weights and measures. Browse through links to images of physicists and Nobel prize winners, research laboratories, and other physics-related servers.

URL:
http://www.nosc.mil/planet_earth/physics.html

PHYSICS: ASTROPHYSICS

Astronomy and Astrophysics

The Web's Virtual Library will connect you to tons of interesting astronomy and astrophysics sites throughout the Web. There are links to documents and other web sites where you can learn about these subjects if you don't know about them already. And if you do, they may be downright useful.

URL:
http://www.w3.org/hypertext/DataSources/
bySubject/astro/astro.html

AstroVR

AstroVR encourages research and study of astronomy and astrophysics by using a multiuser networked environment with access to many astronomical tools and databases. Equipped with the AstroVR software (which can be downloaded for free) and an X-Window display with microphone, users can talk, work together on a whiteboard, share images, make mongo plots, look up astronomical data, and more.

URL:
http://brando.ipac.caltech.edu:8888

Harvard-Smithsonian Center for Astrophysics

Interested in the future of space and astrophysics? Or have you ever wondered what experiments are being conducted on the space shuttle? Here's the place with the answers.

URL:
http://cfa-www.harvard.edu

High Energy Astrophysics Instrumentation

In this comprehensive encyclopedia of satellites and space stations you'll find pages dedicated to the Apollo-Soyuz; Salut-4; Solar Max; Mars Observer; and color images and charts from all high-energy astrophysics missions, including X-ray and gamma-ray missions.

URL:
http://heasarc.gsfc.nasa.gov/docs/heasarc/
missions.html

High Energy Astrophysics Science Archive Research Center

Wow! If you are growing bored with textual documents about astronomy and astrophysics, check out this graphical site created by the people at the NASA Goddard Space Flight Center. Of particular interest are the photos and inline animations of celestial bodies such as supernovas and black holes. This is a very popular site on the Web.

URL:
http://heasarc.gsfc.nasa.gov

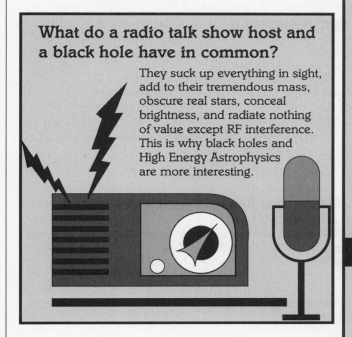

What do a radio talk show host and a black hole have in common?

They suck up everything in sight, add to their tremendous mass, obscure real stars, conceal brightness, and radiate nothing of value except RF interference. This is why black holes and High Energy Astrophysics are more interesting.

High-Energy Antimatter Telescope

The grownups of the High-Energy, Antimatter Telescope (HEAT) experiment must have suffered terrible childhoods. Their preoccupation with matter/antimatter annihilation foretells of some seriously destructive personalities. NASA-supported (amazingly), this program of high-altitude balloon-borne experiments studies antimatter in the primary cosmic radiation. Ah, what good is being grown up if you can't be childish?

URL:
http://tigger.physics.lsa.umich.edu/www/
heat/heat.html

The History of High Energy Astrophysics

This is a fascinating timeline of the history of astrophysics that allows you to discover missions and historical space ventures. This collection of documents will help you trace back to the early days of astrophysics, hundreds of years before humans even dreamed of space flight.

URL:
http://heasarc.gsfc.nasa.gov/docs/heasarc/
headates/heahistory.html

NASA Space Technology Ground-based Solar and Astrophysics

This is a guide to all the observatory resources available to the public on the Internet. It is a long list of national and international observatories of interest to astronomers and those interested in astrophysics.

URL:
http://ranier.oact.hq.nasa.gov/Sensors_page/
GroundObserv.html

Smithsonian Astrophysical Observatory

The Smithsonian Astrophysical Observatory is the home of the Central Bureau for Astronomical Telegrams, which is responsible for disseminating information on transient astronomical events. Issued at regular intervals, announcements are by postcard and Internet postings. Plot a course for this page to learn about current discoveries.

URL:
http://cfa-www.harvard.edu/sao-home.html

Telescopes

There are now so many astrophysical telescopes operating in space that you need this index of them to keep them straight! Focus in on the peepers floating around out there, and find out who controls them and what they're looking for. Also provided are connections to observatories throughout the world.

URL:
http://www.w3.org/hypertext/DataSources/
bySubject/astro/astroweb/yp_telescope.html

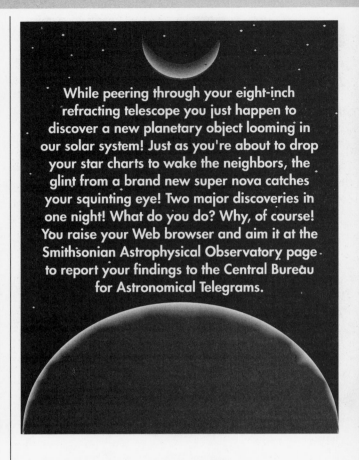

While peering through your eight-inch refracting telescope you just happen to discover a new planetary object looming in our solar system! Just as you're about to drop your star charts to wake the neighbors, the glint from a brand new super nova catches your squinting eye! Two major discoveries in one night! What do you do? Why, of course! You raise your Web browser and aim it at the Smithsonian Astrophysical Observatory page to report your findings to the Central Bureau for Astronomical Telegrams.

WebStars: Astrophysics in Cyberspace

Originally designed for NASA, this page is a list of hyperlinks to sites that relate to space and astrophysics. Other links of interest are cyberspace, data formats, programming information, and articles about astronomy.

URL:
http://www.stars.com/WebStars/

PHYSICS: NUCLEAR

Atomic Physics Links

Here's a long list of atomic physics links throughout the Web. Included are research centers, universities, and direct connections to individual laboratories.

URL:
http://www-phys.llnl.gov/N_Div/atomic.html

Continuous Electron Beam Accelerator Facility

CEBAF studies accelerators, superconducting radio waves, and other properties of strongly interacting matter. A tremendous resource for the classroom, this page includes interactive instructional tours of planets, atoms, and more to help students understand nuclear physics, quantum chromodynamics, and related fields.

URL:

http://www.cebaf.gov:3000/cebaf.html

Fermi National Accelerator Laboratory

Level this baby at the neighbor's dog that keeps you up all night and you'll be chef of the largest outdoor weeny roast. Fermi National Accelerator Laboratory houses the Tevatron, the world's most powerful particle accelerator and microwave oven. See how Fermilab is exploring the most fundamental particles and forces of nature. Please bring mustard and relish.

URL:

http://www.fnal.gov

Fusion Research Center

The Magnetic Fusion Energy Database provides the fusion community with complete sets of experimental data from a variety of machines and discharge conditions. This is an important resource for any Galaxy-class starship.

URL:

http://hagar.ph.utexas.edu/frc/

Lawrence Berkeley National Laboratory

America's national laboratory system started here when Ernest Orlando Lawrence founded this laboratory in 1931. Lawrence invented the cyclotron, which led to a Golden Age of particle physics and revolutionary discoveries about the nature of the universe. This page offers updates on lab news, research, and development, and describes the achievements of LBNL Nobel Laureates.

URL:

http://www.lbl.gov/LBL-PID/LBNL-intro.html

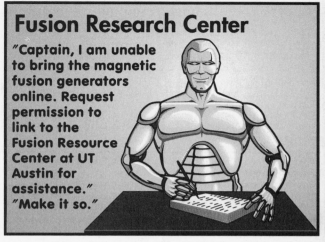

Fusion Research Center

"Captain, I am unable to bring the magnetic fusion generators online. Request permission to link to the Fusion Resource Center at UT Austin for assistance." "Make it so."

Los Alamos National Laboratory

Welcome to the home page of the most famous nuclear physics laboratory in the world. This is where the first atomic bomb was constructed, and it continues to be a mecca for nuclear research. Since World War II, the mission of the laboratory has changed to include biomedical science, environmental protection and cleanup, computational science, and other scientific fields.

URL:

http://www.lanl.gov/Public/Welcome.html

Madison Symmetric Torus

The nice folks at the University of Wisconsin at Madison built this big doughnut for you. This is no ordinary doughnut. It's a toroidal reversed pinch device. But don't ask us what they do with it. We figure it's a kind of "Way Back" machine, or something you'd dunk into a huge cup of java. This page outlines the project, model, and materials involved in the development of this plasma physics and magnetic fusion energy device.

URL:

http://sprott.physics.wisc.edu/mst.htm

Excerpt from the Net...

(from the pages of the Continuous Electron Beam Accelerator Facility)

Jupiter is surrounded by a large radiation belt that tends to fry electronics and people. Some of the satellites are in the radiation belt, so be careful!

A B C D E F G H I J K L M N O P Q R S T U V W X Y Z

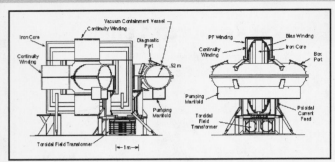

Prepare for lift off. While it looks like a space ship, this view of the Madison Symmetric Torus gives you an inner view of this nifty toroidal reversed field pinch device. Check out the Madison Symmetric Torus.

Office of Fusion Energy

The OFE is connected with the U.S. Department of Energy and strives to make fusion an environmentally attractive, commercially viable, and sustainable energy source for the nation. This site includes maps and lists of laboratories across the country, direct links to research centers across the world, and other general information about fusion and fusion research.

URL:

http://wwwofe.er.doe.gov

Research and Training Reactors

Educational institutions and other organizations seek training and research services available only at particular nuclear reactor sites around the country. Here's a list of research and training reactors around the U.S.

URL:

http://web.fie.com/web/fed/doe/doemnrt.htm

Review of Particle Properties

This is an online book for particle physicists to use as a reference. And what a reference it is! It's also a great source of information for physics students hoping to learn more about quarks, neutrinos, and all the other subatomic particles scientists study.

URL:

http://pdg.lbl.gov/www/rpp/book/contents.html

Table of Nuclides

This is a clickable image map that charts all known nuclides with colored dots showing the known isotopes of each element. Click in a region of the map, or type in the name of your nuclide for more information.

URL:

http://hpngp01.kaeri.re.kr/CoN/

PHYSICS: QUANTUM

Accelerator Physics

This will lead you to one of the last frontiers of physics. Connect yourself to the labs at the forefront of particle research into the atom. There are connections to magazines, articles, laboratories, job listings, and many other useful resources.

URL:

http://beam.slac.stanford.edu/www/library/
w3/alab.htmlx

A Brief Review of Elementary Quantum Chemistry

Quantum mechanics is the foundation of chemistry, and you'll see why with this page that reviews and explains some of its fundamental principles. Among other topics covered are quantization of electronic angular momentum, the Schrödinger equation, and wave-particle duality.

URL:

http://zopyros.ccqc.uga.edu/Docs/Knowledge/
quantrev/quantrev.html

Center for Gravitational Physics and Geometry

The Nittany Lions of Penn State have their paws planted firmly on the ground with this handy page for physicists that supplies electronic preprint archives (xxx.lanl.gov, hep-th new, index, gr-qc new, index, astro-ph new, index, CERN, SPIRES). You'll find a list of publications on new variables in TeX and PostScript, plus information on seminars and conferences associated with this center, and email addresses of physicists worldwide.

URL:

http://vishnu.nirvana.phys.psu.edu

General Relativity Around the World

Relativity beckons at this NCSA Relativity Group page that follows in Einstein's footsteps. The group maintains a list of relativity-related services and lists of resources, including servers for other relativity groups and lots of physics, astrophysics, and astronomy resources.

URL:

http://jean-luc.ncsa.uiuc.edu/World/world.html

"Imagination is more important than knowledge."
Albert Einstein.

Oh, sure. You would say this, too, if you had an IQ of 242. Besides, when it comes to inheriting intelligence, it's all relative. Improve your IQ by studying General Relativity Around the World.

HyperSpace

If space is the final frontier, quantum mechanics is the ultimate frontier! Beam yourself into hyperspace with this set of hypertext-based services for general relativity research. You'll encounter software for finding general relativity mail addresses, a GR news archive, a GR scientific chat, GR-related FTP archives, and other GR groups.

URL:
http://www.maths.qmw.ac.uk/hyperspace/

Many-Worlds Quantum Theory

The next time your cocktail party conversation zings its way toward the interactive exchange of gauge bosons, your very mention of the Many-Worlds theory is sure to start a big bang with the rest of the guests. Get your gravitons together and brush up on what Many-Worlds actually means here at this site that also contains a FAQ list, insightful technical data, and links to related sources.

URL:
http://www.gatech.edu/tsmith/ManyWorlds.html

Matters of Gravity

Gravitate toward this newsletter published by the American Physical Society and you won't go wrong. Gravitation in all its manifestations is the topical interest of this group. The material is published in TeX and PostScript, and is highly technical, but who knows? You may end up really falling for it!

URL:
http://vishnu.nirvana.phys.psu.edu/mog.html

Particle Beam Physics Laboratory Quantum Information

If we are indeed the dreams that stuff are made of, make your dreams come true at this page that features selected papers from "The Mouth of Truth." Enjoy a few path-integral approaches to quantum theory, computation, and complex algorithms with this series of introductory articles.

URL:
http://vesta.physics.ucla.edu/~smolin/

Quantum Computation: A Tutorial

If computers could express their feelings, they'd probably tell you to byte and crunch on this tasty tutorial of how quantum mechanics can be used to improve computation with computers. The challenge is to solve the factors for a very large number with a conventional computer (which is an exponentially difficult problem) and apply the results to a theoretical computer that computes in the twilight zone of Hilbert space. Sound odd and interesting? Read more here.

URL:
http://chemphys.weizmann.ac.il/~schmuel/comp/comp.html

Quantum Magazine

Quantum is published by the National Science Teachers Association (NSTA) and is a lively, handsomely illustrated bimonthly magazine of primarily physics-oriented math and science. If you have Adobe Acrobat, you can view a sampler of what the magazine has to offer, obtain an article index, and view news and contents of the magazine's recent issues.

URL:
http://www.nsta.org/quantum/

Quantum Mechanics Movies

If a picture is worth a thousand words, then what's the value of a movie—especially when it explains quantum mechanics? Download graphic versions of such blockbusters as "Vision, Light, and Energy Levels," "Wave-Particle Duality and Diffraction," "Atomic Orbitals," and "The Schrödinger Equation." Siskel and Ebert give them all 3.1416 thumbs up.

URL:
http://www-wilson.ucsd.edu/education/qm/qm.html

A B C D E F G H I J K L M N O P Q R S T U V W X Y Z

Reliability Analysis Center

Reliability is Job One at the RAC. This site is sponsored by the Defense Technical Information Center and aims at providing data and information to improve the reliability and maintainability of components and systems. You'll find many tools for quality management productivity, including late-breaking news and hot technical topics, online issues of the *RAC Journal*, a calendar of events, and professional society and industry announcements. There's a categorized, cross-referenced list of software tools for reliability and related disciplines, and a bibliographic search capability.

URL:
http://www.iitri.com/RAC/

Sarfatti Lectures on Physics

Take a quantum leap forward with this series of lectures on physics, mathematics, particle theory, and space-time continuum. You'll find detailed, yet accessible, information on quantum mechanics, and some lively photographs and artwork created especially for the page.

URL:
http://www.hia.com/hia/pcr/qmbeynd.html

Transactional Interpretation of Quantum Mechanics

Put your fourth-dimensional thinking cap on and come wormhole your way over to this page on contrafactual definiteness, nonlocality and formalism, the Copenhagen Interpretation, and the new Transactional Interpretation. This paper presents new insights into quantum mechanics and events as an exchange of advanced and retarded waves, coupled with the Wheeler-Feynman Absorber Theory.

URL:
http://mist.npl.washington.edu/npl/int_rep/tiqm/TI_toc.html

Warp Drive Physics

Beam over to this page on elementary relativity and quantum physics that bills itself as, "Multimedia self-paced learning for trekkies." It may actually help the next time you want to put together a full-scale model of the Enterprise, complete with warp-drive engines. Learn about Einstein's theory of special relativity and follow that up with general relativity, causality, tachyons and quantum nonlocality that profoundly impact on real-life starship-design engineering parameters and fundamental limits—or lack thereof.

URL:
http://www.hia.com/hia/pcr/st1.html

Wave Theory of the Field

Catch a wave and you're sitting on top of the world, but catch this page on unitary wave theories and you may end up with a new unified theory for all quantum interactions. The page contains dialectics that attempt to coalesce the concept of space-time continuum and clarify ideas of wave-particle theory, relativity, and discrete models. There is a reference section for further scientific surfing.

URL:
http://www.inet.it/ospiti/cassani/

POLITICAL SCIENCE

The Carter Center

The Carter Center is a nonprofit, nonpartisan public policy institute founded by former President Jimmy Carter. The Center is dedicated to fighting disease, hunger, poverty, conflict, and oppression through collaborative initiatives in the areas of democratization and development, global health, and urban revitalization.

URL:
http://www.emory.edu/CARTER_CENTER/homepage.htm

Cato Policy Reports

The Cato Institute seeks to broaden the parameters of public policy debate to allow consideration of more options that are consistent with the traditional American principles of limited government, individual liberty, and peace. No issue is left undisturbed: censorship, term limits, business subsidies, health care reform, immigration, Social Security privatization, and scads of other hot topics.

URL:
http://www.cato.org/main/home.html

Constitutions, Treaties, and Official Declarations

Ponder this interesting collection of documents that relate to governments and international relations. Read the fine print in complicated treaties such as GATT, and international fishing laws.

URL:
http://www.keele.ac.uk/depts/po/const.htm

Constitutions, Treaties, and Official Declarations

We declare and entreat you to enlighten your constitution by reading this collection of documents that have changed the world— sometimes, even for the better!

European Governments Online

Learn more about monarchies, social affairs, foreign policy, cultures, and many more subjects from European Governments Online. Countries include Belgium, Denmark, France, Germany, Luxembourg, Spain, Sweden, and the U.K.

URL:
http://www.cec.lu/en/gonline.html

European Union

This page of the European Union contains information about the economic unification of Europe and the political agenda of the organization. Connect to specific governments in the Union and read official EU documents.

URL:
http://www.cec.lu/Welcome.html

Foreign Government Resources on the Web

Enjoy this incredible home page that will link you literally with all the world. Compiled by the University of Michigan, this site will connect you with governments in Africa, Asia, the Pacific, Central and South America, Europe, the Middle East, and North Africa. Compare and contrast the various government constitutions that are assembled at this site. You'll find a list of embassies as well as links to other political science resources. Place a bookmark on this page!

URL:
http://www.lib.umich.edu/libhome/
Documents.center/foreign.html

Gallop Organization

How popular is the president this week? The Gallop organization has created this web site to provide the results of recent surveys and polls. Find out what the country is thinking!

URL:
http://www.gallup.com

Index of Nonproliferation Sites

The spread of nuclear weaponry throughout the world is a deadly threat that we should all keep an eye on. To find out more about nukes and the politics behind them, check out this list of links to the top nonproliferation sites on the Web.

URL:
http://www.miis.edu/cns/nrilinks.html

International Affairs Network

Connect to academic institutions, national governments, think tanks, and other bastions of political science. The subject of international affairs is categorized into foreign policy, economic development, international law, peace, and conflict resolution.

URL:
http://www.pitt.edu/~ian/ianres.html

International Monetary Fund

The IMF was created to foster political and economic stability around the world. Here's a collection of documents about the IMF and the financial and monetary studies that it publishes every year.

URL:
gopher://gopher.imf.org

International Security Network

IntSecNet is an information network for security and defense studies, peace and conflict research, and international relations. It offers lists of international security resources by subject and region, and details of institutions active in international security.

URL:
http://www.fsk.ethz.ch/isn/wwwvl_isn.html

A B C D E F G H I J K L M N O P Q R S T U V W X Y Z

Kennedy School Case Studies

The Harvard University Kennedy School of Government makes available over 1,000 case studies. These are accounts of decision-making in public policy and public administration and, as such, represent frontline research about the nature of public sector operations both in the U.S. and abroad. However, these cases are designed not only to help those who use them learn about government, but also to teach about it. Use the Search feature to search the entire case catalog.

URL:

gopher://ksggopher.harvard.edu:70/11/.KSG/
Kennedy%20School%20Case%20Studies

National Archives Information Server

The National Archives have one of the largest collections of political documents in the world. This can be a very useful resource if you're interested in history or politics. However, in some cases, you might have to go to Washington, D.C. to actually view the documents.

URL:

http://www.nara.gov

The National Budget Simulation

Here's your chance to make a difference and save the country from rack and ruin. This simulation gives you a better feel of the trade-offs that citizens and policy makers need to make to balance the budget. For example, by simply reducing military spending by 30 percent, eliminating wasteful Medicare spending by 50 percent, and cutting welfare by half, you can reduce the deficit by $200 billion, and thereby balance the budget while you increase the more-important space, science, and technology spending by 100 percent at the same time! It's fun—try it!

URL:

http://garnet.berkeley.edu:3333/
budget/budget.html

Corrections? Changes? Additions? See our web page at http://www.iypsrt.com.

The Web is the Net's GUI.

National Budget Simulation

Sure the budget's in a mess, so pretend that you're one of America's quality policy makers, roll up your sleeves, sharpen your pencils and your hatchet, and start hacking away! Eliminate wasteful programs to your heart's content, and in no time at all you'll have that darned budget healthy as an ox. Don't forget to mop all the blood off the floor from those axed programs, though!

National Performance Review

The majority of the American public knows that the government wastes much of what we pay in taxes. The National Performance Review is committed to finding a solution to this problem. With over 800 documents online, you can feel better about our government while reading articles like "Putting Customers First: Standards for Serving the American People," and "From Red Tape to Results: Creating Government That Works Better and Costs Less." I feel better now. Don't you feel better? I think we're feeling better.

URL:

http://www.npr.gov

Official Government Web Pages by Country

Although it's not complete, here's an extensive list of links to governments of the world. The more developed the country, the more information you'll find.

URL:

http://www.keele.ac.uk/depts/po/official.htm

Political Parties, Interest Groups, and Other Political Movements

This is a great resource for finding out more about political parties and political movements throughout the world. There's everything from Greenpeace to the League of Women Voters. You'll find data on political parties from 53 countries and an index to interest groups.

URL:

http://www.keele.ac.uk/depts/po/parties.htm

Political Science Resources

Laugh about it, shout about it when you've got to choose, but try this useful site when you begin your search for political science-related links. You'll find documents and data from governments all over the world. There are hundreds of links worth browsing.

URL:

http://www.keele.ac.uk/depts/po/psr.htm

Political Science Resources on the Web

This collection of documents and links that pertain to political science is provided by the University of Michigan. Topics at this site include foreign politics, international relations, current news, theory, statistics, U.S. politics, and periodicals. From this page you can monitor elections and view the home pages of candidates and political parties. There is a variety of links to other political science sites on the Net.

URL:

http://www.lib.umich.edu/libhome/
Documents.center/polisci.html

President Libraries IDEA Network

Connect yourself to the resources available at American presidential libraries around the U.S. Each is a miniature history lesson regarding the particular president's term of office, and all are of value to students of political science.

URL:

http://sunsite.unc.edu/lia/president/

THOMAS

Full text of all versions of House and Senate bills are available on THOMAS, named after Thomas Jefferson. THOMAS receives current legislative information several times daily, plus the text of the Congressional Record once daily when Congress is in session. Take a moment from surfing the Web to see what your elected representatives are up to. Pocket protectors should be exchanged for pork detectors before entering.

URL:

http://thomas.loc.gov

U.S. House of Representatives

The House of Representatives web page contains information about legislation, ongoing debates, house members, phone numbers, addresses, email accounts, laws, and links to other government information services.

URL:

http://www.house.gov

U.S. House of Representatives

What's more fun than a barrel of monkeys? How about a gaggle of congresspersons? Regardless of whether you're a proponent of big government or little government, you still pay Congress' salary, and they still spend your taxes. Pretty good work if you can get it!

United Nations Information Services

The United Nations has some interesting information on the Web. Find out what's happening in the General Assembly, and read about the latest Security Council Resolutions. There is information about human rights, activism, women, U.N. administration, and many other topics of interest.

URL:

http://www.undcp.org/unlinks.html

A B C D E F G H I J K L M N O P Q R S T U V W X Y Z

United Nations Peace-Keeping Missions

Read a fascinating list of current United Nations peace-keeping operations from around the globe that are more numerous than you might think. Each mission is explained in detail, and you can view statistical information about the annual cost in money and lives for a particular conflict. Often, there are links to related resources.

URL:
http://ralph.gmu.edu/cfpa/peace/toc.html

United States Senate

Peruse these resources on the U.S. Senate that include committee documents and the text of speeches. If you would like to contact your senator, there's a complete list of members and their email addresses.

URLs:
gopher://gopher.senate.gov/1/

http://www.senate.gov

PRIMATOLOGY

ASKPRIMATE

If you're hankering to know what the average weight of an adult male orangutan is or whether there are any videotapes that show sexual behavior in chimpanzees, contact ASKPRIMATE. Ask your question and provide your name, city, state, and email address. Be as specific as possible in asking your question. After all, you might learn something from those videotapes.

Mail:
Address: **askprimate@primate.wisc.edu**

California Regional Primate Research Center

Feel up to some primate research, but don't have a lab or zoo handy? The CRPRC at the University of California at Davis can facilitate your primate project without you ever having to leave home. Dream up your idea and see what they have to offer regarding primate fetal and neonatal studies or research on diseases of the aged.

URL:
http://www.crprc.ucdavis.edu/crprc/

Diseases of Non-Human Primates

In case your primates aren't in such great shape, you'll want to consult this list of diseases. This document contains etiological and pathological information, disease transmission means, clinical diagnoses, and treatment procedures.

URL:
http://vetpath.afip.mil/nhp.txt

Duke University Primate Center

Duke's primate page generally focuses on lemurs, those lovable primitive creatures found mainly in Madagascar. Take a tour of the Duke University Primate Center, learn how to adopt a lemur, catch up on DUPC bulletins and recent news, check out Lemurcon '95, and find other online lemur information.

URL:
http://www.best.com/~dupc/

European Primate Resources Network

EUPREN has developed a network that aids the establishment of synergistic relationships to improve the quality of primate-related research in Europe. See how primate research centers from France, Germany, Italy, the U.K., and the Netherlands are collaborating on programs to tackle problems in primate research. While you're here, explore the Census of Primates in Europe and North Africa.

URL:
http://www.dpz.gwdg.de/eupren.htm

Gorilla Conservation News

Find out how mountain and lowland gorillas are faring, especially with the recent civil war in Rwanda and Burundi. This no-frills journal reports and explains recent primatological research projects in Africa.

URL:
http://night.primate.wisc.edu/pin/GCN9.txt

Gorilla Home Page

You'll go ape (you knew that was coming) over this page featuring gorillas, with sections and overviews of their habitats and behaviors. Check out the Gorilla Modeling Project; and while you're at it, you can knuckle down with several other related links.

URL:
http://larch.ukc.ac.uk:2001/gorillas/

Peel on over to the Gorilla Home Page to find out what gorillas really do after dark!

Gorillas

Albeit a commercial site, this resource presents a good deal of detailed information about gorillas' traits and habitats. You'll learn a lot of gorilla information—including how they are scientifically classified and what their habitats and population distributions are, as well as gorilla physical characteristics and special adaptations to their environment. A host of other topics is covered in depth.

URL:
http://www.bev.net/education/SeaWorld/gorilla/
gorillas.html

Guidelines for Acquisition, Care, and Breeding of Primates

Pamper your favorite primate with these thoughtful guidelines on how to take care of them. Learn how to rear young primates and enrich their physical and social environments—while ensuring that you have the necessary expertise to tackle the responsibility in the first place.

URL:
http://netvet.wustl.edu/species/primates/
primguid.txt

International Primate Protection League

This organization, dedicated to the protection of primates, issues conservation alerts concerning environmental and political updates that affect the geoglobal welfare of simians and prosimians. There are sample articles and news from the organization's current newsletter.

URL:
http://www.sims.net/organizations/ippl/
ippl.html

Jane Goodall Institute

Jane Goodall is recognized as the world's foremost chimpanzee expert. Learn what her institute is accomplishing in wildlife research, education, and conservation. Check out online information and progress within African national parks and chimpanzee sanctuaries, including Gombe Stream National Park in Tanzania, and Kibira National Park in Burundi.

URL:
http://gsn.org/gsn/jgi.home.html

Laboratory Primate Newsletter

Read to your rhesus monkey the latest research articles on studies of lab primates. You'll find in-depth articles, news, meeting announcements, available grants, travelers' health notes, workshop postings, positions available or wanted, and notices and reviews of recent books and articles.

URL:
http://www.brown.edu/Research/Primate/

Mad About Marmosets

Meander over to learn more about marmosets and their cousins, the tamarins—to this site about these jungle tree-dwellers whose habitats range from Panama to Brazil. About the size of a rat and weighing less than a pound, marmosets chatter a lot and possess several scent glands that make them somewhat odiferous. Not surprisingly, they are said to be closely related to humans. Find out other fascinating marmoset minutiae here.

URL:
http://loki.ur.utk.edu/ut2kids/primates/
marmosets.html

Marvel at those amazing marmosets and their fun-loving cousins, the tamarinds, by visiting the **Mad About Marmosets** page.

Mountain Gorilla Protection

The purpose of this project at Rutgers University is to provide a digitized database of the mountain gorilla habitat. This database includes information on vegetation patterns, locales where gorillas are found, and the encroachment of humans into gorilla habitat. You'll find shaded relief maps of Africa, sophisticated digital and 3D elevation models, and QuickTime and MPEG movie fly-throughs utilizing radar imagery of subject study areas.

URL:

http://deathstar.rutgers.edu/projects/gorilla/
gorilla.html

Primate Info Net

What do you get when you mix monkeys and gophers? The Primate Info Net. An Internet gopher network for those interested in the field of primatology, PIN offers lists of helpful organizations, educational and employment opportunities, a bibliographic database, and a newsletter.

URLs:

http://night.primate.wisc.edu/pin/
http://uakari.primate.wisc.edu:70/1/pin

Primate Newsletters

Extra, extra! Read all about these wonderful creatures, the primates, with this list of online journals and newsletters from all around the world. Pick your favorite—from the *Chinese Primate Research and Conservation News* to the *Old World Monkey TAG Newsletter*.

URL:

http://night.primate.wisc.edu/pin/newslett.html

Primate Talk Calendar

Give your gibbon a gift by taking it along when you make your plans to attend one of the gatherings listed on this calendar page. Primate meetings from around the world are posted here along with their agendas. The list is updated frequently.

URL:

http://uakari.primate.wisc.edu:70/0/pin/calendar

Seeking stimulating circuit simulations? See Engineering.

Primate-Talk

Ooh, ooh! Subscribe to the general primatology mailing list and check out our evolutionary cousins' progress from the comfort and privacy of your monitor. Lemurs, chimps, orangutans, and apes are all waiting for you to read about them. You'll find news items, meeting announcements, research, conservation and education issues, veterinary information, and much more.

Mailing List:
Address: **primate-talk-request@primate.wisc.edu**
Body of Message: **subscribe primate-talk** *<your name>*

Purchase a primate or prosimian lately? Well, proceed to primp, pamper, and prolong its prospects with this *Primates as Pets* primer page on proper primate protocol and protection.

Primates as Pets

If you're adamant about keeping a primate as a pet (and wouldn't we all enjoy a lemur or loris) consider consulting this page on how to go about doing it. You'll find sections on ethics, opinions, diets, visiting the veterinarian, diseases, additional sources of info for primate pet owners, and a questions and comments area.

URL:

http://uakari.primate.wisc.edu:70/1/pin/vet/pets

PROGRAMMING

Ada 95 Reference Manual

Countess Ada Lovelace would love this page: a hypertext version of the revised international standard (ISO/IEC 8652: 1995): *Information Technology—Programming Languages—Ada.*

URL:

http://lglwww.epfl.ch/Ada/LRM/9X/rm9x/

Hug a tree. Hang out in Forestry.

Ada Tutorials

A collection of tutorials for the Ada programming language and pointers to other Ada-related educational materials are here. These tutorials include an Ada 95 tutorial, a tutorial that allows one to interactively create Ada programs remotely over the Web, the C/C++ Programmers Ada Tutorial, and off-line tutorials for PCs and Macs. There are also slides describing Ada 95, Ada software archives, course material for computer science classes using Ada, bibliographies, and much more.

URL:
>   http://lglwww.epfl.ch/Ada/Tutorials/
>   Lovelace/tutors.html

COBOL FAQ

Got questions about COBOL? Find the answers on the COBOL Frequently Asked Questions page taken from periodic Usenet postings.

URL:
>   http://www.cis.ohio-state.edu/hypertext/faq/
>   usenet/cobol-faq/faq.html

COBOL Syntax Guide

Time for a little quiz. The seventh column of COBOL source listings is reserved for special characters. Do you know what they are and how they're interpreted? What four divisions make up a COBOL program? These and other COBOL coding guidelines are illustrated here.

URL:
>   http://www.cs.indiana.edu/hyplan/mayer/
>   cobol/cobol.html

Common Lisp

You'll find the complete hypertext of the book *Common Lisp the Language*. Lisp is a popular preference for artificial intelligence applications. If you're thick of programming in BATHICK, take a look at Lithp. (You didn't think we'd pass up an opportunity to exercise this tired old jab, did you?)

URL:
>   http://www.cs.cmu.edu/afs/cs.cmu.edu/project/
>   ai-repository/ai/html/cltl/cltl2.html

Common Lisp Archive

Lisp is a programming language used commonly in the area of artificial intelligence. This archive contains many Lisp resources, including a review of the proposed ANSI standard for Lisp, the complete text of an excellent book about Lisp, a Lisp FAQ, source code archives for popular platforms, newsgroup archives, software implementations, papers, Lisp utilities, and other resources relating to Lisp.

URL:
>   http://agent2.lycos.com:8001/lisp/

Common Lisp the Language, 2nd Edition

Here's the complete text of the book *Common Lisp the Language*, 2nd edition, by Guy Steele, the co-developer of the Lisp language. The book is completely in HTML format, so it's a breeze to read online. The book introduces the Lisp language; its data types, type specifiers, and program structure; and, of course, every other detail about Lisp in which you might be interested. This is a must-read for all Lisp programmers. The book is also available in PostScript and DVI formats.

URL:
>   http://www.cs.cmu.edu/Web/Groups/AI/html/cltl/
>   cltl2.html

Excerpt from the Net...

(from Common Lisp the Language on Carnegie Mellon)

"Writing elegant programming code is a really hard task! Designing algorithms and data structures requires such a difficult sort of analytic thought, it's a wonder anyone can do it well, at all. The main barrier to AI is the need to innovate new ways of thinking about programming—such as object-oriented design, and declarative code. (Jargon is the greatest enemy of the required clarity of thought!)" —Jorn

A B C D E F G H I J K L M N O **P** Q R S T U V W X Y Z

Fortran 90 Tutorial

Resembling something of a Fortran program itself, this tutorial has sections explaining and detailing Fortran language elements: expressions and assignments, control statements, procedures, arguments, recursion, array handling, pointers, specification statements, and input/output. Use this resource to learn about Fortran and find out for yourself why, in 1977, Fortran may have been the single largest reason Rick chose not to pursue a career in computer science.

URL:

http://wwwcn.cern.ch/asdoc/f90.html

Functional Logic Programming

Functional logic programming aims to join the two important declarative programming paradigms of functional programming and logic programming. Functional logic languages have more expressive power than pure functional languages because of features like function inversion, partial data structures, existential variables, and non-deterministic searches. This page has sections for language implementations, mailing lists, workshops, conferences, journals, papers, researchers interested in functional logic programming, and related subjects. There are also descriptions and software for several functional logic programming languages, including ALF, Babel, Life, NUE-Prolog, Oz, and RELFUN.

URL:

http://www-i2.informatik.rwth-aachen.de/
hanus/FLP/

Functional Programming

Devoted to all aspects of functional programming, this page offers a large collection of technical papers, many articles, conference proceedings, and a bibliography. There are also separate sections for the languages Haskell, ML, Erlang, Lazy Hope Interpreter, and others; and links to FP projects, newsgroups, and course descriptions.

URL:

ftp://coral.cs.jcu.edu.au/web/FP/home.html

Take a break. Visit a zoo in Zoology.

Find out what's shaken in Seismology.

Haskell

Haskell is one of the first non-strict, purely functional programming languages. Haskell was developed in 1987 by a language-design committee. This page offers a brief introduction to Haskell, details about the Haskell mailing list, pointers to specific implementations and compilers for Haskell for several popular platforms, the draft standard, FAQs, articles, and links to other Haskell sites.

URL:

ftp://coral.cs.jcu.edu.au/web/FP/hcs/
current.html

Check it out, Beave! Eddie creates his own programming language — Haskell — a language developed by committee! Stick some bugs in it!

Home of COBOL

What's both common and business-oriented and speaks an ancient language? Right, medieval data processing analysts. They speak COBOL. Where do they go when they're speaking COBOL? Home, of course . . . the Home of COBOL. This page is sponsored by The COBOL Foundation; the topics include the capabilities of COBOL language, COBOL compiler systems, supporting development tools and utilities, applications written in COBOL, development projects underway, and job opportunities. Isn't it ironic that such a loquacious language should have such a short URL?

URL:

http://www.cobol.org

Learn Ada on the Web

Learn Ada on the Web (LAW) provides software-development education for beginners using the Ada language. The focus of LAW is to teach people who are new to programming—rather than programmers who already know other computer languages. This lengthy and thorough tutorial also addresses providing software engineering tools on the Web, and it features an interesting facility for interactively creating Ada programs on the Web. This material forms the basis of the second edition of the book *Ada: A Developmental Approach* by Fintan Culwin (Prentice Hall, 1992).

URL:
http://www.scism.sbu.ac.uk/law/lawhp.html

Logo for Kids

Learning to program computers is typically left for later school years, or worse, left out entirely. With a good implementation of Logo, children as young as eight can enjoy programming. Move your turtle over to this page for information and support for MicroWorld's Logo Project Builder. These pages are filled with stories, anecdotes, and sample Logo programs.

URL:
http://www.inasec.ca/com/logo/mainpb.htm

Lovelace: Ada 95 Tutorial

Lovelace is an online tutorial for Ada 95—an object-oriented programming language. Lovelace explains the basics of Ada and assumes that you have had some exposure to other programming languages, such as Pascal or C. Lovelace teaches the latest version of Ada. This site also offers a list of other Ada tutorials, a bibliography, details of how the tutorial was created, and an outline for each lesson.

URL:
http://lglwww.epfl.ch/Ada/Tutorials/Lovelace/lovelace.html

Object Technology Resources

Here's a quick reference guide to some of the major resources for object technology on the Web. These information sources include recent books and articles; training organizations; consultants and system integrators; associations and conferences; and pointers to additional sources of information.

URL:
http://www.taligent.com/resources-list.html

Object-Orientation

Information System on Object-Orientation (ISOO) gives an overview of concepts fundamental to object-orientation and of applications of these concepts and paradigms in a wide variety of situations. These include object-oriented programming languages, databases, and operating systems. It gives an overview of different areas of object-orientation, an annotated bibliography, and a large collection of paper and book references.

URL:
http://ravel.ifs.univie.ac.at/ISOO/isoohome.html

Object-Oriented Information Sources

This page brings together pointers to a variety of information sources on the Web related to object-oriented languages and systems. The page is searchable, but pointers available directly on the page include links to software archives, tutorials, FAQs, bibliographies, OO programming languages, companies and their products, research groups, journals, conferences, search engines, newsgroups, and others. You can also register your own Internet resource in this database.

URL:
http://cuiwww.unige.ch/OSG/OOinfo/

OOP at OSU

Looking for a course on object-oriented design and object-oriented software construction? Here's one that presents the concepts of object-oriented analysis and programming. The course focuses on the concepts of OOP rather than any particular object-oriented language; however, it is illustrated with examples from a variety of programming languages. The course is provided in a book-like fashion in HTML, and it covers topics such as classes, methods, messages, inheritance, composition, polymorphism, software-engineering issues, and implementation.

URL:
http://www.cs.orst.edu/~budd/582/info.html

Looking for software? Try Software.

A B C D E F G H I J K L M N O P Q R S T U V W X Y Z

Most common word uttered by programmers:

OOPs!

Programming Language Research

If you're researching programming languages, or even trying to learn one, this page is the mother of all programming resources. It offers overviews of all the popular languages—including some you've never heard of, subject-oriented pages with topics such as linear logic, logic programming, and object-oriented programming. There's also a wealth of links to other programming pages and resources on the Net.

URL:
http://www.cs.cmu.edu/afs/cs.cmu.edu/user/mleone/web/language-research.html

Prolog Logic Programming Archive

Prolog is one of the most widely used programming languages in the area of logic. This archive offers FAQs on Prolog, a list of Prolog books, compilers and software for many platforms, the draft ISO Prolog standard, manuals, user notes, links to research papers, and other resources relating to the Prolog language.

URL:
http://www.comlab.ox.ac.uk/archive/logic-prog.html#Prolog

Python

Python is an object-oriented scripting language. This, the Python home page, provides a good summary of its capabilities, a tutorial on programming with Python, details of where to obtain sources and binaries for Unix, PC and Mac platforms, documentation, a FAQ, a searchable archive of the mailing list, news, and conference/workshop announcements.

URL:
http://www.python.org

Run a Scheme Program

This form-based page allows you to write and run a Scheme program interactively. Put your program in the "Program" box, choose an expression in the "Expression to evaluate" box, click the "Do it" button, and see your results!

URL:
http://www-swiss.ai.mit.edu/~jar/eval.html

Scheme

Uh, well. . . here's the page author's description of Scheme: "Scheme is a statically scoped and properly tail-recursive dialect of the Lisp programming language. A wide variety of programming paradigms, including imperative, functional, and message passing styles, find convenient expression in Scheme." Whew! Well, if you're still interested, the Scheme page offers documentation, including a user's manual, a reference manual, FAQs, and language updates. Some of this is available in PostScript, as well as HTML. There are pointers to Scheme implementation software for many platforms; and, finally, there are pointers to Scheme software, newsgroups, and research projects.

URL:
http://www-swiss.ai.mit.edu/scheme-home.html

What Is Object-Oriented Software?

This short, illustrated introduction to object-oriented software covers topics like classes and inheritance, offers some real-life examples where object orientation is desirable, and contrasts using object-oriented languages with non-object-oriented languages. With the exception on some discussion on C++ and Smalltalk, it's not specific to any particular language.

URL:
http://www.soft-design.com/softinfo/objects.html

PROGRAMMING: BASIC

BASIC 1.0

Here's an interpreter for a Unix implementation of the most common elements of the BASIC language. Its author claims that anyone who has ever written BASIC programs on a C64 should feel at home. You'll find the full distribution (160 Kbytes) in a .tar file that has been compressed and uuencoded. It contains all sources, makefiles, and instructions for building the executable.

URL:

http://www.Uni-Mainz.DE/~ihm/basic.html

BASIC Usenet Groups

Read and contribute BASIC tips and tricks of the trade to these Usenet groups for discussing "the language that would not die."

URLs:

news:alt.lang.basic
news:comp.lang.basic.misc

Business Basic Page

Find out what's new at this page of Business Basic providers (BBx, OpenBasic, Thoroughbred Basic and ProvideX) and apps (Open Systems Accounting, ADD+ON Accounting). There's also a well-written FAQ on the Business Basic programming language explaining its history and development.

URL:

http://www.gmcclel.bossnt.com/

Liberty BASIC

Liberty BASIC could well be the BASIC language that Microsoft neglected to include in the box when they created Windows. If you know nothing about programming, Liberty BASIC comes with a tutorial designed specifically for you. The program is commercial, but you'll find several shareware versions available from the company via FTP.

URL:

http://world.std.com/~carlg/basic.html

Get your doctorate in Ufology.

The BASIC Archives

Some programmers might suggest that BASIC should be permanently archived, but if you're basically a bold holdover who still enjoys this language, drop on by this archive. You'll find a FAQ, an overview of compilers and interpreters, a brief history of BASIC, and comparisons between it, Pascal, C, and other computer languages.

URL:

http://www.fys.ruu.nl/~bergmann/basic.html

Visual BASIC Archives

Why toil over that complicated function when you can cheat and get away with better code? Since someone else has already done all the hard work for you, here are numerous archive sites with thousands of free Visual BASIC routines on everything from creating custom 3D controls to parsing ZIP archives. But WHILE you're checking out these archives, don't forget that you can always RETURN here for more!

URLs:

ftp://ftp.cdrom.com/pub/cica/programr/vbasic
ftp://ftp.cdrom.com/pub/simtel/win3/visbasic
ftp://ftp.halcyon.com/local/mabry/files
ftp://ftp.springsoft.com/pub/springsoft/win16/
  programming

http://coyote.csusm.edu/cwis/winworld/
  vbasic.html
http://www.apexsc.com/vb/ftp1.html

PROGRAMMING: C AND C++

ANSI C Programming

This reference manual summarizes the ANSI C programming language. It's divided into the main sections of history and background, C Programming, decisions, procedures, data representation, arrays and pointers, input-output, the C preprocessor, and idiomatic C. Each section is subdivided into subsections so that you can locate what you're looking for quickly.

URL:

http://www.bath.ac.uk/~maspjw/NOTES/ansi_c/
  ansi_c.html

A B C D E F G H I J K L M N O P Q R S T U V W X Y Z

C Programming Reference

This is an online reference guide for the C programming language that gives answers to programmers' questions quickly and easily. It includes sections on syntax, technique and style, typical programming problems, example programs, functions, creating your own function library, gcc compiler errors, gdb debugger, and a glossary of terms. The master index is very useful to quickly look up specific functions or keywords. You can also download the entire HTML guide to keep on your local web server.

URL:

http://vinny.csd.mu.edu/C_ref/C/c.html

C Resources

This large collection of pointers to C compilers, software, tutorials, tools, and journals includes software for several platforms to compute just about any function, simulation, or other theoretically interesting algorithm. The tutorials include an introduction to computing, an introduction to programming in C, help on C, and a C programming reference.

URL:

http://www.lysator.liu.se/c/c-www.html

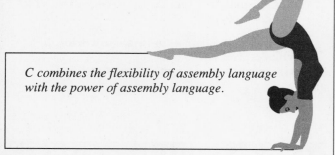

C combines the flexibility of assembly language with the power of assembly language.

C++ Annotations

Intended for helping knowledgeable C programmers make the transition to C++, this tutorial outlines the history of C++; its advantages over C; and the differences between procedural and object-oriented programming and between C and C++. It also explains extensions of C++, such as classes, memory allocation, static data and functions, inheritance, polymorphism, late binding, virtuality, and templates. A number of great examples of programming in C++ are also provided.

URL:

http://www.icce.rug.nl/docs/cplusplus/
cplusplus.html

C++ Programming Language Tutorial Handouts

A collection of handouts developed as part of a series of courses on C++ taught at U.C. Irvine, these documents are in PostScript format. Some of the topics covered are defining abstract data types focusing on classes, templates and exception handling, single and multiple inheritance, dynamic binding, pointer-to-member functions, dynamic memory management, and container classes. Illustrated examples show both the basic and advanced features of C++. There are also links to C++ FAQs, the draft standard, newsgroups, programming libraries, software archives, and other tutorials.

URL:

http://www.cs.wustl.edu/~schmidt/C++/
index.html

The C++ Virtual Library

An excellent resource for both the beginning and advanced C++ programmer, this page offers links to illustrated courses and tutorials in C++, an HTML formatted version of the draft C++ standard, freely available C++ software packages for various platforms, conferences listings, object-oriented programming resources, C++ class libraries, C++ product descriptions, reviews, and newsgroups.

URL:

http://info.desy.de/user/projects/C++.html

The C/C++ Users Journal

Source listings for the print publication *C/C++ Users Journal*. The archive has subdirectories for each year and month that the publication has been in print. The source files are in both .ZIP and tar.Z formats.

URL:

ftp://ftp.mfi.com/pub/cuj/

Introduction to Object-Oriented Programming Using C+

This self-paced online class gives an introduction to C++ programming with an emphasis on object-orientation. The site offers a complete hypertext tutorial, a walk-through exhibit with sample links for visitors, instructions for attending virtual classes, and an explanation of how online consulting is possible using a multi-user system and email. Active and proposed student projects are explained. This course won a Best of the Web award.

URL:

http://uu-gna.mit.edu:8001/uu-gna/text/cc/

Get Virtually Real.

Introduction to OOP Using C++

Introduction to Object-Oriented Programming Using C++ is the first fully virtual course combining a MOO and an HTML hypertext book. The class gives an introduction to C++ programming with an emphasis on object-orientation. With links to other C++ resources as well, this award-winning course is a valuable resource for learning this cool language.

URL:
http://www.desy.de/gna/html/cc/

Learn C/C++ Today

A list of interactive tutorials, software, and aids for learning C and C++, this one goes back to the very origins of C and provides details of the software you need and pointers to get it. There are also several links for other tutorials for C and C++, a FAQ list, book reviews, and other resources.

URL:
http://vinny.csd.mu.edu/learn.html

C++ is the coolest programming language. Oh sure, there are other complex programming languages with arcane focuses, but for day-to-day stuff, nothing beats C++. So derive some class, and learn C++. Then be sure to call the destructors!

Learning C++

Here's a collection of virtual courses and tutorials for C++, including lecture slides, a tutorial for Smalltalk, a transition guide for knowledgeable C programmers, an award-winning Internet C++ course, and pointers to other C++ resources and tutorials.

URL:
http://info.desy.de/user/projects/C++/Learning.html

LEDA

Leda is a C++ library of efficient data types and algorithms. This page provides an overview of some user projects that use LEDA, papers describing LEDA, and a pointer to the software available for the Unix g++ compiler.

URL:
http://www.mpi-sb.mpg.de/LEDA/leda.html

Object-Oriented Software Design and Construction

These class notes describe the C++ language and the object-oriented concepts on which it's based. They also aim to increase programming competence in several ways: by conveying the value of reusable software, by emphasizing the importance of tools and practices, through using an object-oriented library for building GUI-based systems, and by exposure to event-driven systems. The notes are divided into small sections, each designed to match the content of a single 50-minute class meeting.

URL:
http://actor.cs.vt.edu/~kafura/cs2704/

Shortest C Program Contest

Nearly 400 participants competed to write the shortest C program to count from its first argument to its second. Find the results of the competition, along with analysis, feedback, and source code here. The winning solution was a 19-way tie each with 69 bytes of code.

URL:
http://www.unix-ag.uni-kl.de/~conrad/shocc/shocc_en.html

A B C D E F G H I J K L M N O P Q R S T U V W X Y Z

Split C

So you think C and C++ are just too easy? Then you should be moving along to Split C—a variant of C for multiprocessor computers. The folks at Cornell University are developing their own variant of Split C, and they offer some great information and resources about Split C. They also plan to release the source code to their Split C compiler as soon as it's ready. Check it out and be the first Split C programmer on your block!

URL:
http://www.cs.cornell.edu/Info/Projects/Split-C

PROGRAMMING: MARKUP LANGUAGES

First Guide to PostScript

Learn to program in the PostScript page description language with this easily accessible online tutorial. This introduction includes an index of some of PostScript's standard operators and a list of various errors. You should have some experience programming and be familiar with concepts like arrays and variables.

URL:
http://www.cs.indiana.edu/docproject/
programming/postscript/postscript.html

A Gentle Introduction to SGML

Gently introduce yourself to the Standard Generalized Markup Language (SGML, formally ISO 8879). SGML is an international standard for electronic document exchange and the basis of the highly popular HTML Internet standard. Many publishing companies are standardizing on SGML to facilitate integration with more than just the print medium.

URL:
http://www.brainlink.com/~ben/sgml/

HTML Converters

This is an extensive archive of programs to convert from various display and typesetting formats into HTML.

URL:
ftp://src.doc.ic.ac.uk/computing/
information-systems/www/tools/translators/

Feel the squeeze in Compression.

HyperText Markup Language Specification

The foundation for the World Wide Web, HyperText Markup Language (HTML) is a simple markup language used to create hypertext documents that are portable from one platform to another. HTML documents are SGML documents with generic semantics that are appropriate for representing information from a wide range of applications. Here are detailed specifications for programming in HTML.

URL:
http://www.hp.co.uk/people/dsr/html3/
CoverPage.html

PostScript Blue Book Examples

When it comes to learning PostScript, the definitive reference is "the Blue book" from Adobe. This page contains the source of 21 example programs from the Blue book and provides a way to execute them.

URL:
http://www.fwi.uva.nl/~heederik/ps/bluebook/

PostScript People

Where can you go to discuss arcs, strokes, clip paths, fill patterns, and other PostScript tricks? Join **comp.lang.postscript** and talk to PostScript people just like you.

URL:
news:news:comp.lang.postscript

PostScript Resources

This Web page contains FAQs, newsgroup lists, PostScript code examples, utilities, and links to FTP sites that relate to the PostScript print and display language. The utilities include PostScript viewers and a program that converts PostScript into HTML. You can also find fonts and links to other PostScript pages. Code examples here are in shar or HTML format.

URL:
> http://yoyo.cc.monash.edu.au/~wigs/postscript/

Quick Review of HTML 3.0

If standards were never meant to change, they wouldn't call them standards. With a few years of web page evolution behind us, the HTML standard has required revision. HTML 3.0 is a set of extensions to HTML. Like HTML 2.0, it is based on the Standard Generalized Markup Language (SGML), but offers much needed functionality not conceived in earlier specifications.

URL:
> http://www.w3.org/hypertext/WWW/Arena/tour/contents.html

SGML Character Entity Set

Consult this SGML character entity table for declarations of math symbols, Greek symbols, special graphic characters, and publishing and technical symbols.

URL:
> http://www.bbsinc.com/iso8879.txt

SGML Web Page

Embedded links to the SGML archives are integrated with descriptive prose on this extensive page. The links point to other pages, Gopher servers, FTP sites, and important SGML documents archived locally. Included is an annotated and linked bibliography for SGML with over 700 entries.

URL:
> http://www.sil.org/sgml/

Nobody sends you email? Send us a note—we'll answer you.

Weblint

Writing HTML code is like ancient word processing. Remember those old "dot" codes in WordStar? No. That was way before your time. But worse, HTML's arcane typographic codes make it easy to use a second <B> when you really wanted to turn off boldface with </B>. Here's a tool that catches these kinds of mistakes and helps you create correct and efficient HTML. Written in Perl (which means you'll want to run this on a Unix machine), Weblint is a syntax and style checker for HTML. Like the traditional "lint" utility that aids C programmers, Weblint picks the fluff off your HTML pages.

URL:
> http://www.khoros.unm.edu/staff/neilb/weblint.html

PROGRAMMING: PASCAL

Pascal FAQ

Find out the whole skinny about this robust programming language named after the famous mathematician, Blaise Pascal, who created a circulating machine, and thereby, probably the first loop! This is a mammoth question and answer list that will satisfy many queries you may have regarding Pascal.

URL:
> ftp://ftp.csc.cit.ac.nz/pub/faq/ansiiso.txt

Pascal Programming Page

Blaise would be proud of this page on Pascal programming links to FTP sites, newsgroups, other Pascal pages on the web, and other Net resources for programming languages and operating systems.

URL:
> http://www.fiu.edu/~eurzai01/pascal.html

Pascal Usenet Groups

Discuss the intricacies and inventiveness of the Pascal language with other like-minded programmers.

URLs:
> comp.lang.pasca
> comp.lang.pascal.ansi-iso
> comp.lang.pascal.borland
> comp.lang.pascal.mac
> comp.lang.pascal.misc

A
B
C
D
E
F
G
H
I
J
K
L
M
N
O
P
Q
R
S
T
U
V
W
X
Y
Z

Pascal Utility Vendor List

If you're in the need of a vendor for Pascal utilities, consult this large list of worldwide software product purveyors and providers.

URL:

> http://www.wdn.com/ems/lists/pasutil.htm

Turbo Pascal Programmers Page

There's no more thorough a resource for Turbo Pascal programmers than this one. You'll find links to source code, drivers, compilers, manuals, newsgroups, FAQs, and much more.

URL:

> http://www.cs.vu.nl/~jprins/tp.html

PROGRAMMING: PERL

CGI Library

This extensive Perl library offers a rich set of functions for creating World Wide Web fill-out forms on the fly and parsing their contents. Notably, it makes it easy to create forms that remember their "state," a nontrivial task when dealing with stateless HTTP servers. This page is thorough and well documented with interactive examples.

URL:

> http://www-genome.wi.mit.edu/ftp/pub/
> software/WWW/cgi_docs.html

Hypertext Manual for Perl 4

For those slow to upgrade to Perl 5, here is the online hypertext manual for Perl 4, a reorganized version of the hideous Unix manual. The future is bright. For the Perl 5 manual, author Larry Wall handily applied the *split* function.

URL:

> http://www-cgi.cs.cmu.edu/cgi-bin/perl-man

Hypertext Manual for Perl 5

Here's a hypertext version of the Perl 5 "man" page. This was a really good idea. Many people who have ventured into the Unix version become lost and are never found again.

URL:

> http://www.metronet.com/0/perlinfo/perl5/
> manual/perl.html

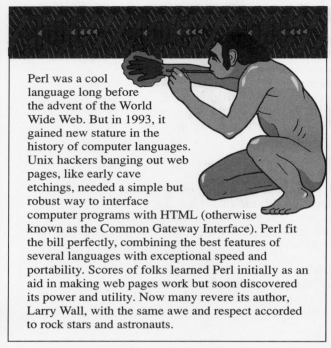

Perl was a cool language long before the advent of the World Wide Web. But in 1993, it gained new stature in the history of computer languages. Unix hackers banging out web pages, like early cave etchings, needed a simple but robust way to interface computer programs with HTML (otherwise known as the Common Gateway Interface). Perl fit the bill perfectly, combining the best features of several languages with exceptional speed and portability. Scores of folks learned Perl initially as an aid in making web pages work but soon discovered its power and utility. Now many revere its author, Larry Wall, with the same awe and respect accorded to rock stars and astronauts.

Perl for Macintosh

Need a complete hypertext manual for the Macintosh version of Perl? These links help you get Perl working in the Macintosh Programmer's Workshop environment.

URLs:

> http://err.ethz.ch/members/neeri/macintosh/
> perl-qa.html
> http://err.ethz.ch/members/neeri/macintosh/
> perlman/perl_toc.html

Perl Reference Materials

Book references, online guides, FAQs, manual pages, newsgroups, Perl code and utilities archives, links to other Perl resource collections, news, Perl5 language descriptions, and more about the Perl programming language are among the many Perl treasures here.

URL:

> http://www.eecs.nwu.edu/perl/perl.html

Perl Support Groups

Face it. You're Just Another Perl Hacker. But it's an affliction that strikes many. So get *associative* with others in the Perl Usenet discussion groups and *hash* it out.

URLs:

> news:comp.lang.perl
> news:comp.lang.perl.announce
> news:comp.lang.perl.misc

Tom's References for Learning Perl

With a focus on the Common Gateway Interface, Tom Phoenix's introductory links to the most useful Perl references on the Internet makes this relatively short page a good starting point for Web developers.

URL:

http://www.teleport.com/~rootbeer/perl.html

Xbase Module

Use the Xbase module to access dBase, FoxBASE, FoxPro, and similarly formatted database files from within Perl. Many traditional dBase functions are available for searching and processing records and fields in Xbase files.

URL:

http://everest.eng.ohio-state.edu/~pereira/
software/xbase/

PSYCHOLOGY

Anger and Grief

Women tend to express their grief through sadness, while a man's expression of grief through anger is very natural. This anger can be directed toward anyone, including God, or even someone who is dead. The release of grief through anger can be beneficial, especially through rituals. This page describes some examples of working with grief and anger in a cathartic manner.

URL:

http://www.dgsys.com/~tgolden/3anger.html

The Arc

The Arc is the largest voluntary organization in the U.S. that is committed to the welfare of children and adults with mental retardation. This site provides a FAQ list and articles on issues such as the rights of persons with disabilities, behavioral supports, advocacy services, guardianship, and research developments. This page publishes announcements on upcoming ARC projects and activities and provides links to related mental retardation and developmental disabilities sites on the Web.

URL:

http://fohnix.metronet.com/~thearc/
welcome.html

Attention Deficit Disorder

A young boy can't sit still or doesn't want to play a game or read a book for more than just a few minutes. He seems to be bouncing off the walls—impulsive and hyperactive. It is possible that you are looking at a child with ADD. This fairly extensive page covers the symptoms, etiology, complications, and treatment of this syndrome.

URL:

http://homepage.seas.upenn.edu/~mengwong/
add/add.faq.html

Autism and Asperger's Syndrome Resources

This page is an organized list of Autism and Asperger's Syndrome resources that are now available on the Net. The topics discussed on this page include a FAQ list on autism; news items; case histories; specific issues about dealing with this disorder; and methods, treatments, and programs that can help. You'll find information on academic and research programs, notices on conferences, links to additional resources, and an autism mailing list.

URL:

http://web.syr.edu/~jmwobus/autism/

Borderline Personality Disorder

This is a common disorder that affects up to 14 percent of the population. A person who has a borderline personality disorder frequently experiences a repetitive pattern of disorganization and fluctuations in self-image, mood, behavior, and personal relationships. These problems can cause difficulties and impairment in friendships and work. This site discusses the etiology, symptoms, and treatment of this disorder.

URL:

http://www-leland.stanford.edu/
~corelli/borderline.html

Carl Gustav Jung

This psychologist—whose pioneering work in the areas of dream analysis, archetypes, and the collective unconscious—will stay forever *jung* with this excellent assortment of arranged quotations and insights from his prolific writings. There's a bibliography of his collected works and links to other psychological material.

URL:

http://miso.wwa.com/~nebcargo/Jung/

You're only as jung as you feel. Get unconscious with the collective selective pithy aphorisms that only Carl Gustav Jung could emit and emote. After all, you're only jung once!

Depression and Treatment

Looking at several methods and points of view concerning the treatment of depression, this article provides general guidelines on the best course to take. The results of psychotherapy are compared to straight pharmacotherapy, as well as the combination of both treatments. The case is made that pharmacotherapy alone is considered to be less efficient in overcoming depression; a combination of both medication and psychotherapy is most helpful, since people also need the support, self-expression, and insight gained through counseling.

URL:
   http://www.coil.com/~grohol/

Down Syndrome

This page provides a FAQ list and a medical checklist on Down Syndrome. There are online articles on DS, pointers to education resources, and information on parent-matching and support groups. You'll find mailing lists, a toy catalog, listings of Down Syndrome organizations worldwide, dates for upcoming conferences, and links to other medical and disability resources on the Net.

URL:
   http://www.nas.com/downsyn/

The Dream Page

Psych yourself up for an interesting and informative exploration of dreams. This page features a collection of people's actual dreams which are submitted anonymously and are publicly available for interpretation. You are welcome to submit a dream of your own or to interpret someone else's.

URL:
   http://www.cs.washington.edu/homes/
   raj/dream.html

Eating Disorders

The most common eating disorders are self-starvation (anorexia nervosa), bingeing and purging (bulimia), and compulsive overeating. You'll find information at this site that describes the symptoms of these disorders and the resulting dangerous medical complications that may develop. Strategies for overcoming food abuse are offered, and other aspects of unbalanced eating habits—such as crash diets, trash diets, and fad dieting—are discussed.

URL:
   http://ccwf.cc.utexas.edu/~bjackson/UTHealth/
   eating.html

Facts for Families

This site seeks to educate parents and families about psychiatric disorders that affect children and adolescents. It lists 46 information sheets that provide information on issues such as eating disorders, child abuse, depression, teen suicide, alcohol and drug problems, step-family difficulties, divorce, grief, bedwetting, anxiety, panic disorders, schizophrenia, and bipolar mood disorders. You will find discussions on knowing when to seek professional help for your child, and information on psychiatric medications for children. These information sheets are available in English, Spanish, and French.

URL:
   http://www.psych.med.umich.edu/web/aacap/
   factsFam/

Take a dream or leave a dream at this Dream Page for you. Analyze yours, or someone else's dream—perhaps it will come true!

Flame Wars in Cyberspace

The point of this page is that happy people make positive judgments, while people who are sad go in a negative direction. The effects of mood on the way people interact becomes very evident with the anonymity of cyberspace. Here, the social norms and rules do not apply. You are not seeing or hearing the other person and the result is an uninhibited mood outlet. Mood will also affect evaluation of others, memory, and attention span. As in the physical world, people need to learn not to be rude through computer-mediated communications.

URL:

    http://www.coil.com/~grohol/storm1.htm

Don't maim your name laying claim to the blame by playing the game that isn't so tame, will not bring you fame, and winds up in shame in the wars known much better as "flame." Quench your burning emotions at the *Flame Wars in Cyberspace* page.

The Flute Player

Imagine being in the belly of a huge snake. It is completely dark. You are totally cut off from the outside world, and the restrictive belly is trying to force you to conform and take control. This is a strong comparison to a person experiencing deep grief. Read the wonderful story at this site of a flute player who combats his grief one piece at a time until he cuts his way out of the belly of the snake. Overcoming grief is a long struggle and it takes many small steps along the way. At the end is the renewal of passion, creativity, and the will to live.

URL:

    http://www.dgsys.com/~tgolden/3flute.html

Graphology

What can you know about someone who does not dot "i"s or cross "t"s? Handwriting analysis, or graphology, will not predict the future, but it can have other varied purposes. The most commonly asked questions on this science are answered on this page. How much will your inner self be revealed through writing?

URL:

    http://www.ntu.ac.sg/~tjlow/grapho.html

Hyperlexia

This Hyperlexia page features articles from the Center for Speech and Language Disorders. The topics discussed include an overview of hyperlexia and how it differs from high-functioning autism and Asperger's syndrome; definitions, diagnoses, prognoses, and school considerations for these disorders; principles for therapy; and recommendations for specific intervention techniques. This site offers a Parent's Toolbox of helpful strategies, links to newsgroups and support groups, a list of non-Net sources of information (books, articles, video and audiotapes), and pointers to related links on the Net.

URL:

    http://www.iac.net/~whaley/gordy.html

Journal of Applied Behavior Analysis

The *Journal of Applied Behavior Analysis* publishes original reports of experimental research on applied behavior analysis and its application to social problems. This page offers indexes of JABA's most recent articles and previous JABA issues, with links provided to abstracts. Hypertext reprints of complete articles from JABA are available.

URL:

    http://www.envmed.rochester.edu/wwwrap/
    behavior/jaba/jabahome.htm

Jungian Newsgroup

Read and correspond with others regarding the wisdom and philosophy of Jungian thought. This newsgroup has a steady flow of insightful posts.

URL:

    news:alt.psychology.jung

Take the Web with you. Learn how in Mobile Computing.

A
B
C
D
E
F
G
H
I
J
K
L
M
N
O
P
Q
R
S
T
U
V
W
X
Y
Z

Excerpt from the Net...

(from alt.psychology.jung)

"Sooner or later, nuclear physics and the psychology of the unconscious will draw closer together as both of them, independently of one another and from opposite directions, push foreward into trancendental territory, the one with the concept of the atom, the other with that of the archetype." C.G. Jung, Axion, 1951

Kingdomality

When knights were bold and dragons were fierce, people had vocations that suited their personalities. They were often named after their job, which was handed down through each generation. The Bakers were bakers, the Smiths were blacksmiths, and so on. Even today, that medieval vocational personality still lurks within us. It can motivate us or set the stage for failure. Should corporations pay more attention to suiting the person to the position? Find out what your medieval personal preference profile is like at this innovative page.

URL:
http://www.cmi-lmi.com/kingdomality.html

Living with ADD

This page features 50 tips for coping with adult Attention Deficit Disorder. The starting point is often regaining hope after suffering embarrassment and humiliation. Treatment of ADD is broken down into five areas, only one of which is through medication. It is the non-medication aspects that this piece focuses on.

URL:
http://homepage.seas.upenn.edu/~mengwong/
add/tips.html

Mood Disorders

This page provides an excellent Depression FAQ list and links to many depression-related mailing lists, support groups, and newsgroups. You'll find information on the symptoms and therapies for Seasonal Affective Disorder (SAD). There are online articles that discuss the many medications now available to treat mood disorders, including tricyclics, MAO inhibitors, and the new selective serotonin-reuptake inhibitors. There are also links to related mood-disorder sites on the Web.

URL:
http://avocado.pc.helsinki.fi/~janne/
mood/mood.html

Multiple Personality Disorder

The new term for Multiple Personality Disorder is Dissociative Identity Disorder (DID). It is now accepted that this psychiatric condition is a common result of extreme trauma early in life, whether physical, emotional, or sexual. Post-Traumatic Stress Disorder (PTSD) is often associated with this illness. An overview of DID is provided on this page, along with descriptions of how the illness can develop and what its symptoms are.

URL:
http://www.access.digex.net/~sidran/didbr.html

Neuro-Linguistic Programming

This is an introduction and discussion of NLP. This method of modeling explores the relationships among linguistics, neurology, and programs of behavior. A background is given on the early work of several people in this field. Neuro-linguistic programming has close ties with hypnosis and is found in much of the work of Milton Erickson.

URL:
http://www.nlp.com/NLP/whats-nlp.html

Panic Attacks, Anxiety Disorders, and Phobias

Anxiety disorders are the most common group of psychiatric disorders in the U.S., affecting nearly 15 percent of the population at some point in their lives. This site describes the symptoms and etiology of panic attacks, anxiety, excessive shyness, and phobias. These problems can be greatly alleviated with a variety of treatments and therapies, but unfortunately the majority of sufferers do not seek appropriate care. At this site you can learn about anxiety disorders and the many therapies that are available to alleviate them.

URL:
http://www.hslib.washington.edu/hsc/newsinfo/
healthbeat/HB1994/panic.html

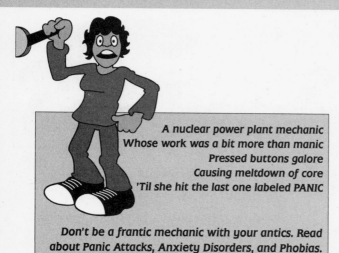

*A nuclear power plant mechanic
Whose work was a bit more than manic
Pressed buttons galore
Causing meltdown of core
'Til she hit the last one labeled PANIC*

Don't be a frantic mechanic with your antics. Read about Panic Attacks, Anxiety Disorders, and Phobias.

Pendulum Resources

Recognizing bipolar affective disorder (manic-depressive illness) is not always easy, but this page may make recognition and diagnosis easier. This resource provides information from the American Psychiatric Association's set of criteria by discussing symptoms for manic and depressive phases. You'll find pointers to the Health Privacy Act, the Neurobiological Disorders Society, and the HEATH Resource Directory. There is information on upcoming bipolar conferences; contacts and useful addresses; links to related disorders; a discussion of patients' rights; and finally, some lighthearted pieces on "What is Normal?" and the emergence of "Cyber Disorders."

URL:
http://www.csn.net/~era/pendulum/

Personality Modeling

Although human personality is far too complex to model exactly, there are definite personality types. This page examines Jung's Theory of Personality Type and the Myers-Briggs Type Indicator. Descriptions of the four personality types and type summaries are offered.

URL:
http://sunsite.unc.edu/personality/faq-mbti.html

The Web will set you free.

Psybernet Carl G. Jung Page

Check out this page for some interesting thoughts on Jung and cyberspace, or psyberspace, in this case. Psybernet's focus is on the psychological aspect of tele-computing. Read more about the Psybernet Conference—an exploration of psychological cyberspace that explores the collective unconscious, symbols, and the value of imagination. There are links to other psychological resources.

URL:
http://vesta.chch.planet.co.nz/~walter/jung.html

Psyche

Psyche is an electronic journal that features scientific studies of consciousness. You'll find articles, overviews of symposia, book reviews, a FAQ list, a discussion list, and archival information. There are also links to conferences, courses, discussion groups, electronic and print journals, bibliographies, and more.

URL:
http://psyche.cs.monash.edu.au/

Psychiatry On-Line

Psychiatry On-Line is an independent, peer-reviewed electronic journal available on the Web; several thousand psychiatrists are now registered as regular readers. This site provides a Psychotherapy On-Line FAQ and articles on a variety of psychiatric issues such as research updates, therapeutic advances, and medication information. There are sections entitled The News Page, Media Review, Letters, and Ask Dr. Ivan. You will also find links to other psychiatric resources on the Web.

URL:
http://www.priory.com/journals/psych.htm

Psychopharmacology Tips

The site is a collection of helpful tips to know for those who take psychotropic medication; these suggestions are tips and are not intended to replace a physician's advice. This page is divided into categories: Focus on Drugs discusses specific medications, their properties and uses, and their interactions; Focus on Problems examines various disorders and discusses which medications are generally used to treat them.

URL:
http://uhs.bsd.uchicago.edu/~bhsiung/tips/
tips.html

A B C D E F G H I J K L M N O P Q R S T U V W X Y Z

Sigmund Freud

Say whatever you want about Sigmund, he's still an influential ghost in the field of psychology. On this page, you'll find a biography of Freud, samples from his major works, literary comment, and a bibliography.

URL:

http://www.iris.brown.edu/iris/freud/
Freud_OV_274.html

There once was a Sigmund named Freud
With half of Vienna he toyed
His thoughts on the ego
Outdid his libido
While the rest of the shrinks were annoyed

SleepNet

This site offers "everything you ever wanted to know about sleep—but are too tired to ask." Approximately 40 million Americans suffer from sleep disorders, and the resulting misery, loss in productivity, and increased likelihood of accidents and illness take a bitter toll. Help is available, and the objective of SleepNet is to provide information on sleep disorders and education about the therapies and medications that are available. You'll find articles on sleep research, information on sleep-related products and equipment, and dozens of links to related sites.

URL:

http://www.sleepnet.com

Society for Quantitative Analyses of Behavior

The Society for Quantitative Analyses of Behavior (SQAB) sponsors conferences and publishes articles on the quantitative analysis of behavior. This site offers abstracts of conference papers, posts job listings, and provides a mailing list of SQAB members.

URL:

http://www.jsu.edu/psychology/sqab.html

Peace of mind is what you'll find
At SleepNet's web page spread
The only place you'll get more rest
Is when you're snug in bed!

Theories and Therapists

The field of psychotherapy involves many types of theoretical techniques and orientations. This page examines four of these: psychodynamic, cognitive-behavioral, humanistic, and eclectic. Are you a product of your parental upbringing, as the psychodynamic theory subscribes? Or are you completely responsible for the choices you make in your life, as the humanistic theory promotes? Find which type is best for you.

URL:

http://www.coil.com/~grohol/therapy.htm

Traumatic Stress Home Page

This site provides information on Traumatic Stress Syndrome, a disorder which can occur from a variety of causes. Scientific and research resources are listed, as well as links to related sites and information.

URL:

http://www.long-beach.va.gov/ptsd/stress.html

Who's Who in Religious Psychology

Five very influential people and their profound effect on the psychology of religion are overviewed on this page. Sigmund Freud, probably the most well-known, saw religion as originating in the father/child relationship. William James attempted to realize a more personal religious experience. The contributions of Carl Jung, Abraham Maslow, and Gordon Allport round out this who's who.

URL:

http://www.gasou.edu/psychweb/psyrelig/
psyrelpr.htm

PUBLISHING (ELECTRONIC)

Adobe Acrobat

What you see is what you get, no matter what type of computer you're using. That's the aim of Adobe's Acrobat document-presentation software. Distribute documents over email, the Web, networks, and CD-ROMs on any PC, Mac, or Unix platform. Acrobat is based on the PostScript page-description language and uses the Portable Document Format (PDF) to preserve Type 1 and TrueType fonts.

URL:

http://www.adobe.com/Acrobat/Acrobat0.html

Adobe PostScript Products and Technology

Check out white papers, overviews, and FAQs about the PostScript page-description language. This page offers additional information for users and developers on Adobe fax technology, and includes a list of recommended books on the PostScript system.

URL:

http://www.adobe.com/PS/PS.html

Agfa Digital Publishing Resources

Agfa-Gevaert is one of the largest manufacturers of photographic materials and products in the world. On their web server, Agfa offers a wealth of information and articles on digital printing hardware, software, and techniques. Topics include digital cameras (see also Photography: Digital), scanners, digital photo-imaging systems, image and plate setters, screening software, film recorders, and Chromapress—Agfa's digital color printing system.

URL:

http://www.agfahome.com/home.html

Architecture for Scholarly Publishing on the Web

The Web is the greatest thing to happen to civilization. 'Tis not. 'Tis too. 'Tis not. Okay, some may argue the point. Take for example this page that discusses the pros and cons of web publishing for members of the scientific community. A case is made for the implementation of extended character sets and more effective interface facilities for inter- and intra-document navigation.

URL:

http://www.oclc.org/oclc/research/publications/
weibel/web_pub_arch/web_pub_arch.html

Clip Art Server

Ah, the joys of free clip art. Now you, too, can plaster electronic pix all over the place thanks to this gigantic linked collection. Most are in GIF format, in color and black-and-white, and there are many different themes and subjects.

URL:

http://www.cs.yale.edu/homes/sjl/clipart.html

Electronic Publishing on the Web

Considering creating your own web page, but don't know how to go about it? Your first choice for guidance should be *The World Wide Web Complete Reference* (Osborne McGraw-Hill). But if you can't wait until the bookstores open, check out this page for a quick primer. Sections explain what the Web is, how to create a document, and how to use Uniform Resource Locators (URLs). Examples of some notable pages are also provided.

URL:

http://www.cch.epas.utoronto.ca:8080/cch/
online_publishing/wwwpub0.html

Electronic Scholarly Publishing and the Web

The formatting and production of scholarly electronic documents via Listserv archives, FTP, Gopher, and the Web are the subject of this white paper. Electronic publishing methods are contrasted with traditional print media, and conclusions are drawn regarding how best to present particular types of scholarly information.

URL:

http://www.deakin.edu.au/people/aet/ausweb95/
ausweb95.html

Guides to Writing HTML Documents

No one likes to look at ugly web pages cluttered with meaningless and illegible text over hideous graphics. (Well, some people do—which is why there are the Worst of the Web awards.) If you don't want to win one of these dubious honors, consult this excellent guide to writing creative and constructive HTML.

URL:

http://union.ncsa.uiuc.edu/HyperNews/get/
www/html/guides.html

Anyone can browse the Web.

A
B
C
D
E
F
G
H
I
J
K
L
M
N
O
P
Q
R
S
T
U
V
W
X
Y
Z

HyperJournal

Read up on the latest trends and developments in the world of electronic publishing with *HyperJournal*. Particular attention is devoted to web page production and publishing, and the journal generally focuses on the aims and needs of professional educators.

URL:
> http://econwpa.wustl.edu/~hyperjrn/hyperj.htm

Hypertext Markup Language

HTML is one of several standardized markup languages, but it has become the *lingua franca* of the Web. Consult this informative page for an overview of the history of HTML, HTML specs and standards, and development and style issues. Like any language, HTML continues to evolve—so feel free to add your own two cents' worth to the ongoing discussions of how HTML and the Web should progress.

URL:
> http://www.w3.org/hypertext/WWW/MarkUp/
> MarkUp.html

HyperZine

HyperZine is an online consumer magazine on publishing in the digital age with an emphasis on digital image reproduction. Commentaries, product information, photo suggestions, new technology, advice about photolabs, creating and editing digital videos, and many more resources are found here.

URL:
> http://www.hyperzine.com

Icons and Images for HTML Documents

Here are tons of transparent GIFs, sorted by category, freely available for you to download and use in all of your web page publishing endeavors. Included are arrows and pointers, bars and banners, stars and symbols, and many more. There are also links to many other icon collections on the Web.

URL:
> http://www.infi.net/~rdralph/icons/

Icons Anyone?

Remember when an icon was a picture of a saint that Russians worshipped? These days, you point your mouse at one and madly click away to launch a program . . . or if you're on the Web, you use them to tastefully embellish your web pages. This site at Cornell references many, many sites where icons abound and are free for the taking.

URL:
> http://www.tc.cornell.edu/Icons/

Journal of Electronic Publishing

Read about electronic publishing issues that range from the philosophical to the mundane in *The Journal of Electronic Publishing*. Find articles about the Internet, electronic libraries, copyright issues, pricing, and economics.

URL:
> http://www.press.umich.edu/jep/

Publishers

If you've just crossed the last *t* in your Great American Novel, you'll need someone to get it into print. At this page, you'll find online publishing companies and selected publishers who probably can't wait to read your *magnum opus*. There's a selection of online bookstores, as well, so you can see where your work will hopefully end up.

URL:
> http://www.comlab.ox.ac.uk/archive/
> publishers.html

Style Guide for Online Hypertext

Considered the Miss Manners Book of HTML, this page illustrates the most effective methods for creating electronic data resources on the World Wide Web. If you create your own web pages, this is a must-read with lots of helpful advice for even the most seasoned HTML hackers.

URL:
> http://www.w3.org/hypertext/WWW/Provider/
> Style/Overview.html

The World Wide Web Consortium

"Oh, what a tangled web we weave. . ." *Marmion* —Sir Walter Scott. The CERN organization in Switzerland was the birthplace of the Web; but today, the W3C at MIT is the organization that sets the future directions of the Web. At W3C, web specifications are developed for the Hypertext Transfer Protocol and HTML. You'll find overviews of the specifications, the history of the Web, technical reports, web software, news updates, and information on how to establish a web site.

URL:
> http://www.w3.org/pub/WWW/

QUALITY MANAGEMENT

The American Productivity and Quality Center

The American Productivity and Quality Center is a resource for businesses that want to pace positive changes in quality management. Although the organization focuses on technology, its primary interest is in human resources. Learn from the experience of this consulting group for productivity and quality.

URL:
http://www.apqc.org

Division of Quality Technology at Linköping University

Find out how these Swedish scientists are developing knowledge and competence in quality management. Receive some basic education on the concepts of QM, and learn more about the university's ongoing research projects at its Center of Excellence. There are also a few quality links to other quality management pages.

URL:
http://galois.ikp.liu.se/

QualiNet

Network with other QM professionals and find sources of information here at QualiNet. You can post messages and your electronic business card, and share ideas and opinions at QualiNet's online conference room. You'll find a special section that pertains to various perspectives of quality management in Silicon Valley.

URL:
http://www.qualinet.com

"Come, give us a taste of your quality," wrote William Shakespeare in Hamlet. So if you're into QM, get thee to QualiNet!

Quality Management Principals

Check out this site for some interesting ideas about improving the quality and productivity of your business. The topics of discussion include leadership, the systems approach to management, the involvement of people, continual improvement, and many other subjects. Links are provided to other quality management resources on the Net.

URL:
http://www.wineasy.se/qmp/

Quality Network

Use this directory of quality management categories to upgrade your productivity and enhance a variety of your quality assurance goals. You'll find ISO 9000 standards and information; sections on environmental, safety, and configuration management; and advice on choosing outside quality management consultants.

URL:
http://www.quality.co.uk/quality/

Quality Resources Online

This may be the mother of all quality management resources on the Web. You'll find an amazing number of excellent links to domestic and international standards institutions; commercial quality organizations and consultants; email discussion lists; and notices of upcoming quality management conferences, seminars, and workshops. Look for some humorous insights from the *Dilbert* comic strip.

URL:
http://vector.casti.com/qc/

The Quality Wave

If quality is your quest, then go for the best with this page of articles and ideas on how to produce premier designs, products, and thought processes. Geared toward quality professionals around the world, you'll find a list of educational programs and resources, as well as quality engineering directories, discussion forums, mailing groups, and a FAQ list.

URL:
http://www.xnet.com/~creacon/Q4Q/

Quality-Related Information Sources

This site features an assortment of institutions, schools, consultants, and other organizations that study the challenges and implement improvements in quality management.

URL:
http://www.qualinet.com/isopage.htm

A B C D E F G H I J K L M N O P Q R S T U V W X Y Z

When Quality is Job #1, surf Quality Wave.

RADIO - AMATEUR AND SHORTWAVE

Amateur Radio Resources

Among the many links of interest on this page for amateur radio enthusiasts are organizations such as the Bavarian Packet Radio Group and Linux for Hams; Windows and Mac software for controlling radio sets, such as the Kantronics TNC and Kenwood TM line; and links to radio equipment dealers and surplus stores.

URL:

http://www.acs.ncsu.edu/scripts/HamRadio/
OtherWebs/3

Amateur Radio World Web Sites

This page provides a long list of amateur radio clubs and links to their home pages. Many of the clubs are associated with a particular university, and most are located in the U.S. or the U.K.

URL:

http://www.cc.columbia.edu/~fuat/cuarc/
www-sites.html

Automatic Packet Position Reporting System

APRS is a shareware program for PCs and Macs that uses the GPS (Global Positioning System) to display the locations of moving stations on your PC. By connecting your PC to a radio network with an inexpensive radio data modem, all stations can see the movements of all the other stations. This web page offers links to FTP sites from which you can download DOS, Windows, and Mac versions of the software. There are also screenshots, maps, user-group details, and links to other packet radio sites.

URL:

http://www.ccnet.com/~rwilkins/aprs.html

Chris Smolinski's Home Page

Macintosh hacker Chris Smolinski writes software for science and radio applications. Download demos of programs that search the FCC's AM, FM, and TV engineering databases and allow your Mac to communicate with Icom radio. Produce an electronic radio schedule, display a real-time map of the Earth, or try out a database package for a complete radio logbook.

URL:

http://www.access.digex.net/~cps/

Directory of R/C Aviation

The Directory of R/C Aviation is a well-rounded list of links to remote control aviation sites on the Web. You'll find links to local and international clubs, commercial sources for equipment, practical information pages, and many other pointers to pages of interest.

URL:

http://maple.nis.net/~rmathes/rc/rc-dirct.htm

Galaxy Amateur Radio

This page is a large index of articles, book reviews, announcements, software, discussion groups, organizations, and other directories relating to amateur radio. There's a sample amateur radio exam, details of swap meets and conventions, opinion polls, information on regional groups, and much more.

URL:

http://galaxy.einet.net/galaxy/
Leisure-and-Recreation/Amateur-Radio.html

Ham Radio Today

Read the latest amateur radio news from around the world, evaluate product reviews, and check out software for popular platforms and links to many other sites with amateur radio information.

URL:

http://www.tcp.co.uk/~slorek/

IRC #HamRadio

The IRC #HamRadio channel is a place for hams to talk about their latest and greatest radios and recent events, discuss the past and future of ham radio, and tell tales of their contacts. Basically, anyone interested in amateur radio can join in and discuss any aspect of ham. There's an information sheet about the channel giving details of the channel bots and how to obtain files from them, channel rules, and operators. There are also links to the operators' home pages, IRC FAQs, and a magnificent collection of pointers to many types of amateur radio resources, including clubs, call-sign servers, software archives, and much more.

URL:

http://wb5fnd.tech.uh.edu/irc/

Forget frequency charts, postpone your packets, terminate your TNC, and retire your rig! With a computer and IRC, you can communicate with more people than you can with a hundred mountaintop repeaters (mainly because there are more people listening in IRC). With IRC, everyone can transmit at the same time without keying over each other. You can transfer data files simultaneously without monitoring sidebands or getting an earful of screeching bits. You can squelch people, not just a signal. Okay, okay, so IRC may never take the place of honest ratchet-jawing, but you have to admit that it has some desirable features. Join the #HamRadio channel on IRC to reminisce about the good old days.

Javiation

Javiation is one of the U.K.'s leading distributors of scanners and associated equipment. Available here is their guide to air band listening, detailed product specifications, and pictures of scanners from most of the leading manufacturers. There are also software and publication descriptions, and links to other radio scanning sites on the Internet.

URL:

http://www.demon.co.uk/javiation/

North American Amateur Radio Callbook

Here's a handy resource if you're looking for your ham radio friends over the Internet. Simply enter the radio call sign and the search engine will scan the North American Amateur Radio Callbook. This procedure will provide you with an email address and perhaps some other useful information.

URL:

http://buarc.bradley.edu/wwwvl-ham-search.html

Packet Radio

A collection of resources of interest to packet radio users, this site includes a FAQ on the subject, software for PCs, newsgroups, links to packet radio user groups, and pointers to other packet radio sites.

URL:

http://www.yahoo.com/Entertainment/
Radio/Amateur_Radio/Packet_Radio/

Packet Radio Home Page

An excellent resource for packet radio containing a packet radio primer and FAQ, a bibliography, abstracts, technical protocol specifications, pointers to organizations and people who produce or have information related to packet radio, newsgroups, software archives, mailing lists, papers, and links to other amateur radio sites.

URL:

http://www.tapr.org/tapr/html/pkthome.html

Packet Radio: An Introduction

Scratching your head trying to figure out just what packet radio is? Here's a great little primer covering the history of packet radio, the elements of a packet station, distance limitations, channel sharing, networks and special protocols, network schemes, and BBS message transfer. The network schemes discussed include Digipeaters, KA-nodes, NET/ROM, ROSE, TCP/IP, and TexNet. This document is available in both plain-text and HTML formats.

URL:

http://www.tapr.org/tapr/html/pktfaq.html

A
B
C
D
E
F
G
H
I
J
K
L
M
N
O
P
Q
R
S
T
U
V
W
X
Y
Z

Remote Control Stuff and Remote Control Flying FAQ

Buzz on over to this collection of documents that will explain to you in detail how to select equipment and construct remote control airplanes and helicopters. Included is information on what to buy, how to hook it up, and helpful recommendations on setting up your remote controls.

URL:
http://www.paranoia.com/~filipg/HTML/ RC/F_RC.html

Remote Control Web Directory

This directory is an excellent collection of links to hobby-related remote control information. The subdivisions are R/C aircraft, R/C cars and trucks, and R/C boats; there are also links to Usenet newsgroups and other sites of interest.

URL:
http://www.towerhobbies.com/rcweb.html

Shortwave Radio Catalog

Shortwave and radio hobbyists find this comprehensive site informative, if not daunting, due to its substantial content with many links to related resources. Timely information on shortwave listening, satellite radio, and related topics make this site a tremendous contribution to the Net.

URL:
http://itre.ncsu.edu/radio/

University and Amateur Radio Clubs

Turn up your gain to this listing of amateur and short wave radio clubs around the country. The signal-to-noise ratio is high and only the faintest of sites are squelched.

URL:
http://www.acs.ncsu.edu/scripts/HamRadio/ OtherWebs/2

What Is Amateur Radio?

Here's a beginners' guide to amateur radio that discusses the ideas and terms of short wave radio. There are also pointers to other amateur and ham sites on the Internet worth browsing. If you are interested in becoming a certified operator, this guide will tell you how, and even give you some of the test questions. If "ham" is your passion, don't miss this page.

URL:
http://www.acs.ncsu.edu/HamRadio/FAQ.html

Excerpt from the Net...
(from the pages of WXYC in Chapel Hill, N.C.)

"There's much more to radio than meets the ear. It's radio waves zapping your leftovers back to edibility every time you pop them in the microwave. It's radio astronomy eating up over 3 billion dollars of U.S. government research money. It's radio satellites ensuring that American children spend as much time listening to Muzak as to their parents. And it's radio-controlled implants making it possible to neutralize a rabid animal at the press of a button. Even your body has a biological radio set, which can be triggered by a seizure of the temporal lobe. Radio knows no boundaries: its signal is as unavoidable as it is unstoppable." —Neil Strauss in Radiotext as found on http://sunsite.unc.edu/wxyc/

World of Internet Amateur Radio

CQ! CQ! One of the most visited places on the Internet for "ham" radio operators. Services include a weekly news page, operating information, FCC rules and regulations, licensing information, software, and directions to other radio web servers.

URL:

http://www.acs.ncsu.edu/HamRadio/

RAIL AND LIGHT RAIL TRANSPORTATION

Amtrak's Station on the World Wide Web

Climb aboard Amtrak's web site for reservations, train schedules, routes, and even vacation promotions and contests. If you are thinking about taking a train, find out if Amtrak runs the route by viewing the National Passenger System Map.

URL:

http://www.amtrak.com

Cyberspace World Railroad

All aboard the Cyberspace World Railroad! In conjunction with the Association of American Railroads, this site takes you to just about every web page with tracks. It also includes numerous photos for antique train lovers. For the more artistic side of railroading, make tracks to CWR.

URL:

http://www.mcs.net/~dsdawdy/cyberoad.html

European Rail Passes

Save money and time when you buy a rail pass through one of several of your favorite European countries. Fill out the online order form and more information is sent to help you book your complete European vacation.

URL:

http://www.eurorail.com

The Eurostar

They said it couldn't be done, but the French and English governments built "The Chunnel" under the English Channel. Read the fascinating stories behind the construction of this new wonder of the world. These pages also include pictures of the trains that run under the channel, their timetables, booking information, exchange rates, and information about other services the Eurostar offers.

URL:

http://mercurio.iet.unipi.it/eurostar/eurostar.html

The Eurostar is the speedster that runs under the English Channel connecting Brussels, Paris, and London.

Penny Bridge

A true commuter's companion, Penny Bridge offers a wealth of information about national and international rail transportation on the Internet. Penny Bridge also provides useful information about local commuter lines in major cities of the world, including subway maps for New York City and London.

URL:

http://bjr.acf.nyu.edu/railInfo/railinfo.html

Railroad Historical Societies and Museums

Travel beyond the information highway crossroads to reach over 30 railroad historical societies and museums throughout the U.S. and England. Many web sites include photos and interesting archives worth browsing.

URL:

http://www.cse.ucsd.edu/users/bowdidge/railroad/rail-http-servers.html

A B C D E F G H I J K L M N O P Q R S T U V W X Y Z

Railroad-related Internet Resources

Get on the right track with this server, which will send you in the direction of any railroad interest you can think of. If this server doesn't have the information, it knows exactly where to look for it.

URL:
http://www-cse.ucsd.edu/users/bowdidge/
railroad/rail-home.html

Subway Navigator

The Subway Navigator is a resource you truly have to see to believe. If you're planning a trip to virtually any large city in the world, chances are you'll find a subway map here for your destination city. But it doesn't stop at maps—choose a starting station and a destination station and the server will give you a detailed itinerary including each stop along the way and the estimated time en route for your ride. This incredible resource will guide you through the subways of 56 major cities around the world.

URL:
http://metro.jussieu.fr:10001/bin/cities/english

Train à Grande Vitesse

Discover the incredible speed of the Train à Grande Vitesse (TGV) bullet trains, some of the fastest vehicles on rails in the world. Visit many interesting places and explore the history of high-speed trains from around the world. Find out which countries currently use high-speed trains and learn how new lines are constructed.

URL:
http://mercurio.iet.unipi.it/tgv/tgvindex.html

Transportation Resources

Princeton engineers a great place to find information about rail and commuter transit. Check train schedules of countries from around the world, or simply find a map of local commuter trains. This site also offers rail traffic reports from major cities in the U.S.

URL:
http://dragon.princeton.edu/~dhb/systems.html

ROBOTICS

Bucknell Robotics Laboratory

Wind yourself up and into the Bucknell Robotics Laboratory at Bucknell University for an update on the latest machinations regarding robotics.

URL:
http://www.eg.bucknell.edu/~robotics/brl.html

Chaos, Complexity, Systems Theory, and Learning

This page from the University of Minnesota touches on all of the above as they relate to robots—especially in chaos, complex systems, and nonlinear systems. It also features robotics want/have ads (old Robbie the Robot or Klaatu parts, anyone?). There is an electronic suppliers list, and links to robotic journals and other related robotics endeavors.

URL:
http://lenti.med.umn.edu/~mwd/robot.html

Corporate Robotics Web Servers

Here's a handy list of 25 robotics companies and their home pages. Whether you're trying to build something fancy for a factory or are just interested in basic robot hobby kits, this is an excellent place to look.

URL:
http://piglet.cs.umass.edu:4321/cgi-bin/
robotics-corporate/

Dante II

This is the resource page for a very interesting and helpful robot that climbs into volcanos and measures their gases and temperature. Dante II will no doubt save many vulcanologists' lives, since there have been many accidents involving these scientists being in the wrong place at the wrong time. Dante II may be the future key to this field of study.

URL:
http://maas-neotek.arc.nasa.gov/dante/
dante.html

Information on Hobby Robots

If you feel like building robots in your spare time, this page provides you with electronics hints and plans and tells you where to buy your equipment and how to put everything together. There are also links to other robot resources.

URL:
http://www.cs.uwa.edu.au/~mafm/robot/

Interesting Robotics Destinations

Arrick Robotics offers this page with many robotics-related devices, services, and information. One of the neatest features is a facility that allows anyone to create a web page describing his or her own robot—complete with a picture and description. Also available here is informative pages describing positioning systems and instruments used in robotics.

URL:

http://robotics.com/robots.html

Jay's Robot Page

Step-by-step instructions guide you in building a remote control toy car. A simple but rewarding introduction into robotics.

URL:

http://challenge.tiac.net/users/jfrancis/
robot.html

JPL Mars Pathfinder

This is a fascinating page for an upcoming NASA project called the JPL Mars Pathfinder. Rocky IV, a robot designed to roam the surface of Mars, will be used to perform experiments and provide us with more data about the planet Mars. All of the engineering and robotic challenges entailed in this project make this a great read.

URL:

http://mpfwww.jpl.nasa.gov

Laboratory for Perceptual Robotics

Clank on over to the University of Massachusetts robotics lab to see what's new on the slab. This UMass program focuses on controlling robots in uncertain and unstructured environments. (Sort of like the Nine Inch Nails' performance at Woodstock '94...) There are lots of MPEG videos of their own robots in action, and you can take a free-form tour of the lab. They also maintain a resource-related page for robotics.

URL:

http://piglet.cs.umass.edu:4321

NASA Ames Intelligent Mechanisms Group

The Intelligent Mechanisms Group creates robots and machines that are capable of learning and adjusting to the external environment. There are links to many projects that will give you a taste of the future.

URL:

http://maas-neotek.arc.nasa.gov

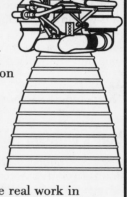

With the tremendous costs and risks of manned space missions (and the shortsightedness of the government), the future of serious planetary exploration belongs to our robotic counterparts. Robotics is going to be a big deal in space. While human cargo jockeys fumble around in 0g over televised shuttle missions on CU-SeeMe, the real work in exploring deep space will be the domain of the machine. JPL's Mars Pathfinder is just the beginning.

POPBUGS—Simulation Environment for Track-driven Robots

The School of Cognitive and Computing Sciences at the University of Sussex provides a program for simulating robot motions in a 2-D world through its downloadable POPBUGS package. The site contains many MPEGs of robotic-simulation demos based on preprogrammed scenarios.

URL:

http://www.cogs.susx.ac.uk/users/christ/
popbugs/intro.html

Robotic Creations

Several home-made robots are described and pictured on this page. Check out the SKIMER, which tests visual navigation and is fairly sophisticated. Then there's the 3 Leg Walker, which was assembled in two hours. A robot named "Thing" looks like a metal spider and is only 11 inches tall.

URL:

http://robotics.com/robomenu.html

Robotic Tele-Excavation

Remotely tele-operate a robot arm over the Web. The system consists of a commercial robot arm positioned over a terrain of fine-grained sediments. View the environment via live images from a CCD camera attached to the arm. A pneumatic system mounted on the arm lets you direct short bursts of compressed air onto the environment at selected points. Using this, excavate regions within the terrain by positioning the arm, delivering a burst of air, and viewing the newly cleared region. Various artifacts are buried in the sediment. Anyone find Jimmy Hoffa yet?

URL:
  http://www.usc.edu/dept/raiders/story/

Robotics and Control Papers

This page is an index to robotics and control papers at Boston University's Aerospace/Mechanical Engineering department. Listed as hypertext links are four primary researchers at Boston. Under each link is a list of the publications each researcher has written.

URL:
  http://robotics.bu.edu/pub/papers/

Robotics at Brown University

Read up on the adventures of Ramona and Rai—two robots constructed by students at Brown University's computer science/robotics and artificial intelligence labs.

URL:
  http://www.cs.brown.edu/research/robotics/

Engineering students at Brown University are shamelessly building robots. With no deference at all to the tin man or Lost in Space, these students ply their specialties and construct creatures with names like Ramona and Rai that wander the halls of the computer science department carrying with them breakfast foods and other gifts meant to endear themselves to us. But don't be tricked by this ploy. Find out about the robotics and AI labs at Brown—before they find you.

Robotics at CalTech

CalTech can do robots, too. Read about their research projects, the friendly staff at the robotics lab, and view pictures of "Snakey," "Snakeboard," and their other creations. (The walking biped is really scary!)

URL:
  http://robby.caltech.edu

Robotics Internet Resources Page

Here's a well-organized list of robotics links created by the computer science department at the University of Massachusetts; here you'll find the nuts and bolts on both academic and corporate robotics. If you're not too serious about robotics, check out the clubs and hobby info.

URL:
  http://piglet.cs.umass.edu:4321/robotics.html

Robots for Space

Robotic systems are an essential part of the NASA Space Shuttle and Space Station Project. The Flight Robotics System Branch attempts to make robots easier to use and increase their potential. One day soon we will have robots that vacuum, sort the laundry, and complete tax returns for us.

URL:
  http://tommy.jsc.nasa.gov/ARSD/app-C/

Spare Parts

Here's a compilation of interesting web sites focusing on the subject of robotics. Included is information about magazines, robot parts, university departments, NASA robotics sites, and much more.

URL:
  http://www.robotics.com/robots.html

USC's Mercury Project

This was one of the first systems to allow web users to manipulate real-world objects with a robot. In a seven-month period, these pages were visited over 2.5 million times. USC has also created a new project that you should look into.

URL:
  http://www.usc.edu/dept/raiders/

SECURITY AND FIREWALLS

AT&T's Internet Security FTP Server

You can bury yourself in this collection of papers and slides on Internet security research from AT&T. Among other rich resources, it includes the preface and table of contents from the book *Firewalls and Internet Security—Repelling the Wily Hacker*. These papers are in PostScript and DVI format.

URL:

> **ftp://ftp.research.att.com/dist/internet_security/**

Authenticated Firewall Traversal Working Group

The Authenticated Firewall Traversal Working Group is an organization working on a protocol specification to address the issue of application-layer support for firewall traversal. The protocol will support both TCP and UDP applications with a general framework for authentication of the firewall traversal. This page contains the charter of the group, its goals and milestones, and related current Internet Drafts—including one on the SOCKS security protocol. Also provided are mailing list information and archives.

URL:

> **http://www.ietf.cnri.reston.va.us/html.charters/aft-charter.html**

CERT Advisories Archive

This is the archive of all the published CERT advisories that give warnings about many common security holes, bugs, and vulnerabilities for computers connected to the Internet. Included are reports on sendmail, ftpd, passwd, telnet, rdist, syslog, xterm, ghostscript, and other services that have been attacked or targeted by intruders. Details of patches and fixes for these vulnerable services are also provided.

URL:

> **ftp://cert.org/pub/cert_advisories/**

CERT Coordination Center Service

The Computer Emergency Response Team (CERT) was established by the Defense Advanced Research Projects Agency (DARPA) to address computer security concerns of researchers on the Internet. This archive contains a collection of documents about security problems, solutions, and resources.

URL:

> **ftp://cert.org**

CERT Technical Tips

The CERT's Technical Tips page offers a collection of documents in ASCII format containing practical advice on topics that include anonymous FTP configurations and packet filtering. It also contains the CERT security checklist, which helps system administrators access and improve the security of their sites.

URL:

> **ftp://cert.org/pub/tech_tips**

COAST Security Archive Index

The COAST Security Archive is a large and detailed collection of security-related tools and documentation on access control, authentication, commercial security packages, email security, encryption, firewalls, intrusion detection, Kerberos, legal issues, network and password security, privacy issues, RFCs, software forensics, security guidelines and policies, trusted systems, and viruses.

URL:

> **http://www.cs.purdue.edu/coast/archive/data/category_index.html**

CERT Advisories

If you're a system administrator, getting mail from CERT is never fun, especially if a vulnerability is being reported that affects your system or network. You might think telling everyone—the entire world—about security holes in your network is a silly thing to do, but it actually makes great sense. It forces administrators to take care of a problem, hopefully before a network intruder can exploit it.

Computer Security Institute's Internet Security Survey

This presentation and discussion of the results of CSI's Internet Security Survey shows that one out of every five sites on the Internet has been attacked at some time, that the firewall market is still wide open, and that many enterprises still do not protect their sites by using firewall technology.

URL:

> **http://all.net/journal/csi/survey95.html**

A B C D E F G H I J K L M N O P Q R **S** T U V W X Y Z

Digital Pathways

A provider of security for remote and Internet access. Digital Pathways' pages include book summaries, an online newsletter, summaries of white papers, and references for articles in trade magazines. There's also an index of frequently asked questions, and a list of resources (including CERT) and how to get in touch with them.

URL:
> http://www.digpath.com

Firewalls FAQ

This FAQ starts by describing what a firewall is, and what one can and cannot protect against. It includes references to books about firewalls and pointers to commercial products. The use of firewalls with the Web, DNS, FTP, telnet, finger, Gopher, Archie, and other services are considered, as well as X-Window traffic. Finally, the FAQ also provides an excellent glossary of firewall-related terms.

URL:
> http://www.cis.ohio-state.edu/hypertext/faq/
> usenet/firewalls-faq/faq.html

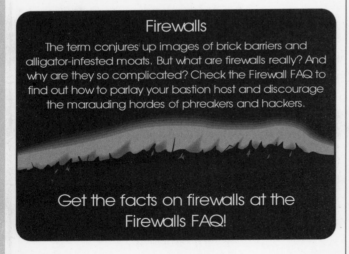

Firewalls

The term conjures up images of brick barriers and alligator-infested moats. But what are firewalls really? And why are they so complicated? Check the Firewall FAQ to find out how to parlay your bastion host and discourage the marauding hordes of phreakers and hackers.

Get the facts on firewalls at the Firewalls FAQ!

Firewalls Mailing List

This mailing list is for discussions of firewall issues, problems you may encounter and their solutions, requests for information, and any ideas or questions you may have. The list is open to the worldwide Unix community, including commercial, educational, and private users.

Mailing List:
> Address: **majordomo@applicom.co.il**
> Body of Message: **subscribe firewall-1**

Firewalls Mailing List Archive

These archives of the firewalls discussion list date back to September of 1992. They contain discussions on all aspects of creating Internet firewalls and the products you need to create those firewalls. The messages are archived in the Unix .Z format—with one for each month, approximately 1MB in size.

URL:
> ftp://ftp.greatcircle.com/pub/firewalls/archive/
> Welcome.html

Internet Security Systems

Internet Security Systems (ISS) specializes in developing network scanning software that detects security vulnerabilities. The ISS web site offers details of their Internet scanner product and scan service, lists of free security seminars, discussions on the threats of hacking, security FAQs, mailings lists, and links to other security sites.

URL:
> http://www.iss.net/

IP Security Protocol Working Group

The IPSEC develops technologies to protect IP client protocols, such as a security protocol in the network layer to provide cryptographic services that will support combinations of authentication, integrity, access control, and confidentiality. On their pages, the working group offers the group charter, its goals and milestones, related current Internet Drafts, and mailing list information and archives.

URL:
> http://www.ietf.cnri.reston.va.us/html.charters/
> ipsec-charter.html

Kerberos

Kerberos is a Unix network authentication system based on a key distribution model for use on physically insecure networks. This page contains announcements about the latest version releases; a FAQ explaining the Kerberos system; the archives for the Kerberos mailing list and **comp.protocols.kerberos** newsgroup; and links to other Kerberos web sites.

URL:
> http://www.mit.edu:8001/people/proven/
> kerberos/kerberos.html

National Computer Security Association

Hackers and crackers got you hopping around changing passwords and updating your filter tables? Who 'ya gonna call? Plug into the National Computer Security Association. (No, this isn't related to the NCSA—the National Center for Supercomputing Applications at the University of Illinois.) This NCSA will gladly fill you in on subjects like computer viruses and computer ethics, and let you use IS/RECON—a searchable database with up-to-the-minute information about hacking/phreaking activities.

URL:
> http://www.ncsa.com

Netscape Data Security

Netscape Communications Corp. developed and maintains one of the most popular Web browsers currently available. Here, Netscape describes the technical security measures and techniques used to secure the data communications between their Web servers and clients. The details of the Secure Sockets Layer (SSL)—which provides data encryption, server authentication, message integrity, and optional client authentication for a TCP/IP connection—are also offered.

URL:
> http://www.netscape.com/info/
> security-doc.html

Network Security

This collection of network security resources includes links to Unix network security tools such as Satan and TCP wrappers, an Internet firewalls tutorial, and links to security technology and software products that meet both government and commercial security requirements for computer systems and networks.

URL:
> http://www.spp.umich.edu/telecom/
> net-security.html

Privacy Enhanced Electronic Mail Charter

The PEM working group has designed a system for sending email that prevents messages from being tampered with, ensures that the sender address cannot be forged, and offers confidentiality of the message content. This page provides details of the group; mailing list information; a collection of related RFCs, including those covering MD2, MD4, and MD5; MIME; and the PEM specifications.

URL:
> http://www.ietf.cnri.reston.va.us/html.charters/
> pem-charter.html

SAIC Security Documents

Another large collection of documents that covers a wide area of topics about computer security, this collection includes papers and FAQs on firewalls, Web security, intrusion detection, Unix security, the Internet Worm, tales of computer attacks and countermeasures, networking, and trusted systems. All of the documents here that aren't FAQs are in PostScript format.

URL:
> http://mls.saic.com/docs.html

SATAN

SATAN is a collection of tools for system security administration. It recognizes several common networking-related security problems, and reports the problems without actually exploiting them. Here, you can find documentation on SATAN, downloadable source locations, a FAQ, and vendor advisories about the release of SATAN.

URLs:
> http://www.fish.com/~zen/satan/satan.html
> http://www.fish.com/dan/satan.html

Satan in Your System
*No, SATAN is not an evil spirit that invades your computer and causes the disk drives to spin backwards, resulting in audible messages from the denizen of doom himself. Nor is it even a computer virus or anything else less than a tremendously useful tool for security administrators. SATAN is the notorious **Security Administrator's Tool for Analyzing Networks** for Unix computers. Turn to the Web before you believe what you hear on the news networks.*

A
B
C
D
E
F
G
H
I
J
K
L
M
N
O
P
Q
R
S
T
U
V
W
X
Y
Z

Secure FileSystem

Secure FileSystem (SFS) is a set of programs that create and manage encrypted disk volumes under DOS and Windows. Each volume appears as a normal DOS drive, but all data stored on it is encrypted at the sector level. This page details SFS features and encryption capabilities, as well as links to the software.

URL:
  http://www.cs.auckland.ac.nz/~pgut01/sfs.html

Secure Shell Remote Login Program

Ssh (Secure Shell) is a free Unix program that you can use to log into other computers on a network, to execute commands on remote machines, and to move files between computers. Ssh features strong user authentication and secure communications, even on insecure channels. Ssh is available for Unix machines, and you can get the source here, as well. Also available on this page is a mailing list for Ssh users, the archives for that mailing list, FTP sites, and other links to sites that pertain to Ssh.

URL:
  http://www.cs.hut.fi/ssh/

SecureWare

SecureWare develops security technologies and software products that meet governmental and commercial security requirements for systems and networks. On their site, SecureWare offers product information; free evaluation copies of their software; technical information on security topics and products; a security glossary; and links to other security information.

URL:
  http://www.secureware.com

Security Mailing Lists

Security mailing lists are important tools for network administrators, network security officers, security consultants, and anyone who needs to keep abreast of the most current security information available. This page offers a list of security-related mailing lists.

URL:
  http://iss.net/iss/maillist.html

Weave your way into the Web.

There's tons of free stuff on the Web.

Security Tools

This FTP site houses Unix software packages for evaluating and maintaining the security of an Internet host. Programs available here include COPS, Crack, Tripwire, TCP wrappers, virus-detection programs, and others.

URL:
  ftp://cert.org/pub/tools

SSL Protocol

Privacy issues are a big deal on the Info Superhighway, and the Secure Sockets Layer (SSL) Protocol is designed to ensure that no third party interferes with or intercepts any communications not intended for them. This Internet Draft explains how the SSL works and provides information on encryption and public and private keys. A glossary of terms and a reference section are provided. From this page, you can also join an email mailing list to keep up with all the latest ideas on SSL.

URL:
  http://home.netscape.com/newsref/std/SSL.html

SSLeay

SSLeay is a free implementation of Netscape's Secure Socket Layer—the software encryption protocol behind the Netsite Secure Server and the Netscape Browser. It supports DES, RSA, RC4, and IDEA encryption algorithms. SSLeay is described here, and a FAQ is provided. Sources for the library, applications for Unix, and a programmer's reference are also given.

URL:
  http://www.psy.uq.oz.au/~ftp/Crypto/

Sun Internet Commerce Group

Sun Microsystem's Internet Commerce Group provides products and implementation services to businesses that want to do business on the Internet. On their web pages, they offer white papers describing network security and cryptography, descriptions of network security products, and certificate and revocation lists for use with some of the products that use cryptographic keys.

URL:
  http://www.incog.com

Technologic

Technologic designs and develops commercial Internet applications ranging from firewalls and network administration tools to Web servers and virtual private networks. Areas of security-related specialization include identification and authentication, access control, privacy enhancement, and network security analysis. Also offered here are details of Internet firewall products, security scanners, and security services.

URL:
http://www.tlogic.com

Web Security Mailing List

The www-security mailing list is a forum for encouraging and stimulating open discussions on the design and development of security technology for the Web. The list encourages the development and implementation of Internet standards for Web security, and it is the official mailing list of the IETF Web Transaction Security Working Group. Currently, this list has over 700 subscribers.

Mailing List:
Address: **majordomo@nsmx.rutgers.edu**
Body of Message: **subscribe www-security**

Yahoo Security and Encryption Directory

The Yahoo directory sports a large collection of links to sites that contain information on cryptography, digital cash, firewalls, hacking, security-related software, viruses, security conferences, security FAQs, and much more. Each area, or security topic, is organized into separate collections for easy access.

URL:
http://www.yahoo.com/Computers/
Security_and_Encryption/

SEISMOLOGY

Center for Monitoring Research

Whoa! What's shakin'? The experts at the Center for Monitoring Research know. Now you can, too, by linking to their technical overview, network maps, seismic bulletins, recent events, and documents.

URL:
http://www.cdidc.org

Earthquake Home Preparedness Guide

Concerned about the safety of your home in earthquake country? Follow this page for information on upgrading and repairing masonry and chimneys, advice on earthquake insurance, preparing for the Big One, developing a family plan for when the inevitable occurs, and what to do after it happens.

URL:
http://www.eqe.com/publications/homeprep/

Don't let your foundation be shaken without first checking the Earthquake Home Preparedness Guide!

Engineering Seismology Group Canada

Consulting services are available to examine natural and induced seismicity through the Engineering Seismology Group Canada. E.S.G. Canada offers expertise in seismology, rock physics, mining, and electrical engineering.

URL:
http://mine.mine.queensu.ca/jaga/esg1.html

Global Shaking

This earthquake bulletin provides almost up-to-the-minute information on earthquakes occurring around the world. It also allows you to view detailed maps of the quake locations.

URL:
http://www.civeng.carleton.ca/cgi-bin/quakes

Hug some trees in the Forestry section.

A B C D E F G H I J K L M N O P Q R S T U V W X Y Z

Excerpt from the Net...

(from What's Lava?)

How is lava made inside the world?

                Stan and Dan

Hi Stan and Dan,

    First, there is a definition I need to make. Just to keep things straight, geologists use the word "magma" for molten rock that is still underground, and the word "lava" for rock that has made it to the surface.

    So, you want to know how magma is made? There is a lot of heat within the earth, and this heat is produced by radioactive decay of naturally-ocurring radioactive elements within the earth. It is the same process that allows a nuclear reactor to generate heat, but in the earth, the radioactive material is much less concentrated. However, because the earth is so much bigger than a nuclear power plant it can produce a lot of heat.

    Anyway, this heat is enough to partially melt rocks in the upper mantle, about 50-100 km below the surface. I say partially melt because the rocks don't completely melt. As you can guess, most rocks are made up of more than one mineral, and these different minerals have different melting temperatures. This means that when the rock starts to melt, some of the minerals get melted to a much greater degree than others. The main reason this is important is that the liquid (magma) that is generated is not just the molten equivalent of the starting rock, but instead something different.

    You could think of making a "rock" out of sugar, butter, and shave ice. Pretend that they are mixed equally so that your rock is 1/3 sugar, 1/3 butter, and 1/3 shave ice. If you start melting this "rock", however, the "magma" that is generated will be highly concentrated in the things that melt more easily, namely the ice (now water) and butter. There will be a little bit of sugar in your magma, but not much.

    Anyway, the most common type of magma produced is basalt (the stuff that is erupted at mid-ocean ridges to make up the ocean floors, as well as the stuff that is erupted here in Hawai'i). Soon after they're formed, little drops of basaltic magma start to work their way upward (their density is slightly less than that of the solid rock), and pretty soon they coalesce with other drops and eventually there is a good flow of basaltic magma towards the surface. If it makes it to the surface it will erupt as basaltic lava.

    Hopefully this will help to answer your question.

Sincerely,

Scott Rowland, University of Hawaii

International Seismic Monitoring

Located in Arlington, Virginia, at the Center for Seismic Studies is the new Experimental International Data Center. Internationally operating, it provides advanced, automated approaches to data collection, analysis, and management.

URL:

http://www.cdidc.org/WebIDC/IDC_Tour/
idc_tour.html

Record of the Day

Tune in to the Record of the Day and find out where the biggest earthquakes took place today—complete with exact location, magnitude, time, seismogram, and a map.

URL:

http://www.gps.caltech.edu/~polet/recofd.html

Solar Seismology

Now, through helioseismology, solar seismic waves can be used to measure the dynamics and internal structure of Sol, our solar system's sun. These studies are being conducted through GONG (Global Oscillation Network Group), which is developing a network of six observation stations around the world.

URL:

http://helios.tuc.noao.edu/gonghome.html

Southern California Earthquake Center

This is the primary distribution center of seismological data recorded by the Southern California Seismic Network, TERRAscope, and various Portable Seismic Instrument Arrays. Also found here is Global Positioning System Data collected by SCEC and contributing organizations such as the U.S. Geological Survey.

URL:

http://scec.gps.caltech.edu

Domestic Science? Of course it's a real science!

What's Lava?

What's the difference between lava and magma? Here are questions from folks like you without school-aged children (because kids know everything about things like dinosaurs and volcanos). Following the questions are answers from scientists and professors (many who learned these answers from their kids) that explain everything you need to know about lava.

URL:

http://volcano.und.nodak.edu/vwdocs/
frequent_questions/group1.html

SOFTWARE

The Consummate Winsock Apps Page

If you're a Windows user, there's only one place on the Net where you can find the latest Internet-related software for your PC. As its name implies, The Consummate Winsock Apps Page is your one-stop resource for finding and keeping your computer up to date. Everything from communications clients to utilities is organized, reviewed, and ranked on a five-star system to help you hunt down the best the Net has to offer.

URL:

http://cwsapps.texas.net/

The Consummate Winsock Apps Page

That's it. Give it up, IBM. Windows 95 is the king of the operating systems. If you don't believe it, pay a visit to the Consummate Winsock Apps Page, and drool over all of the free and shareware software written specifically for Windows 95. There are audio apps, Internet clients of all kinds, utilities, applications, communications programs, virus scanners, and tons of other great code. (Besides, Bill Gates can beat Lou Gerstner in paper football any day.)

A
B
C
D
E
F
G
H
I
J
K
L
M
N
O
P
Q
R
S
T
U
V
W
X
Y
Z

CuteFTP

Sure, WS_FTP is a handy and venerable FTP client for Windows, but it's not as cute as CuteFTP. CuteFTP's best features include intuitive drag and drop, a convenient directory of your favorite archive sites, an extensive icon bar for quick access to popular features, incredible ease of use for novices, and impressive performance for war-torn Net veterans. Shareware versions are available for both 16- and 32-bit Windows systems.

URL:

> http://papa.indstate.edu:8888/CuteFTP/

Say good-bye forever to the command-line Unix FTP program. For that matter, say good-bye forever to the command-line Windows 95 FTP program . . . and hello to CuteFTP! Don't be fooled by its name. CuteFTP is a powerful file transfer utility. Sure, you can use your Web browser to connect to FTP sites and download files, but once you use CuteFTP, you probably won't again. CuteFTP is a shareware program, which means that you can download it and try it out for free. So get CuteFTP today and try it out!

E-mail Notify

E-mail Notify is a native Windows 95 program for notifying you when you have received email. This program integrates tightly with Windows 95—it places an icon in the tray on the taskbar next to the speaker, clock, and schedule icons. E-mail Notify is not an electronic mail program—it simply notifies you when email has arrived. The program works with standard POP servers and requires that TCP/IP be running on your system.

URL:

> http://olympe.polytechnique.fr/~zic/english/
> notify.html

Guide to Available Mathematical Software

This is a gateway to the NIST Guide to Available Mathematical Software, a cross-index and virtual repository of mathematical and statistical software components of use in computational science and engineering.

URL:

> http://gams.cam.nist.gov

Internet Phone

Internet Phone is a Windows program that allows you to use the Internet to speak vocally with any other user anywhere in the world. To use Internet Phone you need the software, a TCP/IP Internet connection, and a Windows-compatible audio device. This web page offers an introductory description of Internet Phone, a FAQ, product specifications, news, a company profile, and a free evaluation copy of Internet Phone.

URL:

> http://www.vocaltec.com

Internet Utility Software

We're not sure why Chrysler Corporation sponsors this page, but it's nice of them! Chrysler brings together onto a single page a collection of the most useful tools for navigating the Internet and the Web. Provided are links to Web browsers, sound software, image viewing and editing tools, QuickTime movie viewers, and other useful utilities. The page is divided into three sections—Unix, Macintosh, and the PC.

URL:

> http://www.chryslercorp.com/help/help.html

Jumbo!

Self-billed as "The biggest, most mind-boggling, most eye-popping, most death-defying conglomeration of freeware and shareware programs on the known web," the Jumbo archive is truly amazing. With well over 20,000 programs, that description may not be far off base. Categories include business; games; home and personal; programming; utilities; and words and graphics. Finding what you want is easy with the excellent organization here.

URL:

> http://www.jumbo.com

Kai's Power Tips and Tricks for Adobe Photoshop

MIT's student newspaper *The Tech* has published this page of Kai Krause's 23 original tips and tricks for Adobe Photoshop. Whether you use Photoshop on a Mac or PC, these tips and tricks directly from the master are sure to have you polishing up your Photoshop techniques and documents.

URL:
http://the-tech.mit.edu/KPT/Tips/

Kaleida Media Player

Multimedia applications written in Kaleida's ScriptX language run on Macintosh or Windows systems and require no platform-specific modifications. Download a free version of the Kaleida Media Player and integrate it with your World Wide Web browser to access ScriptX applications on the Web.

URL:
http://www.kaleida.com

Multimedia Communications Laboratory - Boston University

Researchers in the Multimedia Communications Laboratory (MCL) at Boston University investigate developing general-purpose distributed multimedia information systems (DMISs) such as Video-on-Demand. Here you can download both MPEG playback and encoding software.

URL:
http://hulk.bu.edu

Windows 95 FTP Client

WS_FTP (what a terrible name for a program) is a full 32-bit, native Windows 95 program for retrieving and uploading files to and from remote FTP servers. This is a first-class program that offers many features, including logging, saved configurations, and configurable program and session options.

URL:
ftp://ftp.cdrom.com/pub/win95/inetapps/ ws_ftp32.zip

Find out about your relatives in Primatology . . . uh . . . Genealogy.

Windows 95 FTP Server

WFTPD is a shareware FTP server for Windows 95 systems. This program is a full 32-bit native Windows 95 application that runs quietly in the background.

URL:
ftp://ftp.coast.net/SimTel/win3/winsock/ 32wfd202.zip

Windows 95 IRC Client

Still using an ugly text-based interface to Internet Relay Chat? Now there's no excuse for that! MIRC, the best IRC client, is native to Windows 95—and it's free!

URL:
ftp://ftp.cdrom.com/pub/win95/inetapps/ mirc364.zip

Say good-bye forever to the command-line Unix FTP program. For that matter, say good-bye forever to the command-line Windows 95 FTP program . . . and hello to WS_FTP! Once you get by the name, this is really a great program. Sure, you can use your Web browser to connect to FTP sites and download files, but once you use WS_FTP, you probably won't again. WS_FTP is a shareware program, which means that you can download it and try it out for free. Odds are too, that you fall into one of the categories that can continue to use it free of charge. So get WS_FTP today and try it out! (And after you drag out its icon onto your desktop, rename it "FTP".)

Windows 95 TCP/IP Utilities

WSPING32 is a full 32-bit Windows 95 program offering ping, traceroute, and DNS lookup utilities.

URL:
ftp://ftp.cdrom.com/pub/win95/inetapps/ wsping32.zip

A B C D E F G H I J K L M N O P Q R **S** T U V W X Y Z

TELECOMMUNICATIONS AND CELLULAR

Advanced Communication Technologies and Services

ACTS is a research and development program sponsored by the European Union. This page gives an overview of the ACTS program, examines the situation of the industry sector in Europe today, and provides details on how to make project proposals.

URL:

http://www.lii.unitn.it/EU/ACTS/

Cellular Digital Packet Data

Knock knock. Who's there? Datagram . . . a CDPD datagram, used by your laptop to stay connected to the Internet when you're traveling in your car. (Hey, eyes on the road, not on **www.playboy.com**!) CDPD uses a sophisticated channel-hopping technique that allows both data and voice to share the existing cellular network. Knock knock. Who's there? *Landline shark.*

URLs:

http://www.cdpd.net/rfc-index.html
http://www.pcsi.com/html/cdpd.html

Erlang

Erlang is a programming language, run time system, and a set of support tools developed by the Ericsson company to help design software for its telecommunications products. This page offers an overview of the features and purpose of Erlang, an excellent FAQ on the language, the complete manual pages online in HTML format, and a list of books and papers about Erlang.

URL:

http://www.ericsson.se/cslab/erlang/

On the Internet, you don't need a passport to travel the world.

The Net is your passport. The Web is your magic carpet.

European Telecommunications Standards Institute

ETSI is responsible for setting and publishing European standards for telecommunications and related fields such as broadcasting and office-information technology. This site gives a presentation on the Institute and provides a comprehensive searchable publications catalog along with membership details. There are also certain standards and reports that can be downloaded by registered users.

URL:

http://www.etsi.fr/

Global System for Mobile Communications

GSM, the Global System for Mobile communications, is a digital, cellular communications system that has rapidly gained acceptance and market share worldwide. In addition to digital transmission, GSM incorporates many advanced services and features, including ISDN compatibility and worldwide roaming in other GSM networks. This page provides two excellent overviews of GSM available in HTML or PostScript, as well as some helpful reference points.

URL:

http://ccnga.uwaterloo.ca/~jscouria/gsm.html

Learn how to communicate globally with the *Global System for Mobile Communications.*

RACE

Research and technology development in Advanced Communications technologies in Europe (RACE) is a collaborative European research program that aims to introduce Integrated Broadband Communication (IBC). This page provides details of the many RACE projects, news about telecom developments worldwide, presentations and results from specific projects, and links to other related telecommunications information.

URL:

http://race.analysys.co.uk/race/

Telecom Digest

Telecom Digest is an electronic journal focusing on telecommunications topics. The Telecom pages offer a Telecom FAQ, a collection of articles on future developments, and a link to the mailing list archive. It's also gatewayed to Usenet where it appears as the moderated newsgroup **comp.dcom.telecom**.

URL:

http://www.wiltel.com/telecomd/telecomd.html

Telecom Glossary

A glossary of telecommunications terms and acronyms. It's organized alphabetically, so you can browse the glossary by selecting a letter or by performing a keyword search. You can also add new terms using an interactive form.

URL:

http://www.wiltel.com/glossary/glossary.html

Telecom Information Resources

This page contains references to information sources for technical, economic, public policy, and social aspects of telecommunications. All forms of telecommunication, including voice, data, video, wired, wireless, cable TV, and satellite, are included. It provides nearly a thousand pointers to web resources, with a brief description of each.

URL:

http://www.spp.umich.edu/telecom/
telecom-info.html

> # ISDN is going to digitize your world.

Telecommunications Library

Provided by WorldCom, this page offers resources relating to the technology and business of telecommunications. Included are a weekly newsletter, a listing of telecommunication events and conferences, the *Telecom Digest* FAQ and archives, discussion on the regulation of the telecommunications industry, and a glossary of telecommunications terms and acronyms.

URL:

http://www.wiltel.com/library/library.html

Wireless Source

Here's a bit of insight into what might be the next frontier of telecommunications—wireless. A world without wires would enable us to do our work at the beach or at the top of a mountain. If you are interested in that kind of freedom, take a peek at the future at this easy-to-browse page.

URL:

http://www.wireless-source.com

TELEPHONY

AT&T Internet Toll Free 800 Directory

AT&T might have broken up into smaller pieces, but they still would like you to call, and call often. They've thoughtfully provided a very large list of businesses with toll-free 800 numbers that is searchable by letter, keyword, or company name.

URL:

http://www.tollfree.att.net/dir800/

AT&T Talking Power

This AT&T server provides tutorials on the U.S. telephone network, your phone line, the anatomy of a telephone call, telephone office batteries, sample product manuals, a sample training video in AVI format, and much more about telephones and telephone service. There's also a clickable map to select the topics you're interested in, and a diagram explaining how a normal telephone call is set up and delivered.

URL:

http://www.att.com/talkingpower/

A B C D E F G H I J K L M N O P Q R S T U V W X Y Z

A Brief History of Mobile Telephony

A good introduction to mobile telephony, this paper starts by describing the basics, such as hand-over and locating, and then describes the first real mobile telephone system. It goes on to introduce a bunch of current mobile systems, including GSM as a move toward a world standard. The paper also covers such issues as mobile data communication, packet radio, and security. Finally, the author looks at the future of mobile telephony.

URL:

http://www.cl.cam.ac.uk/users/sgh20/paper/
ttr-372.ps

Digital Telephony Archive

The Digital Telephony Archive is a collection of articles, press releases, testimony, and drafts about the Digital Telephony Proposal in the U.S. and other issues that relate to privacy and digital telephony.

URL:

http://www.eff.org/pub/Privacy/
tDigital_Telephony_FBI/

One ringy dingy . . .
Two ringy dingy . . .

Find out how phone calls actually work at AT&T Talking Power.

Facsimile at the Virtual Library

This web site is a section of the Web Virtual Library for Communications and Telecommunications. It lists links that will take you to anything from modems to fax servers on the Internet. There are also connections to hundreds of hardware and software products related to facsimile.

URL:

http://www.faximum.com/w3vlib/fax

Fax FAQ

If you are interested in integrating your computer with fax machines and fax capability, you should read this FAQ. (That's right, it's a Fax FAQ!) It will make your ride on the information superhighway a little smoother.

URL:

http://www.faximum.com/faqs/fax

The Long Distance Area Decoder

Why let your fingers do the walking when they could be doing clicking instead? Forget those heavy phone books with the fine print, and check out the Area Decoder. Simply type in any city, state, or country to find out its area or country code. The Decoder presents rate information on the cost per minute to call. (Albania's rate is only $1.10 per minute, so call now!)

URL:

http://www.xmission.com/~americom/
taclookup.html

Long-Distance Digest

This weekly newsletter focuses on the long-distance service industry, providing industry information on long-distance resellers, carriers, aggregators, wholesalers, their agents, and others involved in the business of reselling long-distance telephone services for profit. It also profiles prepaid calling cards and other business opportunities in the telecommunications industry. There's also a FAQ and a newsletter archive searchable by date, author, subject, or thread.

URL:

http://www.wiltel.com/ldd/ldd.html

**Unix is the Energizer Bunny
of the Internet.**

Excerpt from the Net...

(from Fax FAQ)

Q.9 How can I share my single phone line with voice, fax, data, etc.

There are a number of devices on the market (suggestions from happy campers welcome) that will try to distinguish between an incoming voice, fax, or data call and route the call appropriately.

These fax switches attach to the phone line and then the other devices (your normal voice phone/answering machine, fax machine, data modem, etc.) are attached to the fax switch).

All devices work on one of two general principles: listening for CNG or voice, or listening for distinctive ring patterns (cadences).

In the first case the device will answer the phone and try to guess what it should do based on what it hears. Some machines play back a sound of a phone ringing so that humans dialing in think the phone is still ringing when in fact the fax switch is listening to see if the call is from a fax machine or a human. If the CNG tone (see Part 1 for a definition of CNG) from the calling fax machine is heard, then the switch connects the call to the fax machine, otherwise the call is deemed to be a voice call and is connected to your phone/answering machine.

A slightly more sophisticated approach is for the fax switch to answer the phone and play a short recorded announcement. If, during the announcement the CNG tone is heard, then the call is switched to the fax machine. If no CNG tone is heard but sound is heard after the announcement, then the call is assumed to be voice and switched appropriately. If nothing is heard then the switch either considers the call a data call and switches it to a modem or considers it a fax call from a machine that does not generate a CNG and switches it to the fax machine.

The other approach relies upon an optional service available from some telcos called "SmartRing", "Distinctive Ring", "RingMaster", "Ident-a-Ring", etc. This feature allows one to have more than one phone number associated with the same phone line. Incoming calls using the different phone numbers can be differentiated by the different ringing patterns (i.e. one long ring, two short rings, three short rings, etc.) The fax switch distributes the call based on the ring cadence it detects.

The advantage of the first approach is that one does not have to send more money to the phone company (or depend upon the availability of the "SmartRing" feature being available). The disadvantage is that it is not always reliable (especially in the face of fax machines that do not generate CNG tones).

The advantage of the second approach is that it is very very reliable. The disadvantage is that it requires the availability of the "SmartRing" feature from one's telco as well as sending more money to the telco every month.

Nautilus

Don't let Big Brother's threats of tapping more phone lines inhibit your phone conversations. Foil them with Nautilus—a program that encrypts voice telephone conversations without any special equipment. Nautilus uses your computer to digitize and compress your speech. Then it encrypts the compressed speech using a variety of encryption algorithms, including Blowfish, Triple DES, or IDEA block ciphers. The digitized, compressed, and encrypted speech is then transmitted over your modem to the recipient's computer. Nautilus is available for PCs running DOS or Windows; Linux; and Sun workstations running SunOS or Solaris—and the source code is distributed with it.

URL:

    ftp://ftp.funet.fi/pub/crypt/utilities/phone/

PGPfone—Pretty Good Privacy Phone

PGPfone is a software package that turns your computer into a secure telephone. It uses speech compression and strong cryptography protocols to give you the ability to have a real-time secure telephone conversation via a modem-to-modem connection. At this site, you will also find the full manual and links to the distribution sites for the software.

URL:

    http://web.mit.edu/network/pgpfone/

Plain Old Telephone Service

A tutorial covering the Plain Old Telephone Service (POTS). It gives a technological background, the history and anatomy of the telephone service, a discussion of telephone deregulation, and an acronym list.

URL:

    http://ganges.cs.tcd.ie/msc-course/tut1/
    tgrp2/start.html

Some Kinda' Phreak?

Ever wonder how to build a "box" to generate tones and manipulate the telephone system? This document describes over 40 different boxes with detailed instructions on how to build them. There are even circuit diagrams for some of them. DTMF, ring, dial tone, busy, fast busy, and quarter tones are provided in .au format. A modem dictionary, CN/A article, and formula for generating certain tones with the software package Mathematica are given. There are also links to the web sites of large telephone companies throughout the world.

URL:

    http://www.cis.ksu.edu/~psiber/fortress/phreak/
    tph2reak.html

Busting Big Bell

Why actually pay for your telephone service when you've already got everything you need to play with the telephone network to your heart's content? You've already got a computer and a modem. All you need are the right digital tones and a little know how! Just say no to those oppressive long-distance bills—even your local service bill if you like! Learn how to be a baby bell's nightmare in just one evening! Tap into *Some Kinda' Phreak?* and find out how to stay off the hook.

Chill out. This is all in fun. Besides, we're betting this page will mysteriously disappear a week after this book ships. Is that a public service or what?

Speak Freely

Speak Freely is a software application for computers running Microsoft Windows, or Sun or Silicon Graphics workstations that allows you to voice-talk over a network. Three forms of compression are provided to conquer slow network connections. To enable secure communications, encryption with DES, IDEA, and/or a key file is available. The full manual with screenshots, and links to the distribution sites are available on this page.

URL:

    ftp://ftp.fourmilab.ch/pub/kelvin/netfone/
    twindows/speak_freely.html

Telecom Regulation

On this mailing list, you can discuss telecommunications regulation on the local, state, and federal levels. Find out about regulatory restrictions surrounding cable, broadcasting, telephony, and data transmissions, as well as the related economic and social issues.

URL:

    http://www.wiltel.com/telecomr/telecomr.html

UFOLOGY

Alberta UFO Research Association

Broaden your UFO info from a Canadian perspective. Peruse this interactive 'zine of UFO news, interconnective journals, sightings, and archival information. One of this site's main highlights is an online form that you can fill out with any data you can offer in reporting UFO sightings.

URL:
    http://ume.med.ucalgary.ca/~watanabe/ufo.html

Alien On Line

View footage of the famous Roswell incident online here at this page that also features extraterrestrial images, movies, MPEGs, and other graphic and textual information regarding UFOs and space aliens. Don't leave home without it!

URL:
    http://www.crs4.it/~mameli/Alien.html

alt.alien.visitors FAQ

Those darned aliens! What are they about, and what do they want from us? Research the queries from this frequently asked questions repository, and you'll know how to greet them the next time you watch them arrive. Before they do, abduct yourself to some terminology and definitions used in UFOlogy, and brace yourself with "facts" on extraterrestrials and their ilk.

URL:
    http://www.iddc.via.at/ufo/ufo-faq.html

Are you plagued by uninvited visitors? Tired of those impromptu trips to the Cassiopeiaen system? Visit the **alt.alien.visitors** FAQ to learn to live with this—one of life's little aggravations.

The Bluebook

These files are part of the famous U.S. Air Force documentation of unexplained UFO reports from the files of Project Blue Book, the originals of which remain under lock and key at the National Archives. There are a total of 585 (minus 13 missing) unexplained cases to read about, including this: "Pilot, Clark, flying a Piper Clipper. Seven delta-shaped objects, 35-55' in span, 20-30' long, 2-5' thick; light colored except for a 12' diameter dark circle at the rear of each. They flew in a tight formation of twos with one behind, and made a perfect, but unbanked, turn. During the ten-minute sighting, they displayed decreasing smooth oscillations. Clark's engine ran rough during the sighting, and upon landing was found to have all its spark plugs burned out."

URL:
    http://www.cis.ksu.edu/~psiber/substand/
    bluebook.html

BUFORA

Explore BUFORA, the home page for the British UFO Research Association, and allegedly the largest UFO research organization within the U.K. They'll bring you regularly updated news, articles, events, and (hold your breath!) up-to-the-minute status reports on UFO cases received by the association as they come in. In the meantime, you can research an extensive case report database of historical information that has been chronicled since 1932 containing over 6,000 entries. There are columns on extraterrestrials, sightings, research progress reports, information on projects being conducted, and a list of contact names and addresses of local UFO groups around the U.K., including UFO group news. This may well be cyber-ufology at its best.

URL:
    http://www.citadel.co.uk/citadel/eclipse/
    futura/bufora/bufora.htm

Biotechnology is the conjunction of Biology and Technology.

A B C D E F G H I J K L M N O P Q R S T U V W X Y Z

Department of Interplanetary Affairs

For ufology from a different tack, try this page of resources flavored with a secret government-conspiracy angle. There are some esoteric articles on UFOs and cosmology, purported alien contact occurrences on the island of Maui, a treatise on NASA's conflicting dis-information about the moon and Mars, and other far-out premises. Expand your mind, but don't believe everything you read, even if it's on the Web.

URL:

    http://www.maui.net/~daryl/enmar.html

The Extraterrestrial Biological Entities

Extraterrestrials unite! If you're out there, dial in to this page of lots of great information about sightings of yourselves. Learn how we humans hold you in both awe and foreboding. You'll be interested to learn about how many spaceships we've seen of yours and how diverse they are. And by the way, would you stop abducting us and just land so that we can take you to our leader? Whoops! If you happen to be a human reader, you'll find lots of UFO documents on Roswell and Groom Lake, and AVI footage to download.

URL:

    http://sloop.ee.fit.edu/users/lpinto/

Fifty Greatest Conspiracies: Jacques Vallee

Stephen Spielberg paid homage to Jacques Vallee in *Close Encounters of the Third Kind* when he based his French scientist character played by Francois Truffaut on this real French UFO theorist. Read an interview with this influential man, and find out about his latest surmisings regarding UFO presence and perception. As Vallee puts it, "I will be disappointed if UFOs turn out to be nothing more than spaceships."

URL:

    http://www.webcom.com/~conspire/val.html

Had enough of real UFO sightings? Link up with http://www.waroftheworlds.com to travel back in time to 1938 to hear Orson Welles' classic broadcast of War of the Worlds by H.G. Wells.

Internet UFO Group

There are many web pages on the Internet dedicated to the research of unidentified flying objects and the creatures that pilot them, but this page stands out for its comprehensiveness toward rich detail and visual appeal. You'll find updates and notices of sightings; UFO primers and controversies; the IUFOG database and media page; government secrets; a library; and links to other pages, sights, and sounds on the Net.

URL:

    http://users.aol.com/iufog/

Monthly UFO Report

Keep up with one man's search for extraterrestrial life via this diary on UFO studies, sightings, and symposia. Read about the latest debates at Oxford and throughout the world regarding the presence of outer-spatial aliens, and keep up with commonly asked question about UFOs. You'll find some interesting Roswell commentary here, and associated sites and publications that pertain to ufology.

URL:
http://medianet.nbnet.nb.ca/medianet/
atlantic/ufo/ufo2.htm

Protree UFO Center

You might need to write a letter to this site for access to this very large archive of various UFO files that also features lots of pictures. Search over 700 files on UFO graphics and materials, and keep up with sightings, experiences, and abductions, courtesy of this resource.

URL:
http://www.protree.com/npt-ufo/

Smitty's UFO Page

Take a virtual trip to the Groom Lake/Area 51 site in Nevada. Read firsthand accounts of alien autopsies, secret flights with Martians in spaceships, extraterrestrial encounters and technology, and purportedly specific information and maps about the base that does not officially exist. Hey, it's still top-secret, and yet the worst-kept secret on the Internet. Why shouldn't you know? See another side of why America has a budget deficit in financing its worst-kept secret—from someone who has allegedly been there. Remember, you didn't hear it here first!

URL:
http://www.schmitzware.com/ufo.html

UFO Gallery

What would UFO research be without a few pictures of them flying around? Here's a gallery of unidentified flying object snapshots collected from all over the world, plus some related information and links. Very cool. . .

URL:
http://www.linknet.it/Spirit/ufo-gallery.html

UFO Information FTP Archive

The compendium of all downloadable compendia regarding unidentifiable flying object information files and their subsequent impact upon human consciousness is to be found here at this Rutgers site. Read and decide for yourself.

URL:
ftp://ftp.rutgers.edu/pub/ufo/

Ufologists

Prominent UFO researchers, witnesses, and skeptics are featured on this database list of significant people in the unidentified flying object realm that includes links to files by or about notable characters, nuts, and/or scientists. You'll find pointers to information on people associated publicly with UFO conspiracy theorists or witnesses. You'll find everyone from Betty Andreasson (an abductee) to Randolph Winters (who claims to be a student of Billy Meier, the famous Swiss contactee and Ufotographer).

URL:
http://www.cris.com/~Psyspy/ufo/people/

Ultimate UFO Page

If what you seek is extraterrestrial, consult this page on ultimate UFO sightings, alien abduction, the Face on Mars, Roswell, UFO crashes, and links to UFO sites and newsgroups. The organization has a response form with which to leave your comments in case you've made a sighting. If you're a beginner, there's also a guide to the mythology of UFOs.

URL:
http://www.serve.com/tufop/

The X-Files

Check out this web page of resources and links to the hit cult television program based upon alien sightings, *The X-Files*. It's a dirty job to put handcuffs on aliens each week and throw them in the slammer, but somebody's gotta do it. Read details of how the cast and producers put the whole thing together, and warp over to other relevant UFO and "X-Files" links.

URL:
http://www.rutgers.edu/x-files.html

A B C D E F G H I J K L M N O P Q R S T **U** V W X Y Z

Don't censor others. Censor yourself.

VIDEO: CONFERENCING

CU-SeeMe

Cornell University's CU-SeeMe is clever in name and famous across the Net for providing real-time delivery of video signals. However, it does not use multicast; so be advised that the bandwidth it consumes increases with each connection, which can be a pain with multiple users. The program is available for Mac and PC users. There are plans to make it compatible with MBONE.

URL:

ftp://gated.cornell.edu/pub/video

CU-SeeMe

(Sung to the Barney tune) "I see you, you see me, we're a happy family . . . with a camera you might really want to stare—back at me in my underwear."

CU-SeeMe Mailing List

Here's a discussion list for the popular freeware video conferencing program CU-SeeMe.

Mailing List:
Address: **listproc@cornell.edu**
Body of Message: **subscribe CU-SeeMene-request** *<your name>*

Biotechnology is the conjunction of Biology and Technology.

CU-SeeMe Schools

Take an electronic field trip with your schoolmates to classrooms around the world with this list of K-12 schools using CU-SeeMe videoconferencing software. Learn about effective videoconferencing and check out postings of upcoming special events. Find out how to participate in live video conferences with scientists; authors; and government, business, and community leaders.

URL:

http://gsn.org/gsn/cuseeme.schools.info.html

Gee, don't the kids of today have things easy? Calculators have replaced slide rules, word processors have done in typewriters, and now **CU-SeeMe Schools** makes it possible for kids to talk to and learn from one another all around the world. Luckily, peanut butter and jelly sandwiches will always stay the same!

Dan's Quick and Dirty Guide to Getting Connected to the MBONE

Dan will show you exactly how to prepare for and implement your MBONE connection so that it's up and running in no time. Many pointers and FTP sites are linked so that you'll get all the information and programs you need to configure your machine(s) for multicasting in a jiff. Thanks, Dan!

URL:

ftp://genome-ftp.stanford.edu/pub/mbone/mbone-connect

Desktop Videoconferencing Products

Check out the latest in video software and merchandise. You'll find a full description of each product, which platform each runs on, system requirements, price, and contact information.

URL:

http://www2.ncsu.edu/eos/service/ece/project/
succeed_info/dtvc_survey/products.html

DT-5

This Virginia Tech desktop video conferencing site evaluates primarily PC- and Unix-based products operating over the Internet and Integrated Services Digital Network (ISDN). You'll find an excellent lengthy introduction to video conferencing, and product links and descriptions.

URL:

http://fiddle.ee.vt.edu/succeed/videoconf.html

Energy Research Videoconferencing Network

Like the rest of us, scientists dislike frequent traveling, especially when *money* is in short supply. If you are a scientist who would like to discuss rocket science with colleagues via computer, check out ERVN—the Energy Research Videoconferencing Network. You can save money while avoiding bland airline food, and you don't have to wear a tie. Read more details about ERVN and its planned evolution in ISDN videoconferencing.

URL:

http://www.hep.net/documents/hepnrc/
ervn-chep94.html

ESnet Video Conferencing Service

If you're a member of the Energy Sciences Network, this page will show you how to schedule video conferences via ISDN with other members of the ESnet community. A user guide provides step-by-step instructions on how to use the video conference reservation-and-scheduling system to set up conference times and retrieve contact information.

URL:

http://www.es.net/hypertext/guides/vcs.html

International Multimedia Teleconferencing Consortium

Setting and maintaining international standards are the goals IMTC is striving for in the world of multipoint document and video teleconferencing. There's a FAQ list and an overview of the ITU-T process, and sections containing white papers, press releases, and reader feedback.

URL:

http://www.csn.net/imtc/

International Telecommunication Union

The ITU has been setting standards since 1865, and this international organization facilitates the global coordination of telecom networks and services between governments and private sectors. One branch of the ITU, the Telecommunication Standardization Sector (ITU-T), creates standards for videoconferencing. On their pages, they offer many databases; press releases; proceedings; publications; and, of course, standards offered by the ITU.

URL:

http://www.itu.ch

International Videoconferencing List

Although it's still under development, the IVL is a formidable service that lists links to videoconferencing providers throughout the world. Pick a university or company near you, check out their equipment and rates, and email them directly to arrange your next video conference.

URL:

http://www.sju.edu/~lees/vc_list.html

Internet Video and Audio Tools

A compendium of MBONE audio- and videoconferencing tools with references for both PCs and Macs awaits discovery at this site. News and articles on videoconferencing include experiments at Loughborough University, multicasts of the campus's computer science lectures, and more.

URL:

http://pipkin.lut.ac.uk/~ben/video/

A
B
C
D
E
F
G
H
I
J
K
L
M
N
O
P
Q
R
S
T
U
V
W
X
Y
Z

Introduction to Videoconferencing and the MBONE

Introduce yourself to the famous Multicast Backbone which provides audio and video across the Internet. Learn what you'll need to have to meet hardware, software, and network requirements. There's a quick-and-dirty guide to getting connected to the MBONE, plus a few downloadable software tools and some tips on how to implement them.

URL:

    http://www.lbl.gov/ctl/vconf-faq.html

ISDN Videoconferencing

Everything you need to know about videoconferencing with the Integrated Services Digital Network can be found right here. Dan Kegel single-handedly organizes ISDN videoconferencing information, site directories, organizations, mailing lists, and vendors linked to the page.

URL:

    http://alumni.caltech.edu/~dank/isdn/
    isdn_ai.html#VIDEO

2B or not 2B, that is the bandwidth question. Channel into the world of Integrated Services Digital Network and bond with Dan Kegel's definitive ISDN page for informative videoconferencing stuff.

IVS Home Page

The INRIA Videoconferencing System is a software package for transmitting audio and video data over the Internet. It uses a sophisticated compression algorithm that produces both low data and low frame rates. The system is freely available, but runs on workstations only. You'll find it here, along with assorted tools, docs, and apps.

URL:

    http://zenon.inria.fr:8003/rodeo/personnel/
    Thierry.Turletti/ivs.html

KVL Video-Link

Similar to the International Videoconferencing List, KVL maintains a large and growing database of worldwide universities and other institutions with videoconferencing capabilities. Access the database by country, check out their rates and equipment, and make a reservation.

URL:

    http://www.videolink.kvl.dk/

Literature Survey: Videoconferencing and Education

Read before you leap. Here's a wealth of articles on practically everything educators need to know about videoconferencing. You'll find academic research papers, training manuals, and other resources designed for both teachers and students. A step-by-step guide shows you how to make the most out of your next videoconferencing experience.

URL:

    http://www.icbl.hw.ac.uk/~cjs/vidconf.area/
    sima1.html

MBONE

Multicast Backbone (MBONE) is an outgrowth of "audiocast" experiments in which live audio and video were multicast to destinations around the world. For the benefit of scientists and rock 'n' roll fans, MBONE has been used to transmit Internet Engineering Task Force meetings and live Rolling Stones concerts. No doubt, the IETF folks had to create MBONE when they couldn't *get no satisfaction* from traditional audiocasting technology. Tune in here to find out how MBONE works.

URL:

    http://www.research.att.com/mbone-faq.html

MBONE Information Web

MBONE stands for the Virtual Internet Backbone for Multicast IP. This site coordinates and provides information on MBONE and related activities. A list of service providers, FTP sites, an array of desktop applications, debugging tools, MBONE documentation, and a section on commercial product vendors can also be found.

URL:

http://www.best.com/~prince/techinfo/ mbone.html

MBONE Mailing List

MBONEs, MBONEs, if they could only talk! Since *we* can talk—over MBONE—subscribe to the MBONE mailing list, and find out everything you've always wanted to know about the Multicast Backbone but were afraid to email.

Mailing List:
Address: **mbone-request@isi.edu**
Body of Message: **subscribe mbone** *<your name>*

MBONE Multimedia Software

You're going to need capable software to take advantage of MBONE, and you'll find a lot of it right here. Links to various FTP sites lead you to applications, tools, and patches.

URL:

http://www.cs.ucl.ac.uk/mice/mbone_soft.html

MBONE Session Agenda

Browse the agenda of material to be broadcast over the MBONE, or book a time slot for your transmission on the 'bone. The schedule is divided into months and all times are given in GMT. A time zone converter is provided.

URL:

http://www.cilea.it/MBone/agenda.html

MBONE, the Multicast BackBONE

Read an informative paper, written lightheartedly for everyone from beginners on up who are interested in the MBONE. Learn more about this virtual network that has been in existence since the beginning of the decade, and find out how it came to be, who uses it, and how to set it up and use it yourself. Included are a lot of other pertinent details, plus a look toward the future of MBONE.

URL:

http://www.cs.ucl.ac.uk/mice/ mbone_review.html

The next time your dog has a birthday party, why not throw him an MBONE? Schedule a videoconferencing session between him and other canine pals he's met on the Net (probably through **alt.canine.woof.woof.woof**). What better way to suck up precious bandwidth than with a bunch of bow-wows telecommunibarking with one another! And even if no one else knows you're a dog on the Net, your best friend still does.

Mr gateway

Find the multicast route between any two hosts with this gateway to mr. Simply type the IP address or hostname of the destination machine, and the gateway will display the route from your machine to that host. (You'll need to be mrouted in order for this service to work.) If you input two addresses, the gateway will display the route from the first to the second.

URL:

http://www.cl.cam.ac.uk/htbin/mr

Multi-Vendor Integration Protocol

MVIP is a standard for integrating diverse telecommunications and applications technology into a single, cohesive package for computers and networking. MVIP supports integration for voice processing, FAX, data communications, video conferencing, and other computer technology requiring connection to telephone networks. Read the MVIP overview statement to see what potential this protocol presents.

URL:

http://www.scii.co.uk/scii/mvip.htm

A B C D E F G H I J K L M N O P Q R S T U V W X Y Z

Multicast Backbone FAQ

How do you join the MBONE? (At the waist?) What's it all about? How do IP multicast tunnels work and how are they configured? Which workstation platforms can support the mrouted program? Where can you get software? Which network providers can you contact? These and many other questions are answered—right here, right now.

URLs:

ftp://venera.isi.edu:/mbone/faq.txt
http://www.cs.ucl.ac.uk/mice/faq.html

Multimedia Integrated Conferencing for Europe

Use your mouse and mosey on over to MICE, the predominant compendium and support center in Europe for videoconferencing. You'll encounter lots of multimedia information including documentation, a searchable index, and a bibliography of nice MICE projects and activities.

URL:

http://www.cs.ucl.ac.uk/mice/mice.html

NERO Project

Read about the efforts of the Network for Engineering and Research in Oregon, a cooperative project to develop high-speed, high-bandwidth data and audio/video connections within and between sites. Detailed information explains how NERO is implementing a high-speed ATM network to support desktop-to-desktop collaboration, distance learning, videoconferencing, and outreach programs.

URL:

http://www.nero.net

Personal Conferencing Work Group

PCWG serves as a discussion forum for the development, delivery, and deployment of innovative, interoperable conferencing and communications products. Here you'll find information on what's new in the telecommunications industry, a list of events, and a section devoted to projects that members of the group are currently working on.

URL:

http://www.gopcwg.org

PicturePhone Direct Videoconferencing

Equipment information, answers to your videoconferencing and ISDN questions, and several free offers await you on PicturePhone Direct's site.

URL:

http://picturephone.com

README file for PictureWindow

Grab yourself a free SparcStation demo of PictureWindow. *pwrx* allows you to connect to any full-featured PictureWindow system and receive video conferences over the Internet from that station. This demo will allow you to talk with the sender, but you will not be able to send video yourself. Hey, it's free, so what are you complaining about?

URL:

http://www2.ncsu.edu/eos/service/ece/project/
succeed_info/dtvc_survey/pwrx_readme.html

ReLaTe Project

If you can relate to Remote Language Teaching over SuperJANET, then you belong here. The folks at this project develop and test application-specific user interface video conferencing software for remote teacher-student interaction. To learn more, read the academic papers produced by the project which address the technical and practical issues of technology and distance education.

URL:

http://www.ex.ac.uk/pallas/relate/

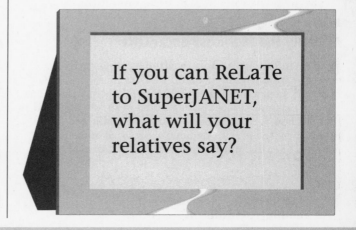

If you can ReLaTe to SuperJANET, what will your relatives say?

The RGO Worldwide Guide to Public Video Conference Centres

Get your whole gang together and go out videoconferencing with this comprehensive list of cities and vendors throughout the planet. Beam your voices and likenesses halfway around the globe to other intelligent beings while marveling at what a wonderful time and world we live in.

URL:
> http://www.ast.cam.ac.uk/~ralf/vcguide/

RTP and the Audio-Video Transport Working Group

Read about RTP—the real-time transport protocol—and the associated RTCP, which provides support for real-time conferencing for large groups on the Internet. These protocols feature support for gateway audio, video bridges, and multicast-to-unicast translators.

URL:
> http://www.fokus.gmd.de/step/hgs/rtp/

Scientist-On-Tap

Here's your chance to learn from some modern-day Galileos and Keplers at the Jet Propulsion Laboratory. The Scientist-On-Tap program demonstrates the remarkable capabilities of distance learning with students and JPL scientists interacting through videoconferencing over the Internet. Make a reservation to sign up your very own scientist, and say that you found him or her here!

Mail:
> Address: scientist@gsh.org

URL:
> http://gsn.org/gsn/sot.home.html

Looking for COBOL resources? Graduate to Fortran.

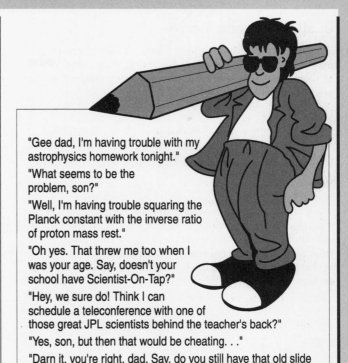

"Gee dad, I'm having trouble with my astrophysics homework tonight."

"What seems to be the problem, son?"

"Well, I'm having trouble squaring the Planck constant with the inverse ratio of proton mass rest."

"Oh yes. That threw me too when I was your age. Say, doesn't your school have Scientist-On-Tap?"

"Hey, we sure do! Think I can schedule a teleconference with one of those great JPL scientists behind the teacher's back?"

"Yes, son, but then that would be cheating. . ."

"Darn it, you're right, dad. Say, do you still have that old slide rule of yours?"

"You bet! Come on, kiddo, let's sit down and I bet we can solve that darned old problem in less than three hours. Hey, I'll race you!"

ShowMe Video

For Sun Solaris Unix workstations only, ShowMe offers a wealth of videoconferencing power and options. It is extremely configurable, provides complete video, audio, and a shared whiteboard, and runs on Internet Protocol. You can order a free 30-day evaluation CD-ROM of the ShowMe system on the page.

URL:
> http://www.sun.com/cgi-bin/
> show?products-n-solutions/sw/ShowMe/
> products/ShowMe_Video.html

UC Berkeley MultiMedia Group Papers

Check out Cal's research program on continuous media (CM) audio and video applications, software toolkits for developing CM applications, and compression/decompression MPEG and wavelet apps. You'll find the aforementioned papers plus the software at their FTP site, and information on their progress toward running parallel systems.

URL:
> http://www.cs.ucl.ac.uk/mice/ucb-mm.html

A B C D E F G H I J K L M N O P Q R S T U V W X Y Z

UCSD Multimedia Archive

Peruse 17 selected, detailed recent multimedia and video research papers and reports from U.C. San Diego's Multimedia Laboratory. Most are quite technical, and all are in compressed PostScript format.

URL:

http://www.cs.ucl.ac.uk/mice/ucsd-mm.html

United States Distance Learning Association

Remember what a pain it was to stand in long lines for registration at college, only to be told that they had just filled the last slot in that last quantum mechanics class you needed? Hey, avoid the hassle and learn from your own home! More and more teachers are using the Internet as a means of remote teaching. USDLA promotes distance learning, and you can check out their endeavors plus get a free copy of *ED*, their *Education at a Distance* magazine and journal—just for taking the time to stop by. So when it comes to crashing classes, remember that a computer is quite adept at crashing.

URL:

http://www.usdla.org

VCN Glossary

Thumb through this great video, compression, and networking glossary, and the next time you're at a teleconferencing cocktail party, you'll not only dazzle 'em with acronyms like ATM, ISDN, HDTV, and DVI, but you'll know what they *mean*, as well. Everything from A:B:C notation to YUV is covered thoroughly and precisely.

URL:

http://www.cs.ucl.ac.uk/mice/vcompgloss.html

Vidconf Mailing List

Audio and video conferencing technology and its uses in daily life are the subjects of interest for Vidconf. Subscribe, and you may just receive your well-deserved 15 minutes of fame.

Mailing List:

Address: **majordomo@pulver.com**
Body of Message: **subscribe vidconf**

Video Conference Technology and Applications

Check out the latest information on Internet and ISDN videoconferencing here first. Post or answer questions, and keep up with the desktop teleconference revolution, both generally and specifically.

URL:

news:comp.dcom.videoconf

Get physical in Physics

Video Conferencing

The concept of video conferencing has actually been around since as early as 1933 in the salad days of Dick Tracy and the comic strips. It's only now that we are finally catching up to it in terms of technological reality. And now that we are, you can catch up as well on its problems and advantages, plus the latest developments, availability, etiquette, links—and, of course, its history.

URL:

http://www.grady.uga.edu/megatech/
videoconferencing/home.html

On the Internet, nobody knows you're a dog. . . . . .as long as you keep the camera turned off.

How will networked, desktop video affect your future?

Video Conferencing Topics

Get the lowdown on setting up, running, and getting the most out of your videoconferencing system. Topics here include hardware considerations, software capabilities, session tools, and online meeting strategies. There are also several anecdotes about successful video conferences.

URL:

http://www.cs.colorado.edu/home/homenii/
videoconf.html

Video Reference

A good source of information on packetized, desktop video conferencing and room-type video conferencing, this page features MBONE FAQs, a map of the major MBONE nodes and links, a survey of desktop videoconferencing products, mailing list archives, and much more to do with video conferencing and the MBONE.

URL:
    http://www.slac.stanford.edu/winters/pub/
    www/net/video.html

Videoconferencing Information and Resources

One of the greatest uses for videoconferencing is in the field of education. This site at San Diego State University educates teachers on the uses and strategies for successful distance learning. Topics include how to set up a videoconferencing center, using compressed video, a review of videoconferencing applications, and lesson plans.

URL:
    http://edweb.sdsu.edu/edfirst/vidconf/
    vidconf.html

Videophone Mailing List

Get all spruced up for that big teleconference by subscribing to the videophone mailing list. You'll encounter lots of videoconferencing issues and answers. The archives for this list are also available.

Mailing List:
    Address: **videophone-request@es.net.com**
    Body of Message: **subscribe videophone** *<your name>*

Web Conference Multicasting Report

Read about the activities, successes, and foibles of the audio/video multicasting portion of the 1994 World Wide Web Conference. This report gives a breakdown of the preparations, setup, arrangements, equipment, sessions, and technical problems that occurred while broadcasting the event.

URL:
    http://www.nlm.nih.gov/reports.dir/
    multicasting.dir/report.html

Don't censor others.
Censor yourself.

Why Do Users Like Video?

If you doubt the power of video in teleconferencing, read these studies undertaken by Sun Microsystems that demonstrate the effectiveness of collaborative technology in enhancing productivity. Analysis shows that video helps mediate human interaction and convey visual communication. Not surprisingly, the study group's use of desktop conferencing dropped dramatically when video capability was removed.

URL:
    http://www.sun.com/smli/technical-reports/
    1992/abstract-5.html

Why Do Users Like Video?

Well, you can watch, hear, and communicate with people from anywhere in the world. Any other questions? Good! Now excuse me, I'm late for my group primal scream teleconference.

X Window Multiplexor

Need an X multiplexor for supporting several networked X servers? XMX is a stand-alone utility for sharing an X Window System session on multiple displays. XMX relies on the X11 protocol and paints the same graphics on all servers, providing a WYSIWIS (What You See Is What I See) environment. It accepts multiple X client connections and displays client graphics on multiple X displays.

URL:
    http://www.cs.brown.edu/software/xmx/

VIDEO: MEDIA AND TELEVISION

C-SPAN

Here's a page for all the fans of C-SPAN—the network that never blinks. Pick through the programming to find out what you'd like to watch. Monitor political campaigns and other interesting government activities here.

URL:
    http://www.c-span.org

A
B
C
D
E
F
G
H
I
J
K
L
M
N
O
P
Q
R
S
T
U
V
W
X
Y
Z

CBS Television

This is the home page for the CBS network and it contains many interesting CBS-related links. Learn about existing shows or get news "up to the minute." There's plenty to explore.

URL:
http://www.cbs.com

ESPNET

A link to the all-sports channel most cable companies carry—get the scores from the NFL, NHL, NBA, and many other sports, and keep up with the latest information on trades and other sports stories.

URL:
http://espnet.sportszone.com

NASA Television

Tune into NASA TV to watch round-the-clock coverage of shuttle missions. This service is available via participating cable TV stations, but many elect to show only portions of shuttle activity. With NASA TV and CU-SeeMe software, you can tune in at any time to watch and hear the events as they happen in real-time. Check this page for a list of NASA Television reflectors and assistance in using CU-SeeMe video software.

URL:
http://btree.lerc.nasa.gov/NASA_TV/

NBC HTTV

The National Broadcasting Company network provides information about their programming, sports events, news, and local NBC affiliates. Also of interest is the NBC Supernet, which offers rich graphics and in-depth news.

URL:
http://www.nbc.com

PowerTV

One day you may be booting up your television if it's connected to a PowerTV set-top box. Check out the PowerTV site to find information on the PowerTV OS— yes, it has an operating system. "Daddy, what does it mean when Mr. Rogers says, 'Segmentation fault: core dumped'?"

URL:
http://www.powertv.com

QuickTime Continuum

Meet the technology that brought video to the personal computer! QuickTime is Apple's cross-platform standard for creating, using, and sharing video, music, text, and animations. For the latest news, developer information, the latest versions available for downloading, and all those QuickTime movie clips—the QuickTime Continuum is the place.

URL:
http://quicktime.apple.com

> **QuickTime was the first multiplatform video format for personal computers to successfully combine synchronized audio and video. Learn more about it at the QuickTime Continuum.**

TNS Technology Demonstrations

Developed by members of the Telemedia, Networks, and Systems Group at the MIT Laboratory for Computer Science, these demonstrations of video processing technologies look great on your monitor. Browse video files captured using video-grabbing hardware, complete a video puzzle, and explore images taken from live video sources.

Note: An X Window terminal is required for most of these demonstrations.

URL:
http://www.tns.lcs.mit.edu/vs/demos.html

TV Net

TV Net gives you easy access to U.S. and International Broadcast and Cable Television addresses and information. This is one hefty site with links to TV-related resources all around the Web. It's an excellent index for locating science and educational shows. (Of course, it's nice to know that "The Simpsons" airs on four dozen networks, too.)

URL:
http://tvnet.com

TV Tonight

Tired of looking through the TV guide? Twizzlers brings you TV tonight with all the important information you need to know about what's on TV. There are brief descriptions of each episode, including late-night talk shows. If you're trying to decide between Letterman and Leno, see what they're up to and pick out your show in advance.

URL:
    http://metaverse.com/vibe/tv.html

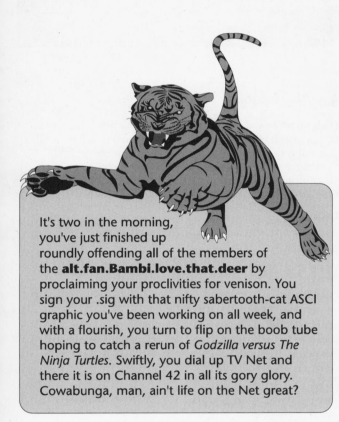

It's two in the morning, you've just finished up roundly offending all of the members of the **alt.fan.Bambi.love.that.deer** by proclaiming your proclivities for venison. You sign your .sig with that nifty sabertooth-cat ASCI graphic you've been working on all week, and with a flourish, you turn to flip on the boob tube hoping to catch a rerun of *Godzilla versus The Ninja Turtles*. Swiftly, you dial up TV Net and there it is on Channel 42 in all its gory glory. Cowabunga, man, ain't life on the Net great?

USA Television Web Links

This is a complete list of television stations with web sites on the Net. Local stations are organized by state and the major networks are listed in a separate section. There's also two sections with lists of cable companies and other sites on the Web.

URL:
    http://www.webovision.com/media/sd/ustv.html

Vidcom

Dal Neitzel may have a bad attitude, but all accounts report that he's a first-rate Video/Multimedia producer with a string of impressive credits to his name. Vidcom is his small production company. On Vidcom's pages, you can read about Dal and his projects, partners, interests, and diversions.

URL:
    http://www.nas.com/~jcofrin/

VIDEO: SATELLITE

Big Dish Satellite Systems

The TVROSAT mailing list is for discussing technical issues relating to large-dish home satellite receiver systems designed to receive television only. The list's archives are available from the list server.

Mailing List:
    Address: **listserv@vm1.nodak.edu**
    Body of Message: **subscribe tvrosat** *<your name>*

Direct Broadcast Satellite Discussions

Beam over to this mailing list discussion of direct broadcast satellite technology.

Mailing List:
    Address: **listserv@vm1.nodak.edu**
    Body of Message: **subscribe dbssat** *<your name>*

European Satellite Information

You'll be beaming with all the European satellite information to be found on this page. There's channel data; TV schedules; satellite broadcasting abbreviations and acronyms; the **alt.satellite.tv.europe** and **decoding pay-TV** FAQs; and online magazines and journals about satellites and telecommunications.

URL:
    http://www.funet.fi/index/esi/

A B C D E F G H I J K L M N O P Q R S T U V W X Y Z

Home Satellite Technology

Join the HOMESAT mailing list to discuss the technical issues related to home satellite receiver systems. Weekly notebooks are kept of issues discussed each week.

Mailing List:
Address: **listserv@vm1.nodak.edu**
Body of Message: **subscribe homesat** *<your name>*

MISCSAT

The MISCSAT mailing list is intended for discussion of miscellaneous technical issues relating to satellite technology. Weekly summaries of the list and full archives are available from the server.

Mailing List:
Address: **listserv@vm1.nodak.edu**
Body of Message: **subscribe miscsat** *<your name>*

Satellite Interpretive Messages

A mailing list for the discussion of all issues and topics that relate to satellite interpretive messages. A notebook is produced each week from the list. Join up and find out where to peruse the weekly notebooks.

Mailing List:
Address: **listserv@vmd.cso.uiuc.edu**
Body of Message: **subscribe wx-stlt** *<your name>*

Satellite Telecom Coordinators Mailing List

If you coordinate satellite telecommunications, you may be interested in conversing with others in your field at the ADEC-STC mailing list.

Mailing List:
Address: **listserv@unlvm.unl.edu**
Body of Message: **subscribe adec-stc** *<your name>*

Satellite TV and Radio Frequency Lists

Figure out where to point your dish once you lug it home with this guide to worldwide telecommunications satellites. You'll find information on each satellite's position, channel, frequency, and whether the signal is scrambled or not. Don't leave your couch without it!

URL:
http://www2.wintermute.co.uk/users/orrock/satidx.htm

Hitch your dish to the skies! Saddle that new satellite receiver of yours so that it's in sync with all those wonderful man-made orbiting transponders that beam old episodes of *Mister Ed* and *Leave it to Beaver* around in space. Heaven help us all if there's other intelligent life watching out there.

Satellite TV Page

Dish up some out-of-this-world treats by orbiting over to this page full of satellite information. You'll find the Satellite Journal International's latest edition and archives, the Scrounger's Guide to Satellite TV, FAQs, satellite radio schedules, and an illustrated satellite chart. There are even links to the home pages of several TV stations.

URL:
http://itre.uncecs.edu/misc/sat.html

TELE Satellit

Get all the news that's fit to broadcast about the satellite world from TELE Satellit. This European resource covers scientific and commercial announcements along with satellite launches, updates, and trends in the industry. Weekly editions in English and daily news reports in German are available.

URL:
http://www.funet.fi/index/esi/TELE-Satellite.html

Video Services

See how video is supporting Western Washington University's academic efforts by utilizing its campus cable system. There is information on how the university provides satellite/off-air recording, teleconferencing, and videotape production and post-production. You can bet your cybershirt that there are links to other video resources.

URL:
http://ra.cc.wwu.edu/WesternOnline/ATUS/Services/VS.html

VIRTUAL REALITY

2Morrow

If you don't feel like forking over the big bucks just yet for an Onyx or a SPARCStation, but you've still got the urge to render and produce VR like the big boys, well, welcome to 2Morrow! This commercial site will be happy to sell you their software suite, which includes an interactive, graphical world builder; a 3D object modeling CAD program; and a runtime module—all for under $100. Don't feel like spending anything at all? Download their tool demo from a variety of sites.

URL:
    http://www.xmission.com/~gastown/imaginative/
    2mro.html

3D Technology Hits the Web

Here's an excellent overview for technically-impaired mere mortals on the marvels of VRML—and wondrous they are, since web surfers will soon be able to fly through 3D worlds. Explore virtual cities and museums; peruse 3D online catalogs; and view other assorted information, such as stock-market trends, in 3D. The article contains links to providers mentioned in the text.

URL:
    http://techweb.wais.com/techweb/ia/12issue/
    12threed.html

Abulafia Gallery

Here's Ohio State's VRML gallery that contains images in high, medium, and low resolutions for all sorts of platforms. A VRML viewer will be necessary for this; Webspace is provided for downloading. There is information on AL, the language used to generate the source code for the environments pictured; the source code itself; and a technical report on representing the textures that you'll be viewing.

URL:
    http://www.cgrg.ohio-state.edu/~mlewis/Gallery/
    gallery.html

ARI MacHTTP Home Page

You might guess that the military is interested in virtual reality—at least we know the U.S. Army is. You can check out their research and read online papers about high-tech Army projects such as those being conducted in synthetic environments and rotary wing aviation. Be all that you can be!

URL:
    http://alex-immersion.army.mil

Avalon Repository

If you're looking for objects, you've come to the right place. Avalon contains a large collection of VR objects in various formats (3DS, Lightwave, BYU, DXF, ENFF, OBJ, OFF, POV, etc.) and a hefty selection of textures (in RGB, WIN, JPEG, and TIFF formats). Thumbnail sketches of object and texture files are in the works, and the site promises soon to become accessible from any Web browser.

URL:
    ftp://avalon.viewpoint.com/avalon

CAVE Lab

Spelunk through the Cave Automatic Virtual Environment (CAVE) at the University of Illinois, Urbana-Champaign. The CAVE is a multi-person, room-sized, high-resolution, 3D video and audio environment. A fractal explorer application enables participants wearing stereo glasses to view and manipulate 3D images using an electronic wand.

URL:
    http://www.mcs.anl.gov/FUTURES_LAB/
    CAVE/cavelab.html

Clemson University Virtual Reality Project

Clemson is actively involved in virtual reality, as you can see for yourself by their ambitious page. They offer their own VR tools, still images, movies, outlines of current research projects, and links to a myriad of VR pages.

URL:
    http://fantasia.eng.clemson.edu/vr/

Encyclopedia of Virtual Environments

The University of Maryland's EVE project is a hypertext compendium of terms that describe virtual reality applications, terms, and technologies. You'll find definitions and examples of system components, active interaction devices, simulators, rendering techniques, and other VR-related applications.

URL:
    http://www.cs.umd.edu/projects/eve/
    eve-main.html

Nanotechnology has some big potential.

A
B
C
D
E
F
G
H
I
J
K
L
M
N
O
P
Q
R
S
T
U
V
W
X
Y
Z

Excerpt from the Net...

(from CAVE Lab)

Argonne CAVE Construction History

On this page you will find a timeline of the original CAVE construction available for historical interest. Updated 7 July 1994

The borrowed graphics engines have been removed, leaving the Argonne CAVE functional with a single front wall active. This will be the state until the CAVE is taken down for shipment and returns from SIGGRAPH.

As of 10:30 PM on Tuesday, 5 July 1994, the Argonne CAVE was functional with two walls and one floor, thanks to the cooperation of the University of Illinois Chicago and Silicon Graphics, Incorporated which kindly loaned us the additional two Reality Engine graphics subsystems to run those extra screens.

As of 6:00 AM on Sunday, 3 July 1994, the CAVE has been moved forward approximately five feet in preparation for a possible installation of ceiling/floor projector. Two mirrors have been placed: one on the left and one on the front. The front projector/mirror set has been aligned.

As of Friday, 1 July 1994, the head tracker and wand are completely working, but not calibrated well. The audio server is an R3000 Indigo, and appears to be working marginally—we tend to get glitches where the CPU can't keep up with the audio data rate required.

As of Wednesday, 29 June 1994, the tracking system is functional but not calibrated or integrated with wand switches. The audio server is on the net and people are bringing up the appropriate software. We have successfully placed one mirror in preparation for moving the CAVE into its final position. It appears that we only need about 10 feet of clearance for the front mirror, even after performing final calibrations.

As of Friday, 24, June 1994, one wall of the CAVE is operational in 3-D. Four audio speakers and one amplifier is in place, and hopefully on Sunday, 26 June 1994, an SGI Indigo will be in place to generate sound under CAVE software control. We have the pieces to implement the tracking system, but ran out of time; work will recommence on Monday, 27 June 1994 along with the delivery of the wand.

Tuesday, 21 June 1994 we assembled unistrut into the CAVE formation so that we could verify the cuts and mark for screen snap holes. Currently we are in the process of drilling those holes.

Friday, 17 June 1994 we took the stainless steel unistrut pieces to central shops to be cut to length tomorrow (Saturday.) Monday the 3/8" holes will be drilled and actual erection of the structure will commence.

Tuesday, 14 June 1994 we pulled Ethernet, FDDI, and HIPPI cabling into the CAVE room from the MCS environment through the ECT communications room. We expect to provide outside-world Internet connectivity at FDDI speeds to the SGI Onyx immediately upon delivery.

On Friday, 10 June 1994 the Electrohome 8000 projectors were powered on and tested with VGA resolution video source from an IBM Thinkpad 350C laptop. The alignment and focusing procedure went as quickly as 45 minutes.

Extending the Web to Support Platform Independent Virtual Reality

It's amazing to think that when we researched the first edition of the original book in 1993, we worked from a cold, hard, Unix command-line prompt. Gopher was the new Internet darling, and Lynx was the only hypertext browser around. Then along came SLIP, PPP, and Mosaic, and nothing's been the same since. These days, people want to make the Web walk, talk, and quack like a virtual duck. So peruse this latest proposal that allows VR environments to be incorporated into the Web.

URL:

http://vrml.wired.com/concepts/raggett.html

Flexible Learning with an Artificial Intelligence Repository

FLAIR is a project of Drexel University that displays information about authors in a 3D cloud VR format. Additionally, *FLAIR World* provides a VR interface to a large artificial intelligence (AI) repository of pictures, movies, and stand-alone applications for a Macintosh platform in multimedia and virtual environments.

URL:

http://christensen.cis.drexel.edu

Georgia Tech School of Civil and Environmental Engineering

Many colleges and universities have full-fledged VR programs within one or more science departments, and Georgia Tech is no exception. Their downloadable demo visualizations include themes such as interactive construction, a lab walkthrough, robotic navigation, networks, and helicopter flight data. Learn more on this web page about their programs for behavior modeling of interactive physical systems.

URL:

http://www.ce.gatech.edu/Projects/IV/iv.html

The Glove

It doesn't matter if you are left- or right-handed with these datagloves. Look into three types of sophisticated datagloves and see which one fits. Spandex is not just for workout clothing.

URL:

http://wombat.doc.ic.ac.uk/foldoc?data+glove

Interactive Origami in VRML

Recycle all your old dot-matrix printer paper by learning the fine art of Japanese folding through VRML. This page on origami at Keio University illustrates several examples of the art, and takes you on a walkthrough of how each one is produced. You can change camera position to see more details and click through frames to view each step-by-step transformation.

URL:

http://www.neuro.sfc.keio.ac.jp/~aly/polygon/vrml/ika/

What could be better than learning origami—the classic Japanese art of folding paper into beautiful shapes—from none other than the master artists themselves? How about by learning how to do it from your home? Save yourself an expensive airline ticket to Tokyo plus all that jet lag by winging your Web browser directly into the Land of the Rising Sun. Talk about the information highway—this is the origami flyway.

Internet VR

Beam your way over to Silicon Graphics for a free copy of WebSpace Navigator, a nifty 3D web viewer. While you're there, you can pick up a few additional WebSpace tools and utilities, and have a peek at SGI's Virtual Reality Modeling Language (VRML) archive. Another link transports you to various spots on the Web for more VRML repositories of demos and information.

URL:

http://webspace.sgi.com

A B C D E F G H I J K L M N O P Q R S T U V W X Y Z

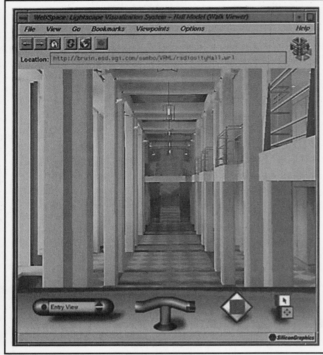

Step into a new virtual reality with WebSpace.

Journal of Virtual Environments

By JOVE, here's a nifty collection of articles on VR and telepresence. Topics covered include Remote Visual Navigation, Air Traffic Control Displays, Gesture Recognition, and Three-Dimensional Zooming. You can read each full article or simply peruse the abstracts.

URL:
> ftp://umd5.umd.edu/pub/vrtp/Jove/
> jove-main.html

Lateiner Dataspace

If you find immersive volume visualization engrossing, visit the Lateiner page for papers and information on discrete physical simulation techniques, distributed datastructures, and simulation environments. This site has Web-based demonstration software available for Windows 95. Learn more about simulation environments as distributed metacomputing engines by reading their online "Interview with a Mathematician about Dataspace."

URL:
> http://www.dataspace.com

LUTCHI

The Loughborough University of Technology Computer-Human Interface (LUTCHI) Research Centre in Leicestershire, UK, has an interesting variety of VR and user-interface projects. Their principal focus is on support and agent-based systems for professional people at work. Learn more about how they design systems to process speech and graphics, and how they manage the interaction or dialogue between users and machines. You can read several of their research reports and visit their multimedia laboratory.

URL:
> http://www.lut.ac.uk/departments/co/lutchi

Mesh Mart

Unless you're a purist, why go to the trouble of designing your own 3D mesh object files when you can pluck so many free ones from the Internet? Mesh Mart provides a resource for mostly 3D Studio (3DS) files and allows you to view thumbnail sketches of rendered object images before downloading. The library contains a large online collection of FAQs and research papers, and there's a useful tools and utilities section.

URL:
> http://www.cic.net/~rtilmann/mm/

Special!
3D Object Files by the Dozen.
Mesh Mart

Montgomery Blair High School Virtual Web Walkthrough

Look out, here come the kids! If you're into adventure games and virtual reality, then come enjoy some fast times with the students at Montgomery Blair High. Take the campus tour—and while you're at it, solve a murder mystery at the school before you are yourself a victim!

URL:
> http://www.mbhs.edu/~dpeck/mwalk/mwalk.html

Movie Samples from the Open Virtual Reality Testbed

Bring along the popcorn for a few QuickTime VR movies of immersion studies, walkthroughs, tours, and experiments in texture mapping.

URL:

> http://www.nist.gov/itl/div878/ovrt/
> OVRThome.html

MR Toolkit

The Minimal Reality Toolkit from this site at the University of Alberta is a robust VR software toolkit with complete documentation, demos, and source code. It consists of a set of subroutine libraries, device drivers, support programs, and a language for describing geometry and behavior. MR is free to individuals and nonprofit organizations, and the package runs on HP, SGI, DEC, and IBM RS6000 workstations.

URL:

> http://web.cs.ualberta.ca/~graphics/
> MRToolkit.html

New College of California: The vrmLab

The Golden State has its share of interesting VR sites, and here's another! This page presents some provocative destinations, but you'll first need a VRML browser application (WorldView or WebSpace), which is available here online. You can cruise an object warehouse and the vrmLab Testbed, then partake of the thought-provoking discussions regarding VR and 3D modeling for the Internet.

URL:

> http://www.newcollege.edu/vrmLab

NPSNET Research Group

OK, wet-nosed scrub, take one look at the high-tech soldier pictured on this page and you'll throw down your weapon and beg for mercy. The Naval Postgraduate School at Monterey, California, focuses on an amazing variety of networked VR defense simulations and capabilities. From terrain modeling and virtual environments to Distributed Interactive Simulation, you'll be captured by their extraordinary research. The days of Pong are long gone.

URL:

> http://www-npsnet.cs.nps.navy.mil/npsnet/

Poor Richard's Virtual Reality Resources

If you want to know about VR, here's a decent jumping-off point. Archives, pointers, FAQs, glossaries, publications, and bibliographies abound at this site. You'll also find a great deal of information regarding HITL, the Human Interface Technology Laboratory at the University of Washington.

URL:

> http://www.iia.org/~rosenr1/sensorium/almanac/
> virtual.html

Presence

Presence is a quarterly journal for serious investigators of teleoperators and virtual environments. Computer scientists, high-tech artists, and electrical engineers will enjoy abstracts of this fascinating journal devoted to computer devices, physics, and teleoperation.

URL:

> http://www-mitpress.mit.edu/jrnls-catalog/
> presence.html

Project Isaac

Isaac is a geometric simulation system supported by the Office of Naval Research. There are demos and 3D movies available for viewing Isaac's capabilities, plus online papers, workshop announcements, and a list of related projects at this Purdue University site.

URL:

> http://www.cs.purdue.edu/homes/vanecek/
> projIsaac/

Toy Scouts

Join the Toy Scouts at the University of Central Florida, and be prepared to have fun exploring new technologies in immersive virtual worlds. Using high-performance graphics workstations and custom software, the Scouts network multiple users into a common virtual environment through a series of games and programs. Fire up your workstation for the likes of Noseball, Ricochet, SyberKnight, Wormhole, and Virtual Darts.

URL:

> http://www.vsl.ist.ucf.edu/~scouts/scouts.html

One bad Apple...

A
B
C
D
E
F
G
H
I
J
K
L
M
N
O
P
Q
R
S
T
U
V
W
X
Y
Z

Experience Virtual Reality.

UK VR-SIG 3D Object Archive

You'll find that the Object Supermarket on this page is one of the premier places to visit on the Web for free VR 3D object files that have been compiled for real-time graphics display. Most of these files are in public domain, and they feature many types of CAD and virtual reality models. The repository also gratefully accepts any VR contributions that you've created, and has links to a variety of VR sources.

URL:

http://www.dcs.ed.ac.uk/home/mxr/objects.html

Virtual Environment Navigation in the Underground Sites

VENUS is a large-scale effort at CERN to use virtual prototyping for its next-generation Large Hadron Collider (LHC) particle accelerator. If you were one of the scientists who invented the Web, of course you would have wanted the world to preview the model in virtual reality, which is precisely what's going on in the VENUS project. See where CAD ends and VR begins with a flight through the LHC prototype. Check out the Visualization Shop and read more about CERN's fascinating modeling programs.

URL:

http://sgvenus.cern.ch/VENUS/

Earn some virtual merit badges for your workstation with the Toy Scouts! You don't even have to pitch a tent or swat a skeeter.

Virtual Environment Technologies for Training Project

The VETT Lab at MIT conducts research in using virtual environments for training and simulation. The multi-modal *Puck* demo integrates visual, auditory, and haptic devices to allow you to play virtual air hockey. *Officer of the Deck* integrates 3D graphics, voice recognition, and spatialized sounds to help train navy officers navigating a submarine into a bay. See these and other projects, plus check out the Virtual Workbench. Don't forget your workstation!

URL:

http://mimsy.mit.edu

Virtual Environment/Interactive Systems Program

Believe it or not, ARPA (the Advanced Research Projects Agency) is still hanging tough after all these years. There are several collaborative projects running at Mississippi State that focus on virtual environment simulation in the areas of education, architecture, oceanography, scientific visualization, and defense.

URL:

http://www.erc.msstate.edu/thrusts/scivi/vr

Virtual Environments Research—Delft University

The Dutch are doing interesting research with VR in such disciplines as physics, psychology, communications, and industrial design. Check out virtual books, VR game design, curing acrophobia, and electronic clay.

URL:

http://dutiws.twi.tudelft.nl/TWI/IS/
vr-overview.html

Virtual Interface Technology

Reality is a virtue unto itself. On this page you'll find a virtual compendium of VR news, articles, magazines, and papers at the University of Washington in Seattle. Recent additions can be encountered in the Virtual Reality Update. This site probably contains the most extensive collection of information about VR presently available on the Internet.

URL:

http://www.hitl.washington.edu/projects/
knowledge_base/meta/

Virtual Reality Alliance of Students and Professionals

Artists, scholars, and professionals are welcome to interact with one another about the beauty, wonder, and potential of virtual reality through VRASP. The informal club meets on Internet Relay Chat every Thursday at 8:00 P.M. EST or 5:00 P.M. PST. At this page you'll find a brochure on the alliance, their membership form, and information on how to participate in the free-for-all discussions.

URL:

http://www.vrasp.org/vrasp

Virtual Reality Modeling Language FAQ in Japanese

This is the Japanese version of the VRML FAQ. You'll need Kanji fonts to view it properly.

URL:

http://www.anchor-net.co.jp/rental/andoh/
vrml/vrmlfaq.html

Virtual Reality Modeling Language Frequently Asked Questions

Did you know that VRML is pronounced "vermel," sort of like vernal? Learn this and a thousand other amazing facts from this FAQ on the language that promises to change the look and feel of the Web in the years to come. There are hypertext links up the ying-yang to many different VRML authorities and sources.

URL:

http://www.oki.com/vrml/VRML_FAQ.html

Virtual Reality Roller Coaster

Why spend money at an expensive theme park when you can ride a stomach-churning roller coaster and puke in the privacy of your own home for free? Strap on your virtual seat belt for a dizzying series of video demos that explore the scope and power of VR. Customize your nausea by choosing where to place tunnels and supports. So have a few hotdogs, and enjoy the ride. Plastic keyboard protectors recommended.

URL:

http://www.ce.gatech.edu/Projects/IV/
coaster.html

Dude,
I'm
Gonna
HURL!
Experience the Virtual Reality Roller Coaster.

Virtual Reality Samples

Flight simulation and virtual reality soar well together. You can see why in this series of airborne scenes featuring highlights from the 1995 Reno National Championship Air Races and views from the cockpit of a Gee Bee R-2 raceplane. You'll need QuickTime VR Player software to view them; but since you're on Apple's page, the link for downloading it is just a click away.

URL:

http://qtvr.quicktime.apple.com/Samples.htm

**On the Internet,
you don't need a passport to
travel the world.**

A
B
C
D
E
F
G
H
I
J
K
L
M
N
O
P
Q
R
S
T
U
V
W
X
Y
Z

Virtual Worlds Project

You'll learn about core research areas such as intelligence entity simulation, advanced human-machine interfaces, graphics, and telerobotics at MIT's Virtual Worlds Project. The Artificial Intelligence Lab offers a number of detailed sections of online publications, presentations, and programs for use in science, engineering, medicine, commerce, and industry.

URL:
> http://www.ai.mit.edu/projects/vworlds/
> vworlds.html

Visual Systems Laboratory at IST

Picture a SWAT team being able to enter a building to rescue hostages within minutes, when none of the team has ever been in the building before. Or picture an intelligence officer walking down a hostile street comparing the scene to the most recent satellite surveillance photos. As the landscape changes before him, he points out the differences, which are instantly recorded in a database. The University of Central Florida is visualizing all this and more. Pay them a visit soon before they come to your neighborhood. And while you're at it, check out their papers, reports, and online dissertations. You never know. . .

URL:
> http://www.vsl.ist.ucf.edu

VR Software

To enter the world of virtual reality, you've got to start somewhere, and that somewhere is usually with a good piece of software. There are many programs here for all platforms, and they are listed, linked, and reviewed by the author, Chris Hand.

URL:
> http://www.cms.dmu.ac.uk/~cph/vrsw.html

VR-386

Psssst. . . Hey buddy, how 'bout a steal on some great VR ware? Yeah, yeah, it'll run smooth on yer PC. Ya at least got a '386, doncha? No, ya don't gotta fork any dough out fer it! OK, here's da deal. Ya ready? Just grab it off da Net! Yep, that's right, Slick. The VR-386 graphics package is freeware and boasts the fastest drawing speed of any VR software in its class. This full-fledged program supports many VR devices, including stereo flicker glasses, HMDs, and PowerGloves. One more thing—tell 'em we sent ya!

URL:
> ftp://psych.toronto.edu/pub/vr-386

VRML 1.0 Specification

Programmers and serious-minded students will want to read this definitive spec regarding Virtual Reality Modeling Language. Topics covered in detail are the mission statement and history of VRML, and the basics of how the system works. There is a separate section on browser considerations and file extensions.

URL:
> http://vrml.wired.com/vrml.tech/vrml10-3.html

VRML Forum

Correspond with others about VRML, the Virtual Reality Modeling Language. The Futures Forum presents the VRML Repository at the San Diego Supercomputer Center, SIGGRAPH '95 notes, and a hypermail archive of the VRML forum mailing list.

URL:
> http://vrml.wired.com

VRML from HELL

This is a dandy page of pointers to the latest technical information, system and software information, and VRML sources for VRML design information, Internet VR viewable files, and online and written material. There are many additional links to other sources.

URL:
> http://www.well.com/user/caferace/vrml.html

VRML Repository

One of four designated "Supercomputing Sites" in the U.S., the San Diego Supercomputing Center maintains the Virtual Reality Modeling Language (VRML) repository for information on this future industry standard. Among the many resources you'll encounter are complete documentation and specifications for VRML, examples of VRML applications, and the repository's FTP site for software and tools related to this innovative modeling language.

URL:
> http://rosebud.sdsc.edu/vrml/

The Net is your passport. The Web is your magic carpet.

Excerpt from the Net...

(from the VRML 1.0 Specification page)

Virtual Reality Modeling Language

Introduction

The Virtual Reality Modeling Language (VRML) is a language for describing
multi-participant interactive simulations — virtual worlds networked via the global
Internet and hyperlinked with the World Wide Web. All aspects of virtual world
display, interaction and internetworking can be specified using VRML. It is the
intention of its designers that VRML become the standard language for interactive
simulation within the World Wide Web.

The first version of VRML allows for the creation of virtual worlds with limited
interactive behavior. These worlds can contain objects which have hyperlinks to
other worlds, HTML documents or other valid MIME types. When the user selects an
object with a hyperlink, the appropriate MIME viewer is launched. When the user
selects a link to a VRML document from within a correctly configured WWW browser, a
VRML viewer is launched. Thus VRML viewers are the perfect companion applications to
standard WWW browsers for navigating and visualizing the Web. Future versions of
VRML will allow for richer behaviors, including animations, motion physics and
real-time multi-user interaction.

VRML-o-Rama

You'd be hard-pressed to come up with a larger collection
of links and personal notes than VRML-o-Rama. The
author devotes extensive coverage to worldwide VRML
history, projects, tools, events, and the real people behind
the virtual reality. There are interesting insights and
links to practically everything VRML on the Web.

URL:

http://www.well.com/user/spidaman/vrml.html

Virtual Reality. Digitally simulated sights, sounds, and even
touch. It's all anyone talks about these days, but it's only half
of reality. The other half is as plain as that honker sitting
between your eyes. This time next year the big deal will be
NRML–Nasal Reality Modeling Language—plug-in modules
for your browser. You read it here first. So until there's
NRML-o-Rama (otherwise known as Smell-o-Rama),
VRML-o-Rama will have to do.

What Is Virtual Reality?

"Virtual Reality is a way for humans to visualize,
manipulate, and interact with computers and extremely
complex data." —*The Silicon Mirage* "Man Bytes
Reality!—News at 6:00." This may not be the definitive
treatise on VR, but it's a great place to begin. The page
lists conventional books, online news groups, BBSs, FTP
sites, local interest groups, and commercial VR
companies.

URL:

http://www.cms.dmu.ac.uk/People/cph/VR/
whatisvr.html

You can be a rocket scientist. Start in Aerospace and Space Technology.

WEB DESIGN, TOOLS, AND RESOURCES

The Backgrounds Archive

Brought to you by Kai's Power Tips Online and maintained by MIT's *The Tech* student newspaper, the Backgrounds Archive is a great collection of background images for web pages, animations, and graphics—anywhere really that you need a seamless background. The collection includes MIT's original archive, plus a wealth of visitor-contributed graphics. Most images are in JPEG format, although there are a few GIFs as well.

URL:
    http://the-tech.mit.edu/KPT/bgs.html

Gamelan—A Directory of Java Resources

Gamelan is an online directory of freely available Java applets. The directory is organized by categories, including (among others) animations, arts, education, finance, special effects, graphics, sound, and utilities. These applets are fascinating—and useful too! For example, a mortgage calculator allows you to calculate your monthly payment for a variety of loan balances and terms, and even displays an annual summary of the amount of principal and interest you pay each year.

URL:
    http://www.gamelan.com

HotJava Web Browser

One day, this will probably be a commercial product. But if you act soon, you, too, can help Sun debug their HotJava browser and learn about the Java technology in the process. Prerelease versions of HotJava are available for the Microsoft Windows (NT and 95) and SPARC Solaris platforms.

URL:
    http://java.sun.com/download.html

Java and HotJava User Documentation

This page at Sun Microsystems is the top of an entire hierarchy of user documentation for Java programming and using the HotJava Web browser. The documentation is well written and starts out slowly with an excellent "Where Do I Start?" primer. Next, you can follow a link with several great overviews of the Java language, HotJava, and security issues. Finally, you can jump right into the details of using HotJava or writing Java programs.

URL:
    http://java.sun.com/doc.html

Java Developer's Kit

We admit it. The Web is already getting boring. Just how many millions of static, lifeless pages can a person endure? Forget graphics, color, sound, and canned video clips. Put some real excitement into your pages with talking heads, animated figures, and rotatable 3-D diagrams and illustrations. All you need to get started is the Java developer's kit from Sun. Kits are available for SPARC Solaris, Windows NT, and Windows 95. And if you really want to get into the internals of Java, the source is available too. The best thing about it? It's all free!

URLs:
    http://java.sun.com/download.html
    http://www.javasoft.com

No, it's not about a new coffee fad. *Java* is a programming language that paves the way for the next great leap in technology for the Web. And *HotJava* is the name of Sun Microsystem's new Web browser that employs Java technology. (Next thing ya know, someone's going to name a competing product Latte.)

Java Discussion Groups

Having trouble motivating your Java applets to dance? If you have questions or problems using HotJava or writing Java applets, take them to the experts on Usenet. This is a very active newsgroup reflecting the hot hot status of Java and HotJava.

URL:
    news:comp.lang.java

Java Home Page

Java is an innovative new programming language for creating hardware- and operating system-independent applets (small applications) that can be embedded in web pages and integrated tightly with HTML. Sun Microsystems is the company behind Java and offers many useful and informative links about Java and Sun's Java-compliant browser, *HotJava*, on the Java home page. Topics include an overview of Java, the Java language environment, a FAQ list, mailing lists, and a handy starter kit to help you start using HotJava and writing your own Java applets.

URL:

> http://java.sun.com

The Java Zone

Welcome to the Java Zone—where nothing is as it seems, and everything is unexpected. But don't even think about visiting this enclave with a browser other than HotJava. This page is chock-full of Java applets that you won't even see with another browser. Check out a scrolling marquee that changes colors before your eyes, text headers that leap onto the page, a live feedback image map, online audio, and the ever-popular waving icons. Oh, there's also a pretty good introductory summary of just what Java is and how it works.

URL:

> http://metro.turnpike.net/S/suen_g/java.html

Mklesson

Mklesson is a tutorial-generating program for the Web available in Perl source code for Unix machines. It was developed to make tutorials easier to create and modify, and the program doesn't require any modifications to the local Web server. This page provides a user guide for Mklesson, the archived software, installation instructions, examples, and pointers to some other tutorial-generating packages.

URL:

> http://lglwww.epfl.ch/Ada/Tutorials/Lovelace/
> userg.html

Perl FAQ

What could be a better resource for the budding Perl programmer than a searchable version of the Perl FAQ in HTML format? This excellent FAQ introduces Perl; provides pointers to Perl software; compares it to other scripting languages; gives numerous information sources for Perl; explains and describes Perl programming aids; covers regular expressions, input/output, and signal catching; describes external program interaction; and much more. You can even make Perl 5 regular expression searches of the FAQ content.

URL:

> http://pubweb.nexor.co.uk/public/perl/faq/
> intro.html

Perl: An Introduction

Here's a great, 21-page introduction to the Perl programming language in PostScript format. It explains Perl's basic syntax, data types, operators and comparators, default arguments, regular expressions, flow control, built-in routines, C library routines, and system calls.

URL:

> http://www.eecs.nwu.edu/perl/SAG-perl.ps

PostScript to HTML Converter

Ps2html (the Sequel) is a Perl program that converts PostScript text into HTML. The Perl code is available in a tar file, and there is additional information about the program on this page. Several translated documents are also presented as examples of the program's capabilities. To use the program you need a PostScript interpreter such as Ghostscript.

URL:

> ftp://bradley.bradley.edu/pub/guru/ps2html/
> ps2html-v2.html

Random Images

How'd dey do dat? Here are detailed instructions on how to make a random image appear on a web page—so that every time you access a page, you get a different image. You'll need write access to the /cgi-bin directory on your Web server in order to get this to work, or you'll need to know someone who can put a program there for you.

URL:

> http://www.iaf.nl/~abigail/HTML/CGI/
> random_images.html

Scanning Photos for the Web

Maximize your equipment to produce the best images possible for the Web. The author of this short article provides a list of hardware that makes digital images look better.

URL:

> http://photo.net/philg/how-to-scan-photos.html

Tom's Tips for Web Designers

Tom Karlo, the photo editor at MIT's *The Tech* student newspaper, has graciously published a collection of tips and tricks for Web designers using Adobe Photoshop on the Mac or PC. Read Tom's tips and become your own graphics expert!

URL:

> http://the-tech.mit.edu/KPT/Toms/

A
B
C
D
E
F
G
H
I
J
K
L
M
N
O
P
Q
R
S
T
U
V
W
X
Y
Z

Excerpt from the Net...

(from Tom's Tips for Web Designers)

Using text in WWW graphics

General Techniques and Principles

The most important thing to remember about
creating graphics that include type is that
they are meant to be read. You may know
perfectly well what that header you spent an
hour on says, but don't expect the guy who
just wandered onto your page to know it as
well. Type is meant to communicate, not just
look pretty. I realize as much as anyone how
much fun it is to break the monotony of normal
text with some graphics, but don't forget that
you are not only trying to impress the viewer
with your Photoshop abilities, but you're also
trying to get a message across. If you just
want to create aesthetically pleasing stuff,
put it in a gallery; that's what art is for.
But please don't torture Web readers with trying
to read it.

There are, of course, techniques that will
allow you to bend but not break these rules.
However, most of the time, the best graphics
can be produced by strictly following the idea
of maximum contrast...

The Tutorial Gateway

The Tutorial Gateway is a program that uses CGI scripts
to make it easier to develop tutorial-style questions and
have them presented by a Web browser. The online
documentation gives instructions on using the program
and examples of generating multiple-choice questions,
true/false questions, single numeric questions, and single
algebraic-expression questions. The Perl source code is
available here, as well.

URL:

http://www.civeng.carleton.ca/~nholtz/tut/
doc/doc.html

WebStone Benchmark

WebStone is a proposed benchmark for evaluating the
relative performances of Web servers. Proposed and
developed by Silicon Graphics, WebStone is enjoying
some acceptance as a tool for comparing HTTP servers.
On this page, you can find the original white paper on the
WebStone proposal, the WebStone benchmark software,
and a FAQ list about, and about using, the WebStone
benchmark.

URL:

http://reality.sgi.com/employees/mblakele_engr/
webstone/

The Wild, Wild World of HotJava

The Wild, Wild World of HotJava is a collection of pages
maintained by Joey Oravec, a sophomore at the University
of Detroit Jesuit High School and a dedicated Java
enthusiast. Joey fills you in on all the latest scoop about
the Java programming language, the HotJava browser,
available releases, and even information about his school
and course work.

URL:

http://www.science.wayne.edu/~joey/java.html

World Wide Web Security

This page offers information on security for the World
Wide Web, the HTTP protocol, the HTML language, and
related software and protocols. It presents an overview of
the issues surrounding this topic; a list of draft proposals
for HTTP security protocols; references to related web
sites, mailing lists, and standards documents; details on
the IETF Web Transaction Security Working Group; and
the archives for the www-security mailing list.

URL:

http://www-ns.rutgers.edu/www-security/

The World Wide Web Security FAQ

The Web Security FAQ describes some of the problems
and answers questions about the security implications
of running a Web server. It includes information on how
to configure and run a secure server, how to protect
confidential documents at your site, how to use CGI and
Perl scripts safely, the uses of server logs, and privacy.
The bibliography lists books covering general security for
Web servers, firewalls, Unix system security, cryptography,
and Perl. There is also a section on Web security from the
client's perspective.

URL:

http://www-genome.wi.mit.edu/WWW/faqs/
www-security-faq

Learn how to protect your web site
from unintended access. Read the
World Wide Web
Security FAQ.

X-RAY TECHNOLOGY

Atomic Scattering Factors

This technical page should be of great interest to X-ray researchers and physicists because of its detailed resources on X-ray interactions with matter. Database and indices provide lookup functions for atomic X-ray scattering factors and calculations of properties such as X-ray transmission, reflectivity, and others.

URL:
   http://www-cxro.lbl.gov/optical_constants/

International XAFS Society

This society's interests represent scientists researching fine structures that are associated with near-edge and extended inner-shell excitation by various X-ray and electron probes. This site maintains several databases of elemental X-ray data. There are sections on XAFS news, minutes of meetings, a list of positions offered, and related resources.

URL:
   http://xafsdb.iit.edu/IXS/

Medical Imaging Resources on the Web

Here's a huge list of imaging-related sites on the Web. You'll find hundreds of links to sites that pertain to X-ray technology, including a useful collection of links to medical imaging vendors.

URL:
   http://pubweb.acns.nwu.edu/~dbk675/
      netsites.html

MRI Tutor

Magnetic Resonance Imaging is the future of X-ray technology because it operates without radiation. Explore the basics of this relatively new technology and learn about instrumentation, artifacts, safety, contrast agents, and many other topics of interest to X-ray technicians and radiologists.

URL:
   http://www.xray.ufl.edu/~rball/mritutor.html

> ## Looking for Earth Science resources? Check Earth Science, Ecology, Geography, and Geology.

Real Time Studies of Crystal Growth with X-rays

Keep up with Cornell University's program on the study of the growth of single-crystal thin films with X-ray diffraction techniques using gallium nitride thin films in a temporary vacuum chamber. You'll also find relevant papers on X-ray and crystalography techniques.

URL:
   http://www.tn.cornell.edu/SIG/RTgrowth/
      RTgrowth.contents.html

Ultranet: Ultrasound Technology

Another form of radiation-free X-ray technology, ultrasound is a popular choice among radiologists. Here are some great links to research projects and institutions involved in ultrasound technology. You'll also find some wonderful digital graphics and explanations of 3D reconstruction using ultrasound images.

URL:
   http://www.quake.net/~xdcrlab/Ultrasound.html

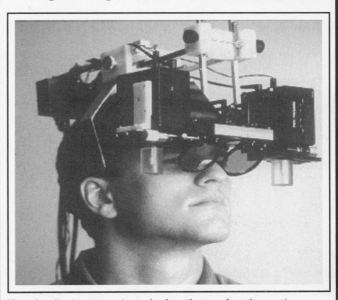

Virtual reality is an emerging technology that envelops the user in a simulated 3D world. This world appears to the user typically via a head-mounted display and is controlled by movements of the user's hand communicated to the display via a specially designed glove or other input device.

A
B
C
D
E
F
G
H
I
J
K
L
M
N
O
P
Q
R
S
T
U
V
W
X
Y
Z

Wilhelm Conrad Roentgen

Meet the man without whom radiologists would be out of a job. A century ago, Professor Roentgen was experimenting with a Hittorf valve when he observed that a few crystals of barium platinocyanide that were accidentally lying on the table gave off a fluorescent light. He had discovered a previously unknown form of radiation that he called "X" because of its mysterious glow. Applying this observation by submitting objects to the glow that captured their images on photographic film was the next logical step. An invention that changed the world was born. Read about the professor (in either English or German), relive his momentous breakthrough, and see the famous first X-ray of Roentgen's wife's hand—complete with a wedding ring on one of her fingers.

URL:
   http://www.fh-wuerzburg.de/roentgen/
   index_e.html

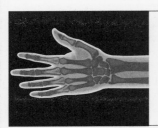

Don't try to impress Wilhelm Conrad Roentgen with your knowledge of X-ray technology. He can see right through you. Become acquainted with the father of X-ray photography.

X-ray Crystallographic Facility

This Chemistry and Biochemistry department at the University of Delaware is concerned with research on characterizations by X-ray diffraction of small-molecule organic and inorganic single crystals. Researchers worldwide may submit samples for analysis, and the site maintains X-ray information files and an online gallery of structures and other graphics. There are links to other related resources.

URL:
   http://www.udel.edu/arcade/x.html

X-ray Library

Wow! This is a truly incredible site that will be of interest to X-ray technicians, radiologists, medical students, and anyone with an interest in what bodies look like beneath the skin. There are hundreds of X-ray images of the human body, categorized by region. There are X-rays of normal and abnormal subjects to make comparisons. Here's a fascinating resource!

URL:
   http://www.njnet.com/~embbs/xray/xr.html

X-ray Properties of the Periodic Table

X-ray your favorite element by using this convenient database. Simply click on any button's associated element name to return a list of its chemical and fluorescent properties. The server allows for inputting different energy values to return X-ray cross-sections based on those values.

URL:
   http://www.csrri.iit.edu/periodic-table.html

X-ray World

This site is maintained by the Department of Physics at Uppsala University in Sweden, and it features the repository of the COREX bibliography and database, the Henke atomic scattering factors, and other information of interest to X-ray spectroscopists. COREX is a database of atomic and molecular core-edge excitation oscillator strengths that can be displayed or downloaded. The site also provides information on upcoming conferences and links to additional X-ray resource information and research sites.

URL:
   http://xray.uu.se/

YOUTH

Amateur Radio Astronomy

Using either imaging or non-imaging techniques, the young radio amateur can look at large areas of the sky for long periods of time. Professional observatories are compared to looking at the universe through a straw. Non-imaging techniques are relatively simple and inexpensive. Discover something new among the stars.

URL:
http://irsociety.com/0c/sara/why.html

Ask Dr. Neutrino

Do you have a physics question? Do you have an answer? "What makes the aurora borealis visible?" Maybe you don't like the answer given and think you can provide a better one. On this page, you can review questions and answers spanning several months or ask a new question of your own.

URL:
http://nike.phy.bris.ac.uk/dr/ask.html

Beakman's Electric Motor

This page provides step-by-step illustrated instructions for building a working electric motor with very few parts, most of which you can find laying around the house, like a D cell battery, a rubber band, paper clips, and a toilet paper tube.

URL:
http://fly.hiwaay.net/~palmer/motor.html

Biosphere 2 Global

Relive the days of yore when Biospherians made headlines by sealing themselves in their glass environment for two years, living off the hydroponics. They're back with an online educational project, the Biosphere 2 global change test bed. Their aim is to link students, teachers, and scientists in collaborative ecological studies. Take a virtual tour while you're there.

URL:
http://www.netspace.org/biosphere2/

The Web will set you free.

The Net is humanity's greatest achievement.

Chicago Academy of Sciences

The Chicago Academy of Sciences was founded in 1857 and is dedicated to science literacy for all. At this site there is information on the museum's collections, exhibits, research, and educational and outreach programs. Here you'll also find information on upcoming scientific meetings, conferences, and events.

URL:
http://www.mcs.com/~cas/home.html

Family Explorer

Your kids can help you learn about the mysteries and wonders of science through Family Explorer. These pages have many interesting and educational projects, and you don't need to be a rocket scientist to understand how they work. Two of our favorite projects were constructing a Star Clock and making green eggs and ham in the kitchen.

URL:
http://www.parentsplace.com/readroom/explorer/

Feet Stink!

If you don't agree, take off your shoes and socks at the end of the day and take a whiff. What causes this and how does it happen? Do everybody's feet smell the same? This is a fun page for learning about, and experimenting with, microbes.

URL:
http://www.nbn.com/youcan/feet/feet-smell.html

Fun Stuff To Do

Family Explorer offers a list of activities and experiments for the young scientist. Learn how to make stained glass, homemade butter, a camera out of an oatmeal box, boomerangs, kitchen chemistry, and a slew of other great projects.

URL:
http://www.parentsplace.com/readroom/explorer/activity.html

A B C D E F G H I J K L M N O P Q R S T U V W X Y Z

Excerpt from the Net...

(from Chris Palmer's Beakman's Electric Motor page)

I saw this on the TV show "Beakman's World" and I was very impressed that you could actually build a working electric motor with so few parts. I built one and brought it to work--it was a big hit with all the engineers around here. This writeup was for a friend of mine who wanted instructions that his son could follow for a science-fair project. So, if you missed the show, here's how to build one. BTW, my friend's son won second place in the school's science fair.

Goats, Goats, and More Goats

The 4-H Clubs of America provide young people with the opportunity to learn about raising farm animals. One of the most fun to raise is a goat—they're loaded with personality and affection, and they'll even allow kids to milk them and trim their hooves. Find out about what's involved in the care of goats, the different types of goats, and how to feed and care for them.

URL:

http://www.ics.uci.edu/~pazzani/4H/Goats.html

Microworlds

Explore the structure of various materials. Designed as an interactive multimedia page for grades 9-12, this Lawrence Livermore Laboratory page is devoted to the practical applications of chemistry. With interesting articles, top-notch graphics, and text geared toward young scientists with inquisitive minds, you can really see what things are made of.

URL:

http://www.lbl.gov/MicroWorlds/
MicroWorlds.html

Before our good friends on Capitol Hill manage to outlaw the Internet entirely, make sure your kids get a chance to see what the Net is really all about. Visit Microworlds now.

Need some great music? Lend your ear to the Music: MIDI section.

Like lizards? Check out Herpetology.

ZOOLOGY

American Association of Zoo Keepers

The AAZK is a volunteer organization of individuals who are dedicated to professional animal care and conservation.

URLs:

ftp://adams.ind.net/pub/aazk

gopher://adams.ind.net/1

http://aazk.ind.net

Animal Information Database

Do you need quick information or fun facts about a terrestrial or aquatic animal? Visit Animal Bytes. Or take the Animal Information Quiz. View movies about killer whales, and find books appropriate for young readers. Sponsored by Sea World's Education Department, this page is just plain fun and informative.

URL:

http://www.bev.net/education/SeaWorld/
infobook.html

Aquatic Animals Mailing Lists

A comprehensive listing of Internet mailing lists on various interests related to keeping or studying aquatic life.

URL:

http://www.actwin.com/fish/lists.html

If you're an aquatic animal, make sure you stay in touch—join the *Aquatic Animals* Mailing List.

Books on Zoology

An index to specialized books on general zoology, agricultural zoology, birds, fishes, invertebrates, marine biology, amphibians, and reptiles. You can order these books by fax, email, and (of course) snail mail.

URL:

http://www.demon.co.uk/ssb/zoomenu.html

Excerpt from the Net...

(from the pages of the American Association of Zoo Keepers)

Zoo Keeping As a Career

"Zoo Keeper" is a term used to describe an individual who cares for animals in zoological parks or aquariums. A career as a Zoo Keeper offers a unique opportunity in the specialized and demanding profession of maintaining captive exotic animals for conservation, research, public education, and recreation.

A B C D E F G H I J K L M N O P Q R S T U V W X Y Z

Excerpt from the Net...

(from the gopher at the University of Delaware)

Several years ago, there was a fad for keeping 'Vietnamese pot-bellied pigs' as house pets. At the time, it wasn't uncommon to hear of bred gilts selling for $10,000 or more. People jumped on the bandwagon, hoping to make a lot of money fast. Unfortunately, the only ones to make money were the first people to get into the business. They did an excellent job of selling their idea, so to speak, getting many more individuals into the business. . . Now we know better. Most people don't want to have a pig in the house.

— PIG TALES by Dr. Richard Barczewski, Cooperative Extension Livestock Specialist, University of Delaware

ChibaZOO

Not pigs in a MUD, but zoos in a MOO. That's right, ChibaZOO is an interactive virtual zoo in an adventure game-like setting. Post a note on the zookeeper's corkboard, then visit Picos of Pavilion, where you can enter the Planetarium, Museum of Science and Industry, EcoSphere, and other rooms that lead to hundreds of links to related Net resources. It's not quite as exciting as fending off grues in Zork, but a clever navigational concept nonetheless.

URL:
http://sensemedia.net/sprawl/

Electronic Zoo

Pick an icon of an animal and click! This entertaining and educational romp through NetVet's Electronic Zoo gives you an interactive Net encyclopedia on amphibians, birds, cats, cows, dogs, exotic animals, ferrets, fish, horses, invertebrates, marine mammals, pigs, primates, rabbits, reptiles, rodents, sheep, and so on. For example, pick pigs, then click on dozens of related links to resources on the Net, such as *Pig Images*, or the favorite of any swine, the *National Pork Producers Council*.

URL:
http://netvet.wustl.edu/ssi.htm

Email Directory

A searchable database of Internet email addresses of zoo and aquarium employees, veterinarians (exotics), administrators, wildlife rehabilitators, educators, and researchers.

URL:
http://www.wcmc.org.uk/infoserv/zoodir.html

Feline Conservation Center

You don't need to drive to Los Angeles to enjoy the variety of wild cat species at the Feline Conservation Center. Over 50 cats, some weighing up to 700 pounds, currently live at the compound. But cat lovers need only use their mice to check out the photos and sounds on this page.

URL:
http://www.cathouse-fcc.org

Gophers at Stanford

There is something too logical about using gopher to browse through zoological resources, wouldn't you say? The Stanford Genome Gopher server provides an abundance of related resources, plus searchable databases.

URL:
gopher://genome-gopher.stanford.edu:70/11/
topic/zoological

A Hunt for Squid

Like a new Tom Clancy novel, "A Hunt for Squid" reveals the exciting and action-packed life of this ocean invertebrate. The squid has played a supporting role in many books and major motion pictures such as, *Moby Dick* and *20,000 Leagues Under the Sea*. If it has *squid* on it, you'll find it here.

URL:
http://sleepless.cs.uiuc.edu/sigsoft/squid.html

A
B
C
D
E
F
G
H
I
J
K
L
M
N
O
P
Q
R
S
T
U
V
W
X
Y
Z

Excerpt from the Net...

(from Moby Dick at Project Gutenburg)

What was it, Sir? said Flask. The great live squid, which they say, few whale-ships ever beheld, and returned to their ports to tell of it. But Ahab said nothing; turning his boat, he sailed back to the vessel; the rest as silently following. Whatever superstitions the sperm whalemen in general have connected with the sight of this object, certain it is, a glimpse of it being so very unusual, that circumstance has gone far to invest it with portentousness.

Invertebrates

Invertebrate Zoology is that branch of science dealing with animals that have no backbones, such as the shrimps, crabs, sponges, worms, jellyfish, snails, squids, and a certain selection of lawyers and politicians. Check out the Smithsonian's Invertebrate Zoology page to find out how to borrow specimens, such parasitic copepods, gorgonians, and leeches, which love to attend Supreme Court nomination hearings.

URL:
> http://nmnhwww.si.edu/departments/
> invert.html

Museum of Vertebrate Zoology

Since 1908, the U.C. Berkeley Museum of Vertebrate Zoology has been putting their backs into maintaining a permanent historical record of specimens and information about the ecology, evolution, and geographic distribution of terrestrial vertebrates. These online catalogs provide information on over a half million bird, reptile, amphibian, and mammalian specimens. The MVZ maintains a frozen tissue collection, DNA extracts, audio tapes of bird songs, collectors' field notebooks, and more.

URL:
> http://www.mip.berkeley.edu/mvz

Nature and Wildlife According to GORP

This Great Outdoor Recreation Page includes many links with descriptions to nature and wildlife resources. You'll find pointers to national parks, conservation centers, the Whale Adoption Project, U.S. Fish and Wildlife Service, and more. Be sure to see *Above the Clouds Trekking*, an in-depth exploration of remote lands and exotic cultures!

URL:
> http://www.gorp.com/gorp/activity/wildlife.htm

Tropical Rain Forest Animals

Where do you find the poison arrow frog with its brightly colored skin that produces some of the strongest natural poison in the world? Or the hoatzin, a bird capable of producing a terrible odor to scare away its predators? In a tropical rain forest, of course. This page includes information on rain forests, and the people and animals who make them their homes.

URL:
> http://www.ran.org/ran/kids_action/
> animals.html

Excerpt from the Net...

(from the Rainforest Action Network)

An average of 35 species become extinct every day in the world's tropical rainforests. The forces of destruction, such as logging, cattle ranching, and overpopulation, have all contributed to the loss of millions of acres of tropical rainforest. Animals and people alike lose their homes when trees are cut down. These animals are given no warning to move--no time to pack their bags--and most die when the forest is destroyed.

— Susan Silber, Rainforest Action Network

Seeking stimulating circuit simulations? See Engineering.

U.S. Fish and Wildlife Service

Before you answer the call of the wild (not to be confused with "when nature calls"), check out the U.S. Fish and Wildlife Service page for information on conservation, endangered species, environmental contaminants, fire management, fisheries, migratory birds, wetlands, wildlife law, and more.

URL:

http://www.fws.gov

Vertebrate Zoology

Vertebrate Zoology is the study of animals with backbones such as fish, amphibians, reptiles, birds, and mammals. The Smithsonian's National Museum of Natural History includes a wealth of information from online databases and documents. One notable item on this site is an interactive metro subway map allowing you to determine the time it takes to get to the museum from anywhere in Washington, D.C. Did you know it takes 10 minutes from the southeast side of the Pentagon? So when you're really in a hurry to check out the National Museum of Natural History, you'll know just how long it'll take to get there.

URL:

http://nmnhwww.si.edu/departments/vert.html

World Species List

The Bible (which you can search in its entirety on the Project Gutenberg page) is filled with commandments telling us what not do. Besides being fruitful and multiplying, the only work the Bible offers us is to *name all the animals*. This has been interpreted by scientists as a directive to catalog everything in nature, from the smallest cells to the most distant stars. That's why the World Species List is so important. It is Holy Work. So stop coveting thy neighbor's computer and get busy by contributing to this noble cause.

URL:

http://www.envirolink.org/species/

Do you know the species of the world? Memorize the World Species List. You'll be tested at 5 o'clock.

Zoo Atlanta

Zoo Atlanta was created out of the bankruptcy of a 19th century traveling circus. Today, Zoo Atlanta is one of the premier zoos in the world—especially on the Web. This unique zoo is on the forefront of conservation and environmentally sound practices. The zoo sponsors volunteer programs for teenagers and participates in other community and environmental awareness activities.

URL:

http://www.gatech.edu/3020/zoo/home-page.html

ZooNet

When you want to know about zoos, zip over to ZooNet. Find information about every zoo in the world, as well as a sampling of pictures, recordings, and video clips of the animals in those zoos.

URL:

http://www.mindspring.com/~zoonet/

The Net is the new medium.

Index

Main subject headings are shown in **bold**

*Main subject headings are shown in **bold***

*Main subject headings are shown in **bold***

*Main subject headings are shown in **bold***

*Main subject headings are shown in **bold***

*Main subject headings are shown in **bold***

*Main subject headings are shown in **bold***

*Main subject headings are shown in **bold***

*Main subject headings are shown in **bold***

Main subject headings are shown in **bold**

*Main subject headings are shown in **bold***